Sustainable Energy Harvesting

Sustainable Energy Harvesting

Edited by **Ted Weyland**

SYRAWOOD
PUBLISHING HOUSE
New York

Published by Syrawood Publishing House,
750 Third Avenue, 9th Floor,
New York, NY 10017, USA
www.syrawoodpublishinghouse.com

Sustainable Energy Harvesting
Edited by Ted Weyland

International Standard Book Number: 978-1-68286-114-1 (Hardback)

Printed in China.

Contents

	Preface	VII
Chapter 1	**Development of a Wall Collector Unit and Phase Change Material (PCM), Air Heat Exchanger for Heating Application in Greenhouses** Liang Hao, Yang Qichang and Luis Alejandro	1
Chapter 2	**Concentrating Solar Power to Be used in Seawater Desalination within the Gulf Cooperation Council** Mohammed Saleh Al. Ansari	10
Chapter 3	**A Multi-Method Evaluation of the Potential for using the Electricity Bill to Encourage Energy Savings in Norwegian Households** Håkon Sælen and Hege Westskog	24
Chapter 4	**Electricity, Uncertainty and the Good Life** ***A Comparison of French and Norwegian Household Responses to Policy Appeals for Sustainable Energy*** Tanja Winther and Sophie Bouly de Lesdain	39
Chapter 5	**The use of the Leopold Matrix in Carrying Out the EIA for Wind Farms in Serbia** Bosko Josimovic, Jasna Petric and Sasa Milijic	53
Chapter 6	**Enhancing Furnace Thermal Efficiency by Adjusting Fuel Temperature** Chien-Li Lee and Chih-Ju G. Jou	65
Chapter 7	**A Study of Energy Conservation Policies at (Primary) Eco-Schools in Istanbul** Kutlu Sevinç Kayıhan and Seda Tönük	71
Chapter 8	**Lipase Catalysed Ethanolysis of Jatropha Oil for Biodiesel Production** Elizabeth Funmilayo Aransiola	85
Chapter 9	**Characterization of Inorganic Elements in Woody Biomass Bottom Ash from a Fixed-Bed Combustion System, a Downdraft Gasifier and a Wood Pellet Burner by Fractionation** Adrian. K. James, Steve S. Helle, Ronald W. Thring, Gurkaran S. Sarohia and P. Michael Rutherford	93
Chapter 10	**Remediation of Nickel from Crude Oil Obtained from Bomu Oil Field using Cassava Waste Water Starch Stabilized Magnetic Nanoparticles** Joshua L. Konne and Kenneth Okpara	103

Chapter 11 **Comparative Study on Charcoal Yield Produced by Traditional and Improved Kilns: A Case Study of Nyaruguru and Nyamagabe Districts in Southern Province of Rwanda** **110**
Alphonse Nahayo, Isaac Ekise and Alphonsine Mukarugwiza

Chapter 12 **Design and Realization of a 300 W Human Power Energy Generator System on a Bicycle** **119**
Ming Chun Hsieh and David King Jair

Chapter 13 **Green Technology for Biodiesel Production from *Mesua Ferrea* L. Seed Oil** **126**
Mohammad Aslam, Prashant Saxena and Anil K. Sarma

Chapter 14 **Large Scale Power Generation: Up-Skilling Welsh Industry** **137**
Elizabeth Locke and Sally Hewlett

Chapter 15 **Oil Extraction From Oil Sludge and TPH Elimination of Solids/Water by Ozonation** **148**
Misri Gozan

Chapter 16 **Transesterification of Soybean Oil with Ethanol using Heterogeneous Catalysts Based on Hydrotalcites** **155**
Evelyn Alves Nunes Simonetti and Gilberto Garcia Cortez

Chapter 17 **Improving Energy Efficiency and Environmental Sustainability of Commercial Insulation** **163**
Celine Manoosingh, James Mihelcic and Manjriker Gunaratne

Chapter 18 **Groundwater Resources Assessment in Tain River Basin of the Black Volta of West Africa** **175**
D. F. Dramani, G. K. Anornu, A. T. Kabo-bah and S. Oduro-Kwarteng

Chapter 19 **Development of a Charge Controller Dedicated to the Small Wind Turbine System** **184**
Ababacar Ndiaye, Cheikh M. F. Kébé, Vincent Sambou and Papa A. Ndiaye

Chapter 20 **CSD Type Family Solar Drying Technology and Its Potential Impact on Rural Development and Nutrition - A Case Study** **194**
Sangeeta Sinha, Anil Roy and S. N. P. Singh

Permissions

List of Contributors

Preface

Energy harvesting refers to the process by which energy is extracted from vast sources and stored for use by low-energy devices. The subjects discussed in this book that address the various aspects which fall under this field are energy conversion and storage, designing energy efficient systems, developing advanced technologies for enhancing extraction of renewable energy, etc. As this field is emerging at a rapid pace, the contents of this book will help the readers understand the modern concepts and applications of the subject. It will prove to be an invaluable resource for academicians and professionals alike.

The information shared in this book is based on empirical researches made by veterans in this field of study. The elaborative information provided in this book will help the readers further their scope of knowledge leading to advancements in this field.

Finally, I would like to thank my fellow researchers who gave constructive feedback and my family members who supported me at every step of my research.

Editor

1

Development of a Wall Collector Unit and Phase Change Material (PCM), Air Heat Exchanger for Heating Application in Greenhouses

Liang Hao[1,2], Yang Qichang[1,2] & Luis Alejandro[1]

[1] Institute of Environmental and Sustainable in Agriculture, Chinese academy of agriculture science, Beijing, China

[2] Key lab. for energy saving and waste disposal of protected Agriculture, ministry of agriculture, Beijing, China

Correspondence: Luis Alejandro, Chinese Academy of Agricultural sciences Center for protected Agriculture and environmental engineering, Room 209, No.12, South Street, Zhong Guan Cun, Beijing 100081, China. E-mail: H.273@live.com

Abstract

Greenhouse heating is essential in the northern region of China during winter season in order to maximize crop production in terms of quality and quantity and thus increase their overall efficiency. But the use of heating systems faces the problem of high running cost and noxious gases emission for heating systems based on fossil fuel and there is also a risk of frost cropping for electrical heating systems due to electrical power failure during heavy weather conditions. In this paper the performance of flat plate collector and a PCM serpentine heat exchange is investigated. An eutectic mixture $CH_3COONa{\cdot}3H_2O\text{-}HCONH_3$ was used as a latent heat storage material, this material possess a melting temperature of 40.5°C, relatively higher compared to commonly used phase change materials for heating applications. The energy capacity of the PCM during charge and discharge phase has been determined based upon the measurement results made in the heating process during February 2012, one of the coldest winter months in Beijing, China. It was observed that the average net energy efficiencies was 45%, therefore the PCM in study, was founded to be suitable for solar thermal storage applications.

Keywords: flat plate collector, latent heat storage, heat exchanger, PCM, Greenhouse

1. Introduction

With a vast land area, China is subject to the most extreme and worst weather conditions, from bitter cold winters to unbearable heat during summer. The winter in the northern region starts from the latest weeks of November and remains until March or April, considering January and February as the coldest months, with temperatures dropping to -40°C. Due to this factor, the use of greenhouses has a special importance for the agricultural sector in the northern region of China. These facilities represent a solution to crop cultivation during the coldest months of winter, allowing crop availability on the local and national markets. Greenhouses are by nature thermal collecting structures, the inside surface collects solar radiation heat during the day and re-radiates it to the Greenhouse inside environment during the night. But usually such heat radiation is not capable to fulfill the crop heat demand adequately. Therefore, the use of heating systems is essential to control the greenhouse indoor temperature and to provide adequate climatic conditions for crop growing during winter season. Among the advantages provided by the heating systems, we can mention: rapid growing and earliness, better control of the inside humidity and danger of disease infestation; reduction on chemical usage with positive consequences on the environment and health in general (Bailey, 2001).

At the same time, the installation and use of heating systems faces the problem of high operational cost, since there is a constant increment in energy prices. As reported by (Bailey, 2006), all types of heating system must increase productivity and sufficient earliness to justify the investment and running costs. Due to this fact, alternative energy sources for greenhouses heating have gained utmost interest in recent years. Some of the most important sources of energy are solar collectors and thermal storage energy (TES); these systems make use of Phase Change Materials (PCM) to store heat energy. The use of solar power is an increasingly attractive alternative as it is a stand-alone energy source, does not require a connection to a power grid, and the sun provides an unlimited and virtually endless supply of energy. In order to convert the solar radiation energy into thermal energy, the use of solar collectors has been widely use, the collector unit is placed outside the facilities and serves to heat the facilities in interest. In a greenhouse, the use of solar thermal energy covers one part of the crop heating demand needed during the day, as solar radiation is available during day only; its usage during night time required an efficient storage medium, so the excess of heat is stored for later use during the night. The solar thermal energy can be stored as sensible heat or latent heat. To store the same amount of energy, the sensible heat storage required significantly large quantities of storage medium, compared to the latent heat storage. In most sensible storage systems, the energy is stored in water or rock, which occupied larges spaces. A clear example of the difference for heat storage between sensible and latent heat storage can be founded in concrete blocks, which heat capacitance is approximately 1.0 kJ/KgK, while for hydrated salts, its capacitance during store and release is 190 kJ/KgK. In latent heat storage, the use of a PCM is considered as storage medium. PCM's are chemical substances, which possess a high fusion heat, which melts and solidifies at a certain temperature, they are capable of storing and releasing large amounts of energy compared to sensible heat storage materials. The process of absorbing or releasing occurs when the material changes from solid to liquid and vice versa.

Basically, PCM's can be classified into four types depending on its application: solid–solid, solid–liquid, solid–gas and liquid–gas phase change. However, the most common phase change material used for space heating application is the solid–liquid. Among the most used solid-liquid change PCM are Salt Hydrates, Fatty Acids, Esters, and a large quantity of Paraffin, some of these are considered products of oil refining or polymerization and are available in large amounts.

Various research works about solar thermal collectors and its application and combination with different storage mediums have been carried out in the past decade. Husseyin and Aydin (2009), conducted a study on the performance analysis of a latent heat storage system. In their study, a 10 piece flat plate air solar collector was used to heat up a greenhouse located in Turkey. A corrugated plate, reverse corrugated plate, trapeze plate, reverse trapeze plate and a flat absorption plate in combination with a phase change material were used for this work. They reported that such system is capable to provide 18-23% of the total daily thermal energy of a greenhouse for about 3-4 hours in comparison to conventional heating devices. And the difference in temperatures between indoor and outdoor was around 6-9 °C. In a more recent study, Laurent and Annabelle (2011) developed a small scale composite solar wall integrated with a phase change material. Their work focuses on the solar wall performance analysis, principally on the delay between absorption of solar radiation and released energy to the environment. Their study makes use of a mixture of calcium chloride potassium chloride and additives that possess a melting temperature of 27°C. They reported, composite wall have the advantage of heat release delay. This wall installation is rather to be used in premises occupied during day.

In the same manner, Phase Change Materials had been extensively studied for space heating applications, due to their capacity to store large amounts of heat when they change from solid to liquid. Studies about the application and utilization of PCM dates back to 1980, when Kern and Aldrich (1979), use calcium chloride hexahydrated as a PCM to investigate the heat energy storage capacity for a 36m^2 greenhouse that use fiberglass as insulation material. They reported that the thermal performance for this PCM higher compared to sensible heat storage mediums. In a recent study Husseyin and Aydin (2008) developed a hybrid system that make use of a GHPS integrated PCM been $CaCl_2 \cdot H_2O$ the PCM used as thermal storage material because of its high latent fusion per unit mass, chemical stability, availability in large quantities and low price. To fulfill the heat energy demand during winters in Turkey, their results showed that this system was capable to increase the greenhouse indoor temperature by 5-10°C and the PCM increases by 1-3 °C depending on the environment temperature, preventing in this way the risk of crop frost during winter nights. In other study, El-Bassuani et al. (2003) conduct a research where they combined urea sodium acetate trihydrated, urea sodium acetate trihydrated and lead acetate trihydrated as phase change material for energy storage purposes. They reported that such system composed by a mixture of urea sodium and acetate lead has the capacity to store 286kJ/kg. As a remark from this study, the composition mentioned before can be modified by adding other organic or inorganic compounds, rising the melting temperature and making this PCM suitable for applications in hot climatic regions. This paper presents the analysis performance of a north wall collector unit with a PCM heat accumulator. The main purpose is to develop an air-type solar collector with high thermal performance that is suitable for heating applications, and can be set up inside a greenhouse in order to optimize the space utilization. The literature review shows us that small works on greenhouse heating in the northern region of China has been carried out, focusing on solar collectors in combination with PCM's. Therefore, this paper will provide simplicity to solar collector's developers and information on the heating performance of the PCM used on greenhouses. As well as this system represents a potential application, there is a special focus in remote areas where the electric supply is non-existent or non-constant.

2. Experimental Setup

In this study, a north wall collector unit with a PCM heat accumulator was developed. This system represents an alternative solution to greenhouse heating demands for remote areas in the northern region of China, where an electrical supply is not available, and also can be used in combination with conventional heating systems in order to reduce the energy consumption. The schematic diagram of the system is given in Figure 1, and its main components are: PCM containers, Air ducts, Absorption plate, Serpentine PCM air heat exchanger, Air inlet valve, Air Outlet valve, and air separating valve. The hollowed stainless steel bars used to contain the PCM are shown in Figure 2. The facilities used for testing are shown in Figure 3a, b and c. The experimental set up consists of a collector unit equipped with a serpentine heat exchanger at the top side of the absorption plate; this unit was set up over the north wall inside a Chinese solar greenhouse with an angle of 90 degrees, respect to the horizontal. The collector unit collects the solar radiation and stores it to be used during the night by using a PCM heat exchanger. The absorption plate was set inside a PVC box at distance of 10mm measured form the glazing plate. A galvanized steel sheet with a thickness of 0.5 mm and black chrome selective coating was used as an absorption plate. The PVC used to construct the solar collector box has a thickness of 2mm, a length of 102mm and a width of 102mm. Its exterior is covered by polystyrene foam with a thickness of 20mm and thermal conductivity of 0.042W/mK. A single polycarbonate rigid film was used as a glazed material to maximize solar radiation on the absorption plate. Some of the polycarbonate properties are: good transmittance, high impact resistance, good temperature capability (270F), light weight, good life and low transmittance in the long IR, reducing heat loss. Two radial fans, 110v, 60Hz, 1.3A, 12x12 cm in size and a CFM of 110 were set inside the serpentine heat exchanger chamber to facilitate air circulation, inside the heat exchanger chamber and towards the adsorption plate. The air circulates through PVC ducts with a 30mm diameter covered by industrial Floormate foam insulation with a 20mm thickness and a thermal conductivity of 0.028W/mK. 3 ball valves were used at the end and mid-way of the air duct to prevent reversibility during the discharging process. The PCM was sealed inside the 12 stainless steel tubes with a 100mm length, positioned horizontally inside the serpentine heat exchanger chamber, 12 thermocouples (T-Type cooper constant) were used to measure the thermal behavior of the PCM inside the tubes. A Binary eutectic composition ($CH_3COONa{\cdot}3H_2O$-$HCONH_3$)) was used as phase change material in a ratio of 3:1. This eutectic PCM was developed by Takahiro Wada (1983), and possess a melting temperature of 40.5°C, with a fusion heat of 255KJ/kg. 99% composition purity was used in this work. The PCM charging and discharging process period occurs as follows: when the absorption plate is heated up by the solar radiation, the electric fans are activated pulling the heated air from the upper duct into the PCM heat exchanger chamber, at this moment, the valve C is closed to procure a direct air flow into the PCM heat exchanger. Then, the hot air circulates through the serpentine PCM chamber and the 12 stainless steel bars that contain the eutectic composition, until it reaches the left side of the collector. At this moment, the valve B is closed and the air is redirected to the absorption plate to be circled again.

The release process is merely a circuit between the greenhouse internal air and the PCM heat exchanger and it occurs in a similar way like in the collecting process. After the total amount of PCM contained in the stainless steel tube change its phase (solid-liquid), the three valves are closed and reopened during the night time when the heat stored is need for greenhouse heating. Then, the valves C and B are opened; the electrical fan pulls air into the chamber and exits by the valve C to the greenhouse environment. Three more thermocouples (T-type) were set at the absorption plate inlet and outlet to measure their respective temperatures. A data acquisition system was used to record the different parameters during collection and release cycles with 20 min intervals, the properties recorded were: ambient air temperature, humidity, inlet and outlet temperature and solar intensity over the absorption plate surface.

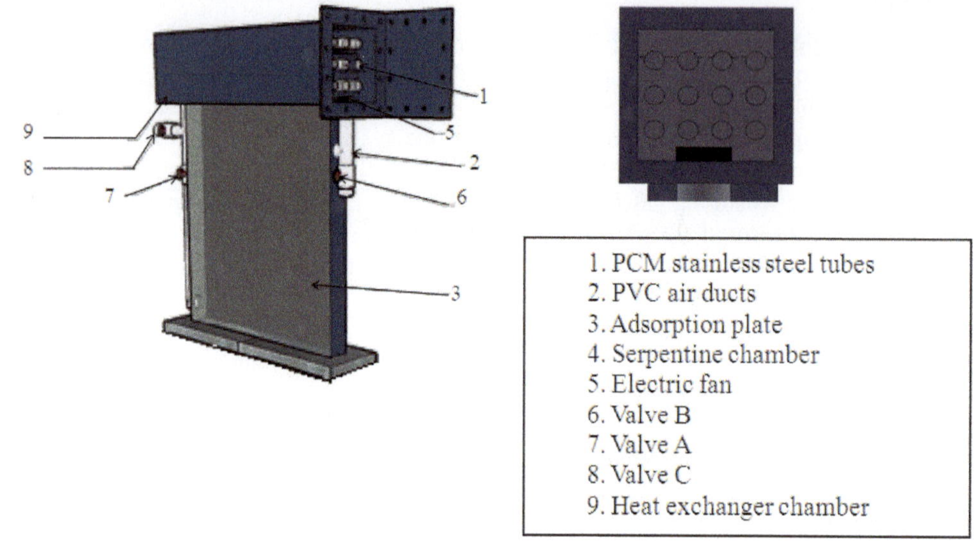

Figure 1. Solar COLLECtor

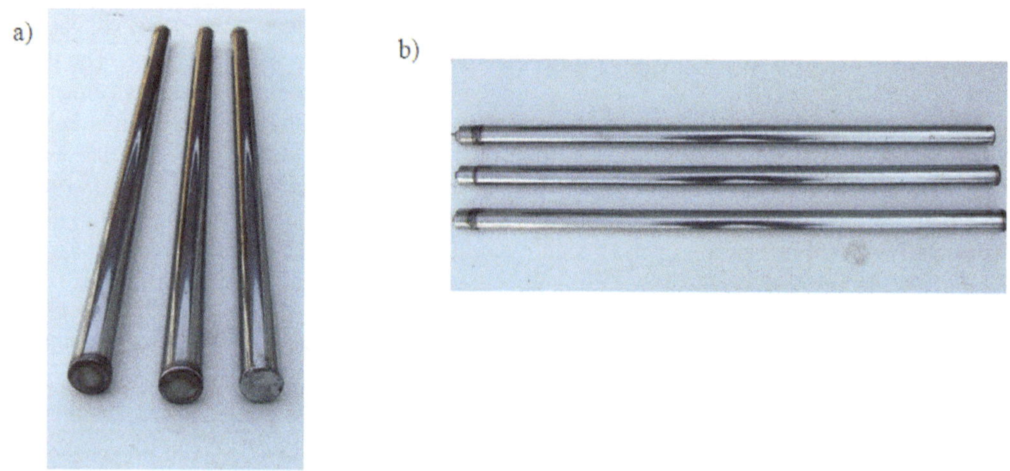

Figure 2. PCM containers (stainless steel)

3. Measurements

Experiments were conducted during the months of January and February of 2012 in the city of Beijing, China. The heating system was designed to be set over a north wall, located inside a Chinese solar green house in order to act in a similar way as a thermal wall. A Chinese solar greenhouse positioned towards south, with a base dimension of 8 meters, a north wall height of 2.8 meters and a width of 80 meters, was subdivided into polyethylene plastic film separated sections for system testing. Figure 3a and 3c shows the inside facilities and the north wall used in this experiment. Experiments were carried out to study the thermal behavior of a eutectic composition ($CH_3COONa{\cdot}3H_2O$-$HCONH_3$) and the possibility of heating application for Chinese solar green houses. The experiments were carried out at the same time period from 9:00am-15:00pm during test days. A data acquisition system (CR1000 data logger, Campbell, USA CMP3, kipp and zonen, USA) was used to record the different parameters during the heating and realizing cycles at an interval of 20 min. The properties recorded were: ambient air temperature, humidity, inlet and outlet temperatures and solar intensity over the adsorption plate. Different air flow rates of 0.2kg/min, 0.6kg/min and 1.0kg/min were tested as well. The temperature variation at the heat exchanger inlet and outlet with different air circulation rate and solar radiation measurement for different times during the days are showed in Figure 4. There was a similar finding on solar radiation and solar irradiance during test days from 10:30am-13:30pm equal to 600W/m^2.

a)

b)

c)

Figure 3. Chinese solar greenhouse

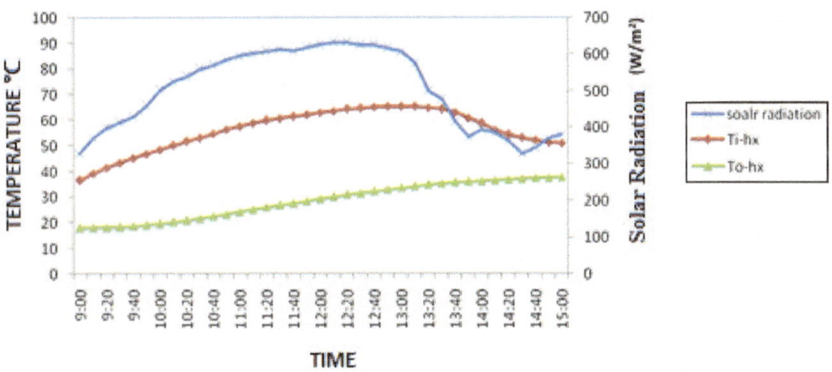

Figure 4A. Air temperature variation for different air flow rate, 0.2 kg/min

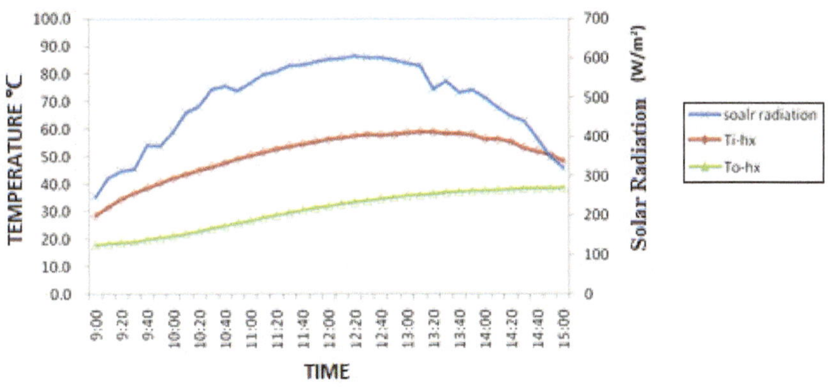

Figure 4B. Air temperature variation for different air flow rate, 0.6 kg/min

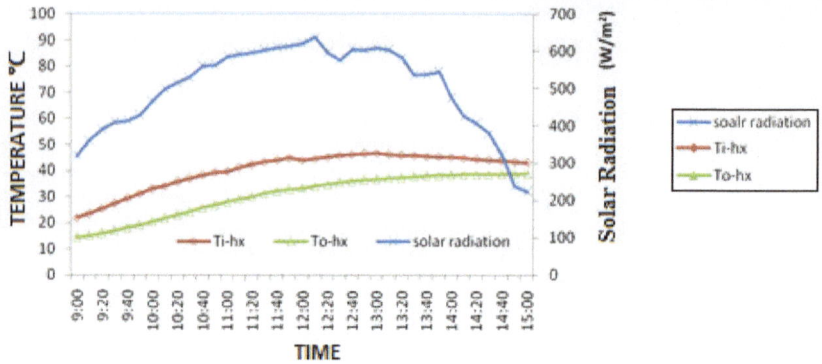

Figure 4C. Air temperature variation for different air flow rate, 1.0 kg/min

4. Energy Analysis

4.1 PCM Analysis

The collector unit and the PCM were tested to study its thermal performance and efficiency. The data collected was used to draw the efficiency curve of the heat exchanger and the collector unit at different mass flow rates.

Energy efficiency during the charging period is defined as the ratio of heat stored in the PCM to the total solar energy received.

$$n = \frac{stored\ heat\ in\ PCM}{Solar\ energy\ input}$$

The total energy efficiency obtained from the absorption plate during charging period $n(t)$ can be described by following equation:

$$n(t) = \frac{Q_S(t)}{Q_t(t)}(100) \quad (1)$$

A heat exchanger is a piece of equipment built to make an efficient heat transfer from one medium to another, as heat is transferred by convection form, there are PCM bars with air circulating inside the surroundings of the serpentine chamber. The heat supplied by the heat storage unit during the discharging period can be calculated by the following equation:

$$Q_r(t) = V\rho c_p\left[T_o(t) - T_i(t)\right] \quad (2)$$

The system COP during a discharging period can be described as the heat supplied by the heat exchanger to the energy consumption of the fan.

$$COP = \frac{Q_r(t)}{Q_d} \quad (3)$$

5. Results and Discussion

Few studies have been carried out on the application of solar concentrators and PCM heat exchangers for greenhouse air heating systems, as the literature review demonstrates, there is not much information available, especially for application in cold regions. A low mass air flow can get a lower outlet air temperature and a bigger temperature difference between inlet and out let. As was demonstrated when the air flow rate was increased from 2.0 kg/min to 0.6 kg/min, the heat collection of the PCM heat exchanger was less, and a much better performance was achieved when it was reduced from 1kg/min to 0.2kg/min. Figure 5 shows the hourly amount of heat collected in accumulator for different air flow rates. Due to the efficient exchange in the PCM heat exchanger chamber, the temperature recorded at the outlet was always lower during the heat collection cycle than at the inlet, what means that the PCM could easily collect the heat from the air. Also, it was founded that the average highest net efficiency happens with a mass flow rate of 0.6 kg/min. Figure 6 shows the net efficiency of PCM heat exchanger during test time at different mass flow rates.

Latent Heat Storage (LHS) has the advantage to store the heat energy collected during the day; this energy is released during night when heat energy is required to maintain an optimum indoor temperature. But in order to complete this process, first, it is necessary to convert the solar radiation into heat energy; such task can be accomplished by using a solar collector unit. In this work the solar collector unit was designed to be held over the north wall inside the Chinese solar greenhouse, during the day it helps covering a percentage of the heating requirement, and supplies the heat energy that needs to be stored into the LHS unit, being its main function during night time when the heating on demand was high. It was founded that the performance of this solar collector is greatly reduced during cloudy days, when is required to make use of alternative heating sources. The absorption plate design present a good performance but a better design could help to improve the overall efficiency of the system. The LHT unit and insulation system provides heat for about 4 hours. Although, it is recommended to be used in combination with conventional heating sources in order to achieve an appropriate temperature. Table 1 show the variations of indoor temperature of the green house and the temperature of PCM.

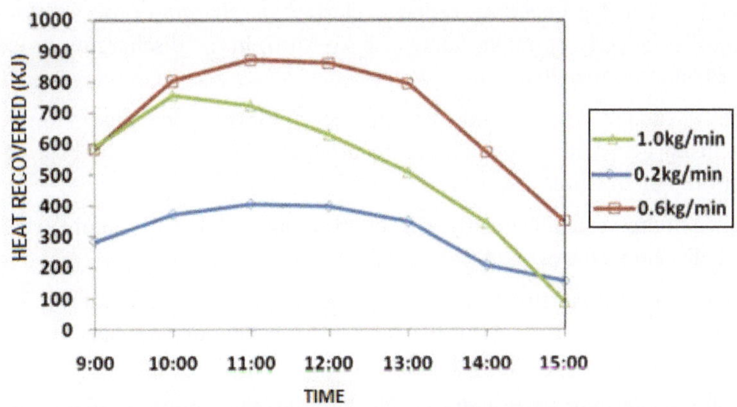

Fig 5, Variations of hourly heat storage in heat exchanger for different Air flow rate

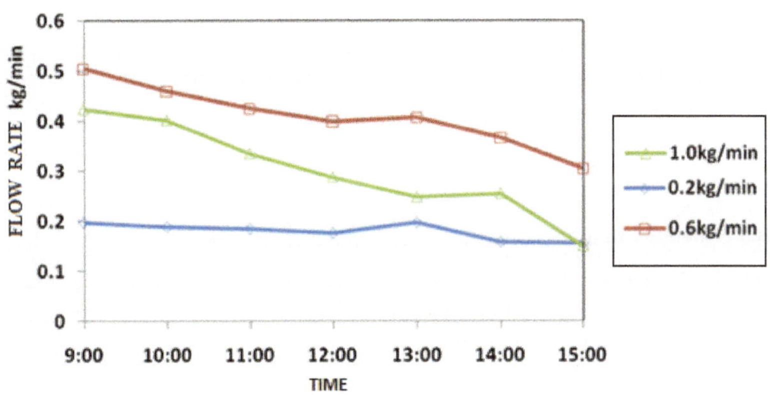

Figure 6. Variations of hourly heat efficiency for different air flow rate

Table 1. Amount of heat recovered

Mass flow tare	Recovered heat from the accumulator during the discharging period(KJ)	Total power consumption (KJ/24h)	COP
0.2kg/Min	1340.0	432.0	3.1
0.6kg/Min	2272.0	1296.0	1.8
1.0kg/Min	2070.0	2074.0	1.0

The amount of heat contribution and heat recover by the PCM heat exchanger of this system was calculated by using the Equation (2).

6. Conclusion

Experiments were carried out to investigate the performance of an integrated PCM solar wall model unit during the charging and discharging processes, this system represents an alternative solution to farmers in remote areas. The following conclusion may be drawn within the parameters investigated in the present study:

- $CH_3COONa{\cdot}3H_2O$-$HCONH_3$, an organic/inorganic eutectic mixture, is suitable for latent heat storage, which is another choice besides the well know calcium chloride hexahydrated and paraffin wax.
- The results of the present experimental work agree well with the one reported by Abdel-Monem and El-Bassuoni et al., who found that the pseudo-binary system with a ratio of 3:1 forms a eutectic mixture at 40.5 °C accompanied by large heat of fusion.
- Air flow rate is a key factor in the operating process. During the charging process, the system stored a maximum solar gain and most efficiently optimum value (0.6 kg/Min). In the discharging process, the duration of the Output Power depends on this parameter.
- The solar collector was capable to supply 4 hours of heat energy for the greenhouse. However, it is strongly recommended to combine this system with a conventional heating system to fulfill the heating demand during winter season in the northern region of China.
- The thermal performance of this system is greatly affected by cloudy days, and is more efficient in days with high solar radiation and high air temperatures.
- Material and design involved in the construction of solar collector plays an important role in its performance. It is suggested to separate the absorption plate from the bottom to let the air circulate over both faces of the collector, resulting in a better performance and air heat collection.
- Future studies should focus on: interaction between the large areas of built-in solar wall and greenhouse environment. This could help to increase the thermal performance of this unit.

Acknowledgements

The authors gratefully acknowledge the financial support from the Center for Protected Agriculture and environmental Engineering (CAAS). Also, special thanks to all the working staff in the green house array who collaborated in the construction of the system in the suburbs of the city of Beijing, China.

References

Beijing. (2008). *The Columbia Encyclopedia* (6th ed.).

Benli, H., & Durmus, A. (2009). Performance analysis of a latent heat storage system with phase change material for new designed solar collectors in greenhouse heating. *Solar Energy, 83*, 2109-2119.

Boulard, T., Razafinjohany, E., Baille, A., Jaffrin, A., & Fabre, B. (2003). Performance of greenhouse heating system with a phase change material. *Agricultural and Forest Meteorology, 52*, 303-318. http://dx.doi.org/10.1016/0168-1923(90)90088-N

Christian, V. Z. (2011). *Integrated greenhouse system for mild climates*. Springer.

El-Bassuani, A. M. A., Tayeb, A. M., Helwa, N. H., & Fathy, A. M. (2003). Modification of urea–sodium acetate trihydrate mixture for solar energy storage. *Renew Energy, 28*, 1629-1643. http://dx.doi.org/10.1016/S0960-1481(02)00052-6

Eman-Bellah, S., Mettaweea, & Ghazy, M. R. (2006). Assassab Experimental study of a compact PCM solar collector. *Energy, 31*, 2958-2968. http://dx.doi.org/10.1016/j.energy.2005.11.019

Ghoneim, A. A., Klein, S. A., & Duffie, J. A. (1991). Analysis of collector-storage building walls using phase-change materials. *Solar energy, 47*(3), 237-242. http://dx.doi.org/10.1016/0038-092X(91)90084-A

Hamid, E. Q. (2009). Numerical analysis of a coupled solar collector latent heat storage unit using various phase change materials for heating the water. *Energy Conversion and Management, 50*, 247-254. http://dx.doi.org/10.1016/j.enconman..2008.09.038

Huseyin, B., & Aydin, D. (2008). Evaluation of ground-source heat pump combiened latent heat storage system performance in greenhouse heating. *Energy and Buildings, 41*(2), 220-228. http://dx.doi.org/10.1016/j.enbulding.2008.09.004

Huseyin, B., & Aydin, D. (2009). Performance analysis of latent heat storage system with phase change material for new designed solar collector in greenhouse heating. *Solar Energy, 83*(12), 2109-2119. http://dx.doi.org/10.1016/j.solener.2009.07.05

Kern, M., & Aldrich, R. A. (1979). Phase change energy storage in a greenhouse solar heating system. Paper presented summer Meeting of ASAE and CSAE. University of Manitoba, Winnipeng, Canada.

Koca, A., Oztop, H. F., Koyun, T. K., & Varol, Y. (2008). Energy and exergy analysis of a latent heat storage system with phase change material for a solar collector. *Renewable Energy, 33*, 567-574. http://dx.doi.org/10.1016/j.renene.2007.03.012

Laurent, Z., Annabelle, J., Stephane, L., Yvan, D., & Daniel, R. (2012). Experimentakl study of small wall integrating phase change material. *Solar Energy, 86*(1), 208-219 http://dx.doi.org/10.1016/j.solener.2011.09.026

Levav, N., & Zamir, N. (n.d.) Phase change material for heat storage in greenhouse in: REUR technical series L: greenhouse heating with solar energy. CNRE, pp. 163-166.

Li, J. H., Zhang, G., & Wang, J. Y. (1991). Investigation of eutectic mixture of sodium acetate trihydrate and urea as latent heat storage. *Solar Energy, 47*(6), 443-445. http://dx.doi.org/10.1016/0038-092X(91)90112-A

Mahmud, M. A., Sopian, K., Alghoul, M. A., Sohif, M., & Ruslan, M. H. (2011). Review of solar air collectors with thermal storage units. *Renewable and Sustainable Energy Reviews, 15*, 1476-1490. http://dx.doi.org/10.1016/j.rser.2010.10.019

Öztürk, H. H., & Başçetinçelik, A. (2003). Energy and Exergy Efficiency of a Packed-bed Heat Storage Unit for Greenhouse Heating. *Biosystems Engineering, 86*(2), 231-245. http://dx.doi.org/10.1016/S1537-5110(03)00134-X

Sethi, V. P., & Sharma, S. K. (2008). Survey and evaluation of heating technologies for worldwide agricultural greenhouse applications. *Solar Energy, 82*, 832-859. http://dx.doi.org/10.1016/j.solener.2008.02.010

Wada, T., Kimura, F., & Yamamoto, R. (1983). Studies on salt hydrate for latent heat storage.Ⅲ. pseado-binary system, CH3CO2Na`3H2O-HCONH2. *The Chemical Society of Japan, 56*, 1575-1576.

Zalba, B., Jose, M. M., Luisa, F. C., & Harald, M. (2003). Review on thermal energy storage with phase change: materials, heat transfer analysis and applications. *Appl Therm Eng, 23*, 251-283. http://dx.doi.org/10.1016/S1359-4311(02)00192-8

Zalewski, L., Joulina, A., Lassuea, S., Dutilc, Y., & Roussec, D. (2011). Experimental study of small-scale solar wall integrating phase change material. *Solar Energy, 86*(1), 208-219. http://dx.doi.org/10.1016/j.solener.2011.09.026

2

Concentrating Solar Power to Be Used in Seawater Desalination within the Gulf Cooperation Council

Mohammed Saleh Al. Ansari[1]

[1] Department of Chemical Engineering, College of Engineering, University of Bahrain, Kingdom of Bahrain

Correspondence: Mohammed Saleh Al. Ansari, Department of Chemical Engineering, College of Engineering, University of Bahrain, P. O. Box 32038, Sukhair Campus, Kingdom of Bahrain. E-mail: malansari@uob.edu.bh

Abstract

The research discussed in this paper is intended to shed some light on the potential that exists when concentrating technology of solar thermal energy for use in Gulf Cooperation Council Countries, or most commonly referred to as GCC. The research provides the opportunity to study the technologies related to the concentration of solar power technologies along with water desalination demands and fresh water availability, which in turn leads to the jeoparadization of water resources that are derived from the ground. Desalination plants that are solar powered, within the Gulf Cooperation Council countries comprise of Qatar, Saudi Arabia, Kuwait, Oman, UAE, and Bahrain. The study recognizes that a political structure is required for introducing the premier foundation of solar power and desalination as well as ensuring comprehension of the proposal.

The study evaluates solar energy accessible resources and the costs of integrating this alternative energy source, desalination of water, long standing scenarios of integrating power production technologies into the water sectors, and quantifying the socio-economic and environmental impacts of this alternative energy notion.

Keywords: Concentrating Solar Power (CSP), Gulf Cooperation Council (GCC), power, water, desalination plants

1. Introduction

The practice of solar desalination has received a massive amount of attention from the engineers that operate within small scales supplying water resources to remote areas of the world, specifically, those areas that are considered to be rural, based on their distance away from other sectors of society. Desalination is best described as expansive technologies whose sole intent is to produce energy intensive technologies and fresh water resources. Desalination has environmental consequences that are rendered in various forms, pollutant emissions that are derived from energy consumption contributes to the dangers associated with global warming along with brine additive discharges and chemicals from outfall culverts into the seas (Rubin ES, 2007). Concentrating solar power, commonly referred to by its acronym term CSP, is a specific technology that supplies energy for seawater desalinations within ranges of an expected solution. The increasing world population steadily helps with the increasing demand for water resources depletion of naturally attainable groundwater sources within the arid global regions needing remedies that are affordable, secure, and match the surroundings, creating a sense of sustainability. The remedies should have the capabilities of dealing with the magnitude of demand with the foundation being accessible along the establishment of expertise, as further approaches avail to remove the vagueness from methodological advances-in case further solutions are not ascertained in time-endangering entire areas (Rubin ES, 2007).

2. Global Horizontal Irradiance within GCC Countries

According to recent studies, there is enough solar irradiance meeting the specifications of green based electrical needs within multiple GCC countries. According to Rubin, ES, R & D is a requirement for developing technological needs (Rubin ES, 2007). There are plethora of motivations for large-scale implementation of CSP desalination systems, known as the basis of energy storage and bio-fuel hybrid operations. Concentrating solar power plants will avail continuous firm capacities that are fitting for scale desalination through thermal processes or membrane processes. Desalination plants that integrate technologies with regards to CSP will be recognized by their ability to produce large units that reaches to 1000000 m^3/ per day. The energy produced by the solar based thermal power stations will be more to economically priced electrical production cost. Efficiency is another factor that is yielded with the perception of managing water use, improving distribution and irrigation systems, reusing wastewater responsibly keeping sustainability measures in mind, should avert 50% of the long term GHC region deficits. Solar desalination and proficient amounts of water use in large scales, forces groundwater over-exploitation within the GCC regions to be forced to come to an end. Figure 1 shows The annual global horizontal irradiance, global irradiance in GCC countries.

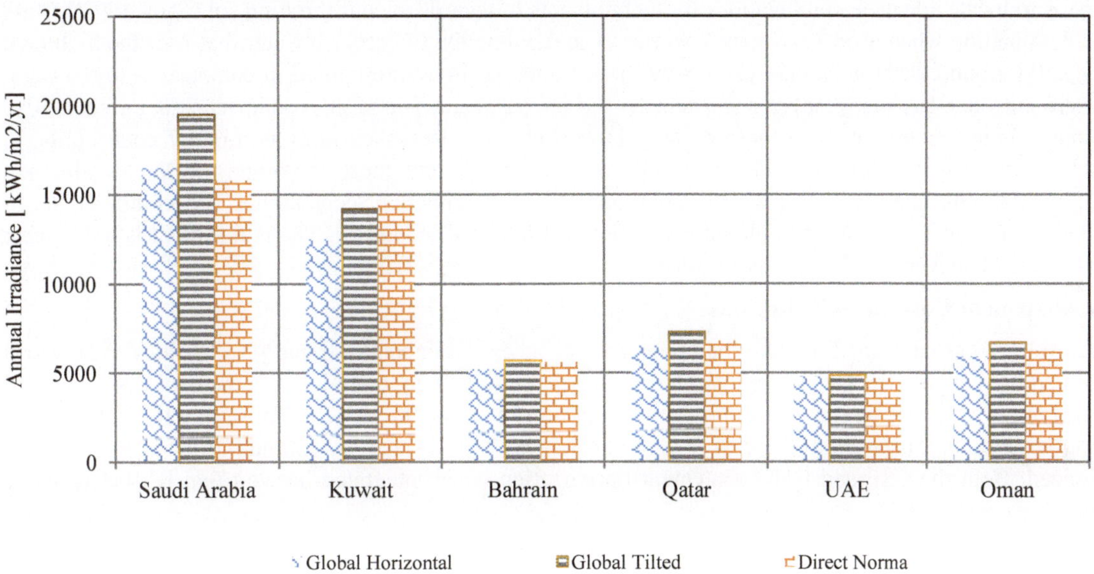

(a)

Bahrain	Area (Km^2)	711
	Population	1,323,535
	horizontal irradiance ($kwhr/m^2/yr$)	5500
Saudi Arabia	Area (Km^2)	2,149,690
	Population	28,082,541
	horizontal irradiance ($kwhr/m^2/yr$)	20000
Kuwait	Area (Km^2)	17,820
	Population	2,818,042
	horizontal irradiance ($kwhr/m^2/yr$)	14000
UAE	Area (Km^2)	83,600
	Population	7,890,924
	horizontal irradiance ($kwhr/m^2/yr$)	5000
Oman	Area (Km^2)	309,500
	Population	2,846,145
	horizontal irradiance ($kwhr/m^2/yr$)	6000
Qatar	Area (Km^2)	11,440
	Population	1,870,041
	horizontal irradiance ($kwhr/m^2/yr$)	6100

(b)

Figure 1. (a) The annual global horizontal irradiance, global irradiance in GCC countries; (b) Map of GCC countries (Kingdom of Saudi Arabia, Kingdom of Bahrain, State of Kuwait, State of Qatar, United Arab Emirates and Sultanates of Oman with the Area of the country and population 2010 and horizontal irradiance yearly

Recent developments suggest that utilizing superior solar powered desalination techniques through the utilization of horizontal drains seabed intakes and nano-filtration shall avert the majority of environmental consequences that are a direct result of desalination. With the support of Europe, the countries included in the GCC nations should commence to establish constructive political conditions and legal frames for introducing the market of solar power technology concentration for electricity and seawater desalination (Rubin ES, 2007). For this reason, this paper is composed to shed additional light on indicates that move towards a sustainable remedy to the threatening water crisis within the GCC region, illustrating a means to attain balanced, affordable, and safe framework for water supply to coming generations that are typically overlooked in modern strategic evaluations.

The scope is to locate adequate combinations for technologies in regards to concentrating solar power (CSP) and seawater desalination when used as an energy source. The desalination of brackish water that uses the high total dissolved solid ground water is another choice, but because its resources are limited in comparison to seawater, utilizing groundwater has many adverse impacts on the environment. Brackish groundwater as an alternative energy source is not an option that should be ignored entirely, but it is taken to be non-major contribution to sustainable fresh water supply systems. Seawater desalination is the most important of the desalination technologies, defining and evaluating several combinations of desalination technologies and combinations of CSP under diverse environmental and climatic situations within GCC countries. The main intent is to evaluate GHG as COseq whenever CSP has an involvement.

3. GHG Abatement Costs in GCC Regions

The cost of GHG was calculated, finding the importance of its environmentally friendly technology to be worth noting. When the cost of GHG abatement is referred to per ton of CO_2eq, there is a five percent cost reduction that applies to the behavior on the operation, following a trend to move negatively, but saving the yield. When CSP along with the optimization of the cost of GHG abatement is integrated, a good output for technology can be recognized. Both the CSP and GHG abatement optimization when integrated, move to an R^2 that is above ninety eight percent.

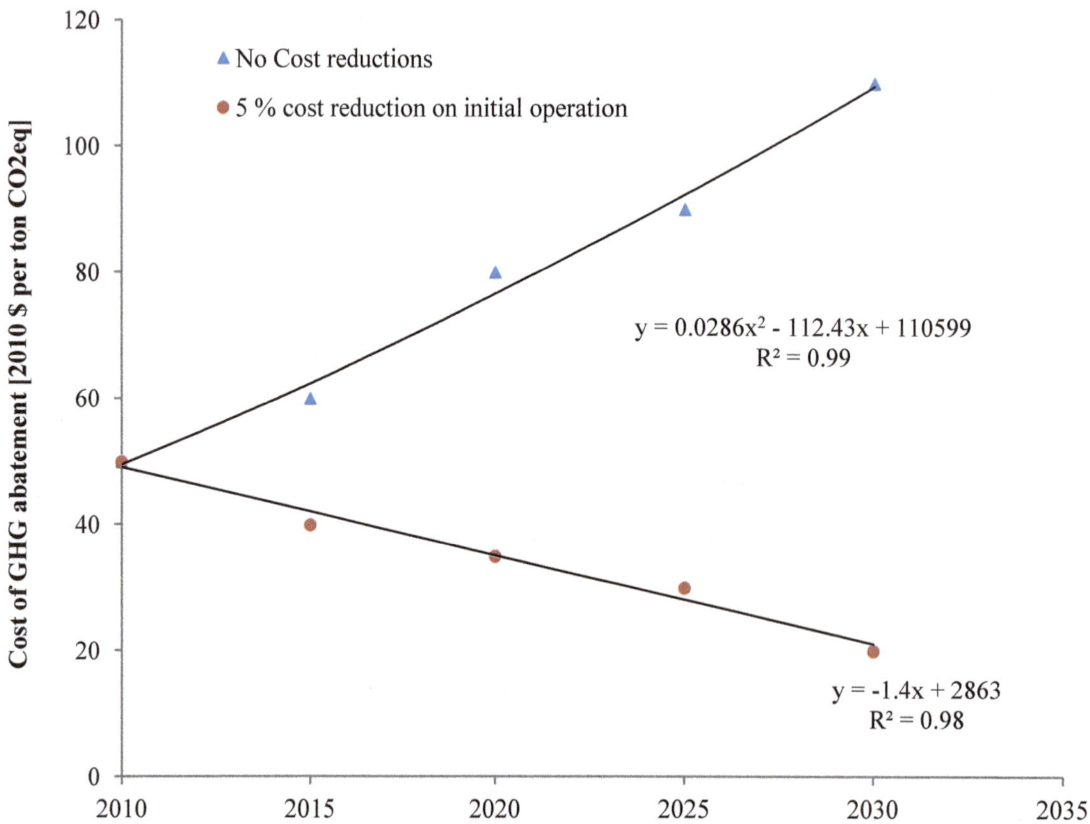

Figure 2. Cost of GHG abatment in GCC

Utilizing a five percent reduction on the initial costs would help benefit GHG abatement by twenty percent, if the policies were fixed towards utilizing CSP technologies by the year 2030 as per Figure 2 that shows the cost of GHG abatement in GCC. The implementation of one type of renewable energy source would have an influential output on the cost of this technological solution, which in turn would also benefit the environment.

4. The Pre-Selection Process of Desalination Technologies in Regards to Seawater Implementation

There are many different types of desalination technologies that are available and utilized within GCC regions. The GCC countries are respectively located in an area where seventy three percent of the world's desalination plants are housed. Some of these plants are applied in large scales, specifically for this region.

Thermal desalination techniques are used to evaporate seawater by utilizing heat that is produced from surrounding power plants. Other desalination techniques use a mechanical approach where filtration is done through membranes. Compression technologies that utilize vaporization are done in conjunction with thermal distillation to increase the productivity of desalination processes.

Multi-Stage Flash Desalination, commonly referred to by its acronym term MSF, is a specific type of thermal desalination unit that comprises of condensing and evaporating product water. The condensation and evaporation phenomena's take place when they are coupled with each other during various stages where latent heat is recovered to be reused for preheating seawater when its concentration levels exceed 48000 ppm. Within what is referred to as a brine heater, incoming feeds of water are heated to their maximum temperatures, through condensing steam that is saturated from the cold ends of steam cycled power plants from other heat sources. The heated seawater will then flow into the initial evaporation stage, where pressures will be set to a lower standard. The abrupt introduction of the heated water within the chamber combined with low pressures forces the water to boil at a quick rate, forcing an explosion, or flashing of steam. During this process, there is only a minute percentage of water that is converted into vapours, based on the amount of pressure that has managed to remain maintained during this stage. Further boiling will persist until the water has the chance to cool down to an equal boiling state, supplying the required heat for vaporization.

Vapours that are generated through flashing will become condensed within tubes of heat exchange that run throughout the upper section of the stage. The tubes will be cooled by the feed water as it comes in and then transferred into the brine heaters, which will preheat the water as well as recover part of the energy released, used for the evaporation within the first stage. The process gets repeated until forty different stages are done, but twenty stages are normally the most that are employed. To ensure that energy recovery and maximization of water is employed, every stage within the MSF unit should operate at lower pressurizations.

Another system, known as the Multi-Effect Desalination, or MED for short, is another type of thermal desalination process, but it utilizes different configurations. During this process, feed water is dispersed onto the tubes that are located in the evaporator, shortly after the evaporator has gone through a preheating session that takes place along the top section of each of its chambers.

When applied to the cold ends of steam cycled power plants, MED plants that do not contain TVC will have a heat consumption that comprises of 0.2-0.4 MJ/kg that comes in the form of steam process that is less than 0.35 -0.38 bars when it is withdrawn from steam turbines.

Reverse Osmosis, or RO for short, is a desalination unit that utilizes membrane separation processes to recover water that has been permeated from pressurized seawater to points that are greater than the osmotic pressures of the seawater, and can reach up to eighty bar pressure. The way that this process works is actually considerably straight forward. The membrane filter will hold all of the salt ions that are present in the pressurized solution back; this allows the additional water to pass through. The RO membranes do have sensitivities when it comes to oxidizers, pH, algae, organics, bacteria, and fouling depositions.

Pre-treatments of the water feeds is a vital processing step that can have an impact on energy costs and consumption in RO processes. This is especially true due to the fact that every portion of feed water that is utilized, even if the water will later be discharged at the end of the process, will need to be treated before it is able to pass through the membranes. Micro, nano, and ultra filtration processes have been proposed as alternative methods to avoid chemical pre-treatments of raw water, avoiding contamination of the additives within the seawater from affecting any plants surrounding. The pre-treatment of reverse osmosis involves the removal of all dissolved gases as well as ensuring that the pH has been stabilized in addition the salts and other dangerous substances must be removed from the brine as well.

MSF and MED are two primary contenders in terms of electricity consumption and energy consumption. MSF has a lower cost, but MED showcases better performance whenever primary energy is the topic of debate. The temperature that MED operates at is lower when steam at lower pressures within steam cycle thermal power plants hydrides with co-generation. Due to this fact, the combination of MED desalination with CSP is more productive and effective than a MSF and CSP combination, whenever the raw feed water being used is seawater.

5. Thermal Vapour Compressions

Vapor compressions have an added effect on distillers to help improve their efficiency. Compression processes that utilize vapors rely on reusing vapours that are produced by distillers for heating steam after the second stage of compression has commenced. Vapours that are produced during the first stage will be recompressed partially again inside of a compressor, and utilized to apply heat onto the initial cell. Vapors are compressed in two different fashions. They can be compressed by utilizing mechanical compressors, or they can be compressed with steam ejectors. In regards to thermal vapor compressions, steam is used at high pressures and then withdrawn from other processes to produce industrialized processed steam.

Thermal vapour compression can be used to help increase the efficiency of the MED desalination process; however it does require steam to be saturated or wet steam at higher pressures if it is being connected to steam powered cycle. Reverse osmosis requires lower consumptions of electrical energy and the costs are done per units product in terms of brackish water being consumed, this of course, is in comparison to the mechanical vapour compression method. Whenever low grade energy is being used as a type of useful energy, thermal technologies for desalination would positively move towards an adoption of MED and MSF. CSP, otherwise known by its extended name, concentrating solar power technologies has grown to become the hub whenever concentrating solar thermal power generation is being discussed or applied, because this method is the most reliable and abundant renewable energy resource within the GCC region. Additional desalination concepts that are based on geothermal energy, wind energy, plus biomass and other sources are additional promising sources that contribute to the supply of freshwater from seawater. Due to the fundamental properties of CSP, it is believed, this technology will provide large scale desalination seawater plants for individuals residing within the GCC countries.

The energy harvesting that is taking place each year on GCC land is producing a massive amount of solar energy that is equal to 1.5-1.8 million barrels of crude oil, per each km^2. One CSP plant that covers square kilometer of the desert land can deliver power and energy to desalinate 6.9 km^3/hr; this is equivalent to the consumption of one enormous desalination plant. The parameters of design for solar power in regards to renewable energy will construct CSP, a key technological source that is required in the future of the renewable energy circuit. CSP is also a key resource of energy in regards to desalination of seawater in GCC countries. CSP has the ability to satisfy the demand for power capacity; its natural resources are easily obtainable and never show obvious signs of restraint. Economically, CSP requirements will be governed by the demand of the market, within electrical sectors of GCC countries that possess high to moderate solar irradiance.

6. Concentration on Solar Power Used for Thermal Turbines

The time series of parabolic plant design utilizes synthetic oils to transfer energy into steam generators within the cycle of a power plant. The shade that is provided by the Fresnel segments are a valuable additional service that is granted by the plant. The plants provide cover for stores, buildings, common lots, and protecting certain crops from an extensive amount of sunshine, as well as reducing water consumption that is required for irrigation. The series of time that is represented in the following figure, reaches a point of over 110000 m^2, which equates to a total of 48 MW from solar heat, along with 1245 W/m^2 of normal direct irradiance. This is all transported from linear Fresnals from parabolic trough collector fields located over the areas mentioned above. To positively generate electricity, fluids flowing throughout the absorber tubes-using either synthetic oils or steamed water-will transfer the heat to conventional steam powered turbines (Tamme et al., 2005).

The storage of heat can consist of either the utilization of two large style tanks, each that contain a mixture of molten salt nitrate as a medium for stage with the ability to sustain heat capacities for hours when full load operations are being performed on the turbine. The heat will then be transferred either from or to the transfer fluids that collect via an additional heat exchanger mechanism. The liquid salt will be pumped directly through the heat exchanger within the cold tanks into a hot that during the charging periods and the opposite will occur during the periods of discharge. The concentration of solar thermal powered generation is the most abundantly utilized and reliable sources of renewable energy within the GCC countries. But, other types of desalination should not be dismissed, simply because this method is seen to be more favorable.

The concentration of technologies that are solar thermal powered is based on the existing concept that when solar radiation is concentrated at higher temperatures to produce heat that electricity can be generated without the need for conventional power cycles. For this to occur, a lot of the systems will employ the use of glass mirrors that will help to track the position that the sun is presently at. Contradictory to most common beliefs, annual sums of solar irradiance that is delivered directly along the tracking surface of the sun within desert regions is typically higher than those globally, that are located along fixed surfaces that possess either a tilt southward or horizontally.

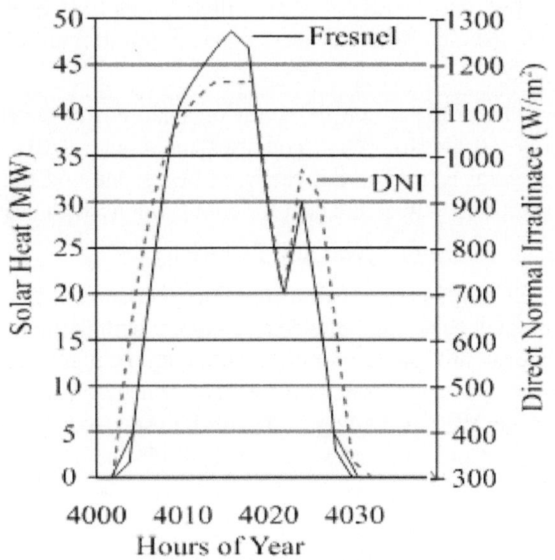

Figure 3. Time Series of solar heat as MW and Direct Normal Irradiance as W/m^2

In instances where CSP is being brought further into the picture, sunlight will be focused on receivers that are designed to help reduce the amount of heat that is lost during its capture. Fluids flowing throughout the receivers will help take the heat that is being emitted from the sun away, forcing it to set along the thermal power cycles; this is where high pressures and steam that is of high temperatures can be generated to help the turbine drive. Additional transfer fluids such as water, air, oil, as well as molten salt can be utilized during the transfer.

7. CSP for GAS Powered Turbines

Before the year 2030, GCC countries should anticipate that the amount of ground water will be depleted, therefore, all efforts to use waste water that has been treated should be integrated as a cost saving technology. Engineering and economically based technologies make water consumption utilizing CSP technologies the main target for GCC countries. Solar towards utilize large fields of two-axis tracking mirrors known as heliostats to reflect sunlight to central receives located along the top sections of a tower. The CSP is then converted into high temperatures of heat. The optical emblematic concentration factors range from 200 to 1000. Plant sizes range from five to one hundred fifty MW or electrical production, with these figures being feasible. High solar fluxes impose the receive, permitting work at elevated temps exceeding 1000 degrees Celsius, combining thermal energies into steam cycles then into gas powered turbines and combining cycles. Solar towers that possess central receiving systems are also compromised, which integrate fossil plants for hybrid operations in wide varieties of options, possessing the potential to proactively generate electricity with increased annual capacity factors when thermal storage is used. Higher efficiencies can be reached when gas turbines that are solar-heated are used. This adoption has the possibility of increasing supplementary within cycle processes that are combined. The quota of persons residing in Arab and GCC countries averages 320 m3 per person per year and the Arab nation is 1000 m^3 capita. These figures are stressed by water consumption areas that are lacking renewable resources for agricultural output; this could in turn lead to a food threat in regards to harvests within the area. There is a gap between supply and demand within the area; desalination is the only way that this gap can be filled sufficiently.

Solar towers can be utilized to help achieve extremely high operating temperatures that can exceed 1000-1200 °C, this helps to enable them to produce heated air that can be utilized for the operations performed by gas turbines. These gas powered turbines can be combined with additional cycles that will yield extremely high efficiencies of conversion within the thermal cycles that equate to more than fifty percent.

Concentrated solar energy that is high temperature can be used for generating electricity and also assisting with processing heat. In this type of scenario, the amounts of energy inputs can be used with an efficiency that totals eighty five percent; Additional applications that combine production of industrialized heat, direct cooling measures, along with seawater desalination.

All of the CSP desalination concepts that exist, aside from one have the same exact perspective to further expand the amount of time that solar operation base loads need, by utilizing a thermal energy storage source and also larger collection fields. To be able to generate what equates to one megawatt hour of solar powered electricity, a land space of 4 to 13 m^2 is a requirement.

The primary characteristics that will make the concentration of solar power a primary technological future advance will require that key energy resources for GCC countries are able to: deliver enough power capacity as demand requires, ensure that the natural resources can easily be accessible and that the amounts of them are unlimited, can be combined with additional power and heat for cooling along with desalination.

8. Concentration of Solar Power for Combined Heat and Electricity

With price fluctuations in fossil fuels, a requirement for a feasibility study of different alternative power plant concepts for producing on-site combined generation of electricity along with thermal for MED desalination units and absorption cooling became apparent. Further worth mentioning is that absorption chillers have been exploited for base load operations, while compression chillers are primarily used for intermittent demands of desalination and electricity. Cold water district cooling grids will be used to distribute cooling power from the plant to the end users. This integrated process makes less than a 35% fuel input, because of the enhanced efficiency of solar fuel and the combined generation of it.

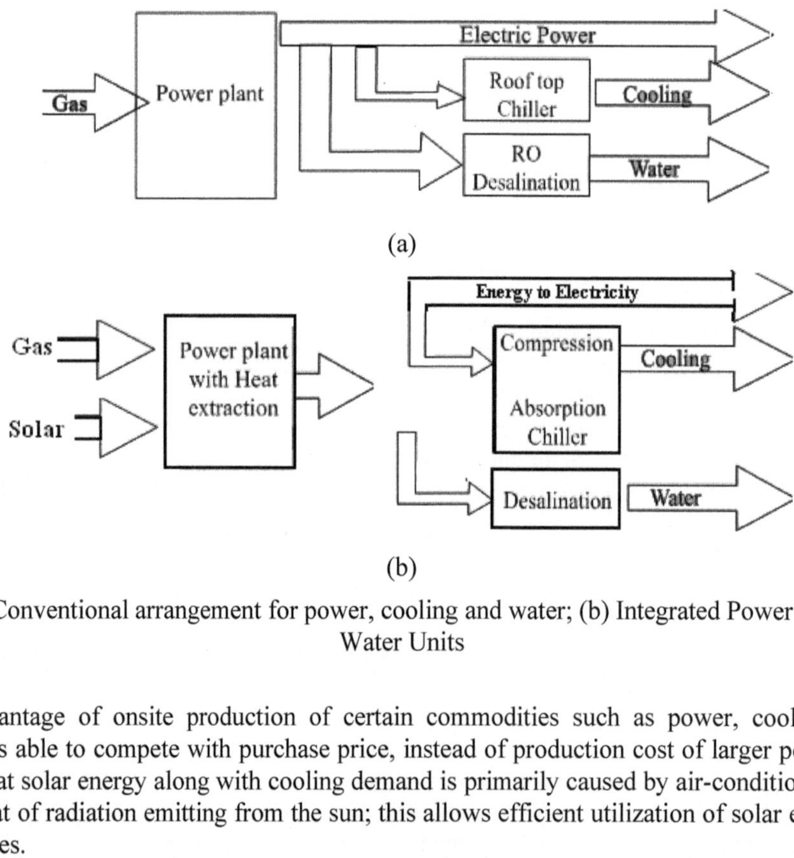

Figure 4. (a) Conventional arrangement for power, cooling and water; (b) Integrated Power & Cooling and Water Units

An obvious advantage of onsite production of certain commodities such as power, cooling, and water is production cost is able to compete with purchase price, instead of production cost of larger power plants. There is coincidence that solar energy along with cooling demand is primarily caused by air-conditioning, which is due to the intense heat of radiation emitting from the sun; this allows efficient utilization of solar energy, saving fuel for peak load times.

9. Concentration of Solar Power for Large Seawater Desalination

Concentrating solar power plants, CSP, can generate energy that can be used in membrane desalination, which takes place during reverse osmosis or RO either with seawater or brackish. The challenge that arises is ensuring smooth operations of seawater RO, which can guarantee continuous production of desalinated or permeated water; Thermal desalination methods, such as MSF or MED are potential hybrids powered by CSP that are directly connected with co-generating power and electricity.

In order to supply constant energy and power, desalination plants that use PV electricity or wind would need to be coupled with an electricity grid for added backup. Discontinuous operations that are allowed for high power capacities of win and PV energy would need to be installed and allowed to produce the same amounts of water and electricity. Renewable shares provided by CSP are 91%, PV renewable shares are 25% and wind powers renewable shares are 37%. Based on the conditions within different GCC locations, these numbers are considered to be the average performance for said systems.

CSP plants are able to save fuel when they are installed to run at capacity, in comparison to other types of renewable sources of energy, such as wind and PV for desalination. Instead of utilizing a backup power technology, the electricity that is generated by these systems can be stored in the form of batteries, hydrogen energy storage, or hydro-pumps, thus providing a continuous flow of power to desalination. If this is the case, additional electrical storage capacity that is required by CSP would be minor, the additional storage that is required for wind and PV power, would steadily increase system costs.

Further concentration will be given to CSP technology in regards to membrane desalination and thermal technologies, descriptions of economic and technical performance of larger scaled systems of this caliber combined with generated power from seawater that has been desalted will be further described in comparing the technical performance.

10. Comparing Technical Performance of Desalination Technologies

In order to be able to correctly compare MED and RO in combination with CSP in regards to different sites, both of the systems were designed to request an identical demand of water and electricity. By calculating the required amount of thermal energy input and the size that the solar collector field would need to respectively be, evaluations of the differences that exist in the systems performance can be further documented.

The vital design parameters are different for each site, due in part by the different nominal performances produced by the solar field, this varies based on the solar incidence angle, which is defined by the fields latitude, the different salinity found in the desalted seawater strongly influences how RO performs. Additional parameters such as the relative humidity and ambient temperature are also core influences of plant performance, having a specific effect on internal energy requirements and efficiency that are required for plant operation. The influence introduced by relative humidity and ambient temperatures do not have as much influence on the systems performance as that of the salinity of seawater and solar irradiance.

Both of the systems should deliver the product water while satisfying the WHO standards, allowing the salinity maximums of 200 ppm. This will require a reverse osmosis plant that utilizes a multi-pass approach, due to the fact that the single stage reverse osmosis process does not produce adequate final salinity figures. However, typical MED plants deliver water that is 10 ppm. The end product is not potable, requiring adequate proportioning of salts and minerals for human consumption. Electrical energy is considered to be a necessary by-product. Plants are designed identically, where the water output and the net power are the same, but the size of collector fields and fuel consumption will vary according to other requirements.

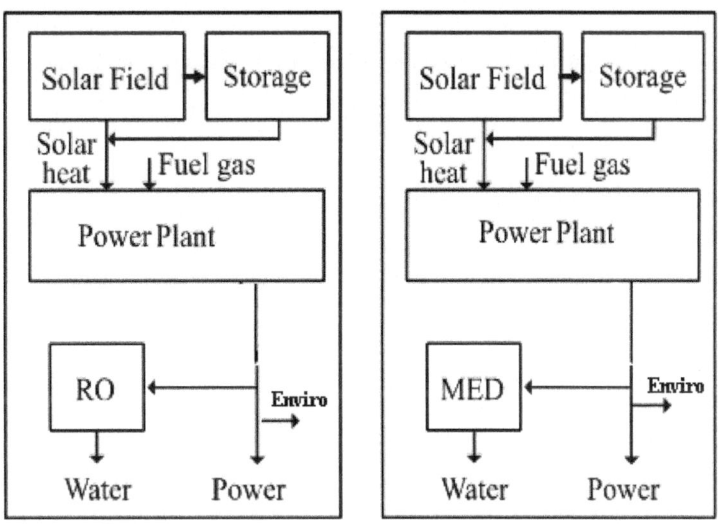

Figure 5. Different arrangements for desalination powered by CSP

There has been a massive amount of literature composed that compares thermal desalination and Reverse Osmosis, commonly referred to as RO. Most of the compositions come to the same conclusion that the thermal desalination process is not as energy efficient as the RO process. A further evaluation performed on this principal study brings up an opposing conclusion in regards to combining full scale CSP plants. In seven different cases performed on the GCC region, when CSP and MED processes are combined, four and eleven percent less input energy is required in comparison to when CSP and RO processes are combined.

The reason why there is a contradiction in the studies is due to the fundamental differences of the design target of solar power and conventional systems: with reliability being given to finite, polluting and expensive fossil energy sources, the conventional system is designed to yield optimal energy conversion from fuels into energy that can be used, which maximizes the electrical output of plants with their given fuel inputs (El-Nashar, 2002). Systems that are powered by the sun are made to maximize solar shares of given energy services. The efficiency of conversion is secondary; competiveness and economic performance are equivalent when both systems are concerned. Considerable reductions of global fuel consumption-a primary target of sustainability- is minimally achieved by increasing the efficiency of conversion, but is fully achieved when the shares of renewable energy sources are increased to 100%; The efficiency of the combined generation, producing two products that are extremely valuable, such as water and power-is extremely high.

Desalination plants are normally operated at constant loads. Part loads can cause problems in regards to fouling and scale performance, this also reduces economic attractiveness of respective projects. Solar energy can only be captured during the daytime and is reduced significantly on days that clouds are present in the sky. Directly coupling MED plants with CSP will require storing thermal energy that is taken from solar inputs or hybrid operations with bio or fossil fuels to allow continuous operations to commence. A reverse osmosis plant that is connected to grids can help compensate for the fluctuations of solar input levels by pulling electricity from other sources that are accessible on the grid.

Electrical transfers from remote CSP plants to RO plants located along the shore can produce a loss in electricity. Putting the CSP plants in hot desert climates will require dry cooling procedures to be carried out, this could lead to colder end temperatures along the steam cycles, which would emulate MED and CSP combined processes. In this type of case, heat would be dispersed into the desert, instead of being reused for desalination. Instead of losing the efficiency of processes by ineffectively cooling, its favorable to place CSP plants near the shores even though lower solar irradiance can occur.

Directly coupling MED with CSP has specific benefits: primary energy consumption will be drastically reduced, with this reduction, environmental impacts caused by the plant decrease. Integrating plants in this fashion is extremely attractive to large consumers, due to the fact that onsite operations are extremely competitive amongst water and power purchase prices of external sources.

RO and MED process design are currently in phases of dynamic development in regards to efficiency gains, cost reduction, reduction of environmental impact, and material enhancement (Alarcon et al., 2007). Combining desalination technologies along with CSP is in a premature stage within the feasibility analysis; there are no plants that fit this description that are presently operable at the time.

All of the desalination technologies showcase a lot of technical learning effects, considerable reduction being given to investment costs within the past years are anticipated to pursue, solar collectors prices will decrease overtime, thus compensating the increased cost of additional measures that are required for controlling population which will be indispensable for implementing these technologies in the large scale within GCC countries.

11. Demand for Desalination Plants in GCC

The GCC area will continue to be the largest segmented market for desalination technologies, where combinations of the increasing population, polluted groundwater sources and capacity of retirement that was built during the booms of oil production in the 1970s and early 1980s will cause a doubling in the capacity requirements.

Growth within the region is moving forward with added emphasis being given to Saudi Arabia's financial stability to help maintain capacity within the timeframe. Desalination is an environmental issue that involves brine discharges and energy consumption. This technology must be reasonable against other types of environmental alternatives and impacts. The desalination industry is not string, with a large portion of small and medium sized players governing operations in different places. It appears that desalination growth worldwide will escalate to 25%, in GCC countries demands for desalination increase by 35%. Saudi Arabia is a primary shareholder within the desalination market, receiving the highest amount of solar irradiance per square meter per year.

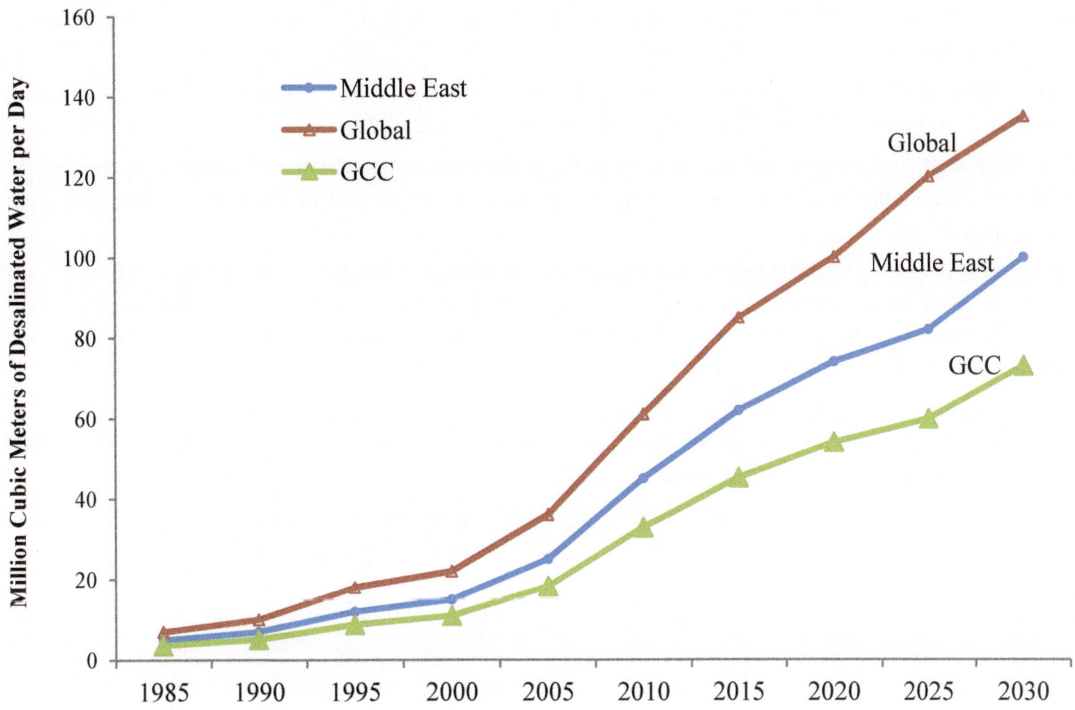

Figure 6. The demand for desalination plants Global, Middle East and GCC

CSP will be required to contend with wind-powered PV systems plus RO-systems and solar thermal collector non-concentrating systems. Aside from these contenders, linear Fresnel and parabolic trough systems are also likely to be competitive within this market segment. These two options offer low costs and probability of energy storage with hot water temperatures falling below 100 degrees Celsius.

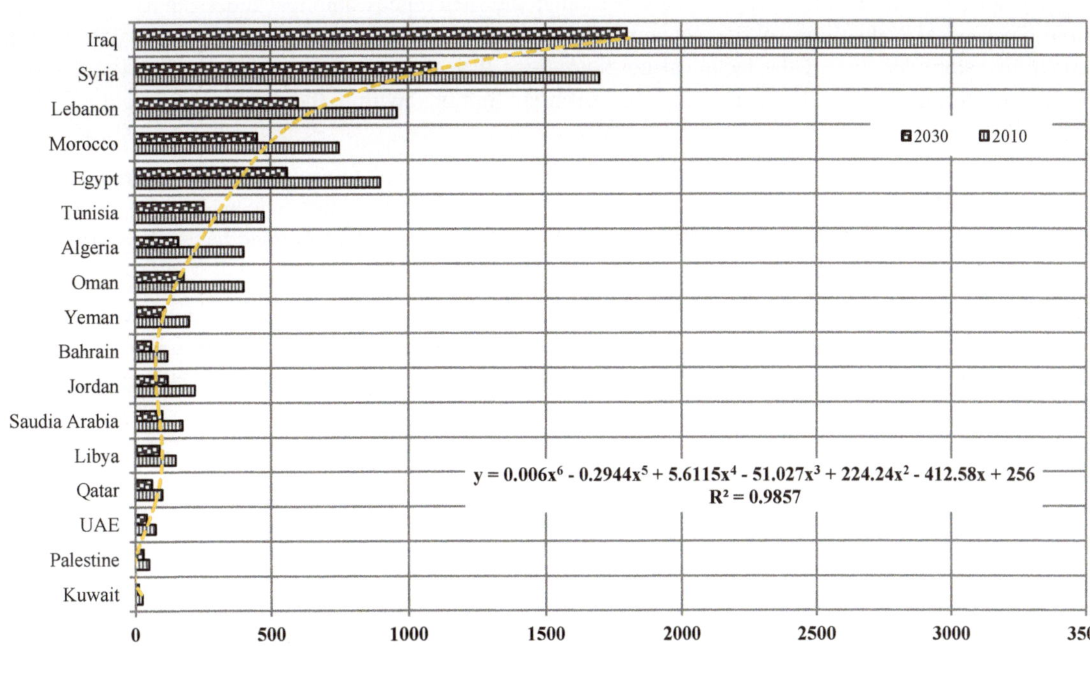

Figure 7. Total available renewable freshwater sources available per capita in the region of the year 2030

12. Socio-economic Impacts of Desalination on GCC

The economic performance of the referenced plants have been done for a site that has solar irradiance that measures 2400 kWh/y and salinity seawater of 40,000 ppm. Taking these conditions into consideration, both plants achieve a yearly solar share of 19% utilizing solar fiends that are designed to provide a nominal power capacity without engaging in thermal energy storing.

The yearly cost on capital is calculated from reviewing the discount rate of five percent, and taking economic plant life of twenty five years, this equates to annually fixed charged rates of 7.1%.

Additional costs related items are maintenance cost and operational costs, which are assumed to be 2% of investments along with annual insurance costs that are equivalent to one percent of investments made per year for both types of plants.

Plants will be operated in fossil modes/hybrid solar modes additional fuel inputs of natural gas will be emitted. Average life cycle costs of fuel is assumed being 25 €/MWH. Related yearly costs and fuel consumption will depend on the total amount of solar share yearly, which will vary with investment and size of the solar field and amount of thermal energy that is stored.

Replacements of membranes required for reverse osmosis will be required to take place once every five years, thus adding twenty percent additionally to the initial investments made on membranes to the yearly cost of operation when a CSP/RO process system is used.

Economic performance of desalted water and combined generated electricity is compared to the fixed sales price for electricity, which equates to 0.07-0.084 €/kWh this would equate to the production costs of cycle powered stations utilizing gas-powered cycles subtracting annual electrical revenue from the yearly expenditure figures. Remaining yearly cost is charged to the yearly production of desalted water, which yields an average cost done on per cubic meter of desalted water.

The CSP/MED process configuration reflects lower costs in water in comparison to the CSP/RO processes. Because CSP/MED systems have increased technical performance abilities, average fuel consumption is ten percent lower than the amount of fuel consumed in CSP/RO processes. An addition in regards to replacing the RO membranes every five years also brings added costs. The cost of these items makes the CSP/MED system appear economically better, even though the initial investment cost is slightly higher than with CSP/RO process systems.

This result is contradictory to the presumed assumption that RO systems are cheaper than MED systems. Even though this assumption is true in regards to initial investments, when both processes are combined with CSP/desalination plants, the end result is opposite, even though the differences amongst the systems is not large. In-depth, project wise evaluations of the economical performance and technical performance will lead to well-founded decisions for one of the technical configurations that are appropriate CSP-desalination systems; competition will further define the shares that the different options within the future desalination market generate.

The considerably high lifetime costs of fuel and the costs of solar collector fields have been taken into consideration. Considerations in the high initial investment costs of MED plants, because of the rising costs associated with raw materials required to build the plants such as steel and copper have also been given considerable thought.

The cost of water is high in comparison to the costs associated with desalted water, which was taken from literary works, even though it still is within the range of values reported. The cost of water from RO and MED is reportedly in the range of 0.40-2.00 €/m3., representing large varieties of sites, product and sea water qualities as well as economic frame conditions (Abu Arabi, 2005).

The cost analysis appears fairly different when considering the learning curve in regards to CSP until the year 2020. Figuratively, during that time, collector costs will have decreased, power blocks will be well-adapted for solar fields, increased enhancements for environmental protection pursuits will be taken, costs of MED will have changes because of the increasing production of commodities such as steel and copper. The efficiency gains that will take place within the solar field will increase solar shares of reference plants to twenty five percent without the requirement for thermal storage to ninety five percent utilizing full scales of storage capacities. The cost for fuel will increase respectively. The revenue for electrical sales are assumed to remain constant, this may be a conservative estimation. Under these proceeding conditions, associated costs of water will decrease for CSP desalination processes that do not require thermal storage.

13. Available Freshwater Renewable Sources per Capita in the Gcc Region Circa Year 2030

Non-renewable resources such as the large fossilized groundwater reservoirs that are located beneath the Sahara desert can partially be utilized, if acceptable time-spans to be able to supply several generations with the valuable resource can be guaranteed. Additional measures, allowing the reutilization of waste water that has been treated and managed with respects given to accountability towards moving forward the distribution systems along with new unconventional water sources are imperative to help avoid the probability of disintegrate water supplies within the region (Gueymard, 2011).

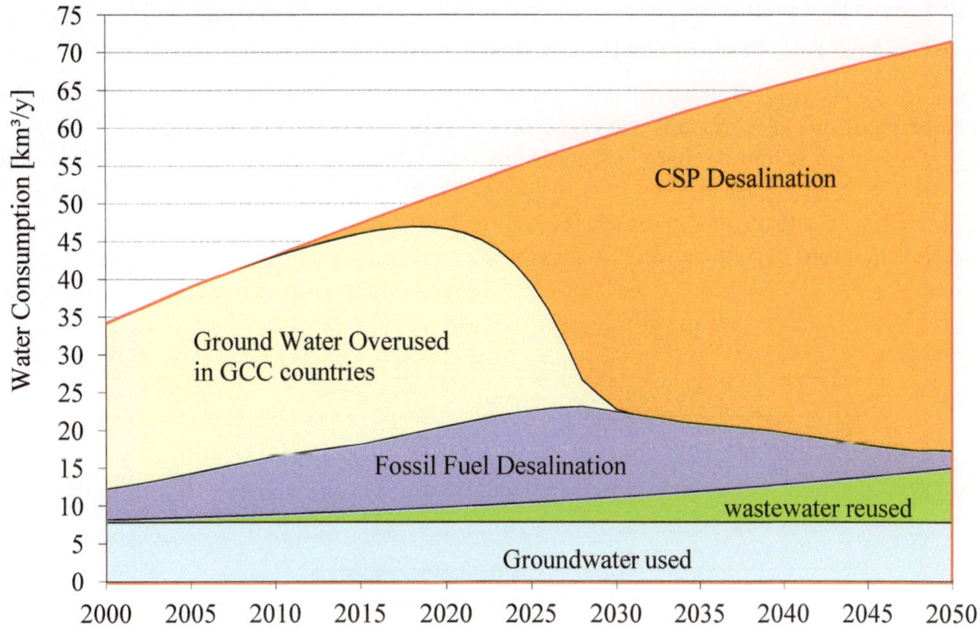

Figure 8. Concentrating Solar Power for Gas Turbines

From Figure 8, GCC countries should anticipate ground water depletion before 2030, so integration of all efforts to be collectively to use both treated waste water and any cost saving technology. Economical and engineering make water consumption using CSP became one of lthe target in Figure 4. Solar towers use a large field of two-axis tracking mirrors (heliostats) that reflect the sunlight to a central receiver on top of a tower. The concentrated solar energy is converted to high temperature heat. The emblematic optical concentration factor ranges from 200 to 1000. The plant sizes ranges from 5 to 150 MW of electricity production and these are feasible. The high solar fluxes imposing on the receiver (average values between 300 and 1MW/m^2) permit working at elevated temperatures over 1000 °C and to combine thermal energy into steam cycles into gas turbines and combined cycles. It is also comprises solar towers with central receiver systems which can be integrated in fossil plants for hybrid operation in a wide variety of options and have the potential to generate electricity with high annual capacity factors by using thermal storage. High efficiencies may be reached with solar-heated gas turbines. This may increased supplementary in combined cycle processes. The quota of the person in GCC and Arab countries were averaged to come with 320 m^3 per head per year and for Arab nation's 1000 m^3/capita/annum. Those figures are still with the stressed water consumption areas that the lack of renewable resources and accordingly agricultural output which eventually means the food strategically is a threat to yield and harvest in the area. There is a big gap between demand and supply in the area and desalination must fill this gap.

14. Probable Environmental Impacts

The primary impacts in regards to desalination of seawater will depend on the idea that taking seawater for desalination processes and cooling the system may result in entertainment and impingement of airborne pollutant emissions and organisms. Carbon dioxide that is caused by the generated electricity and heat that is required to power the desalination plants, biological control agents and chemical additives can be used to help avoid foaming of the raw water resources, biological fouling, corrosion of the systems and scale precipitation of desalination plants can appear finally in brine, and discharges of the hot brine along with highly concentrated salt from the sea can affect the local species of the region (Gueymard, 2011).

The impacts on the environment that are due to emissions from power plants and plume rise regarding life-cycle fundamentals combined with constructing the plant, maintenance and operation as well as de-commissioning of CSP/RO orientation and CSP/MED process plants, and the impact of the said locations have been analyzed in conventional desalination schemes. Because of the direct impacts that desalination plants have on the costal environment, performing impact analysis should be taken into account for accurate evaluations and every prior case up to the erection of large scaled desalination plants to further avoid deterioration of sensitive species that could vanish due to the plants erection. The only sites that should be chosen for erection should allow proper and effective quick dilution tactics of concentrated brine disposal to the seawater to avoid overheating local areas and high concentrated amounts of salinization long with low demands of oxygen (Gueymard, 2011).

15. Greenhouse Gas Emissions Produced by Desalination Types

Greenhouse gas emissions emitted from hybrids within desalination plants are taking origin for lifecycle assessments combining of the GCC regions will increase renewable carve ups within the system. It appears that MSF is amongst the highest within its share in regards to GHG, while the CSP/MED process is in competition and could very well be a contender to hybrid combinations in the near future. Extra efforts are put on R&D within this field. MED is a setback, that most are recovering from, with manufacturers deciding to take a more conventional approach toward implementing MSF processes.

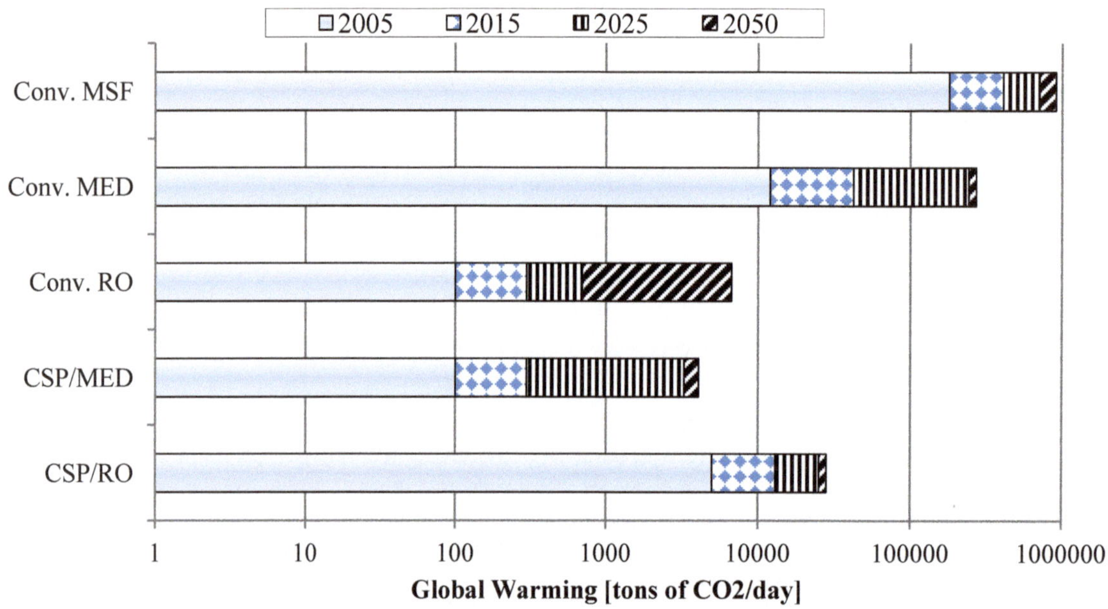

Figure 9. Greenhouse Gas Emissions (GHG) from different type of desalination

Figure 9 suggest that the greenhouse gas emission from hybrids of desalination plant taking as origin for lifecycle assessment the power combine of the GCC countries with increasing renewable carve ups of the system. It looks that MSF is the highest in its share to the GHG while CSP/MED is competing and could be contender to any hybrid combination in the future. Here it could be suggested to run extra efforts on R&D in this field. It came a setback for MED in the past and now is recovering when most manufacturer started to move towards conventional MSF.

16. Conclusion

Apart from the conclusions that other strategic evaluations of the GCC water sectors have proposed, desalination plants that utilize seawater have a quota on the supply of freshwater that is priced reasonably for GCC countries, support is given to domestic energy sources and will propose extreme impacts on the environment if CSP/ concentrating solar power is taken as the primary base of energy supply. Sterilized and clean desalination plants are imperative to help rectify the water dilemma that is taking place in GCC countries. If chemical additives are to be substituted within proper design intakes and the filtration of seawater on the opposing side would need to have its own reflections on energy. Concentrating solar power/ CSP is to be emission-free and the GCC region means will need to be large enough to deal with the added strain on demand. Based on the current available technology, and economics within a fifteen to twenty five year time frame, CSP will be a great solution that will be a contender amongst fossil fuel desalination services.

CSP will increase the way that the water distribution system is practiced in an efficient manner, market introduction for incorporating CSP for seawater desalination and power should start soon and adequate economic and political frameworks will need to sustain GCC countries to help implement initial pilot plants that will pledge to quickly expand this impressive technology in GCC countries that require its implementation. Any setbacks for the strategies will heavily jeopardize the further depletion of groundwater resources and have an adverse effect on social and economic development.

Looking at the concept of GHG abatement along with cost reductions of five percent of initial operations, utilizing solar irradiance will reach the average of 10 MWhr/ per year in GCC countries. The high demand for desalination plants within GCC countries is equal to seventy three percent of the total amounts of desalination plants, putting the extremely high costs of fossil fuels into consideration, all give proper support that CSP/MED processes will be successful, perform highly, and ascertain economically reasonable yields in favor of such technologies.

References

Bockamp, S., Griestop, T., Fruth, M., Ewert, M., Lerchenmüller, H., Mertins, M., & Dersch, J. (2003). Solar Thermal Power Generation (Fresnel), Power Gen.

Garcia-Rodriguez, L. (2003). Renewable energy applications in desalination: state of the art. *Solar Energy, 75*, 381-393. http://dx.doi.org/10.1016/j.solener.2003.08.005

Garcia, I., Alvarez, J., & Blanco, D. (2011). Performance model for parabolic trough solar thermal power plants with thermal storage: comparison to operating plant data. *Solar Energy, 85*, 2443-2460. http://dx.doi.org/10.1016/j.solener.2011.07.002

Gueymard, C., & Wilcox, S. (2011). Assessment of spatial and temporal variability in the US solar resource from radiometric measurements & predictions from models using ground-based or satellite data. *Solar Energy, 85*, 1068-1084. http://dx.doi.org/10.1016/j.solener.2011.02.030

IEA. (2012). *Whitepaper of Technology Roadmap Solar Heating and Cooling*. Retrieved October 26, 2012, from http://www.iea.org/publications/freepublications/publication/2012_SolarHeatingCooling_Roadmap_FINAL_WEB.pdf

Reiche, D. (2010). Renewable Energy Policies in the Gulf countries: a case study of the carbon-neutral "Masdar City" in Abu Dhabi. *Energy Policy, 38*, 378-382. http://dx.doi.org/10.1016/j.enpol.2009.09.028

Rubin, E. S., Chen, C., & Rao, A. B. (2007). Cost and Performance of Fossil Fuel Power Plants with CO_2 Capture and Storage. *Energy Policy, 35*, 4444-4454. http://dx.doi.org/10.1016/j.enpol.2007.03.009

Stoffel, T., Renne, D., Myers, D., Wilcox, S., Sergupta, M., George, R., & Turchi, C. (2010). Concentrating solar power: best practices handbook for the collection and use of solar resource data. Technical Report NREL/TP-550-47465. Retrieved October 23, 2012, from http://www.nrel.gov/docs/fy10osti/47465.pdf

3

A Multi-Method Evaluation of the Potential for Using the Electricity Bill to Encourage Energy Savings in Norwegian Households

Håkon Sælen[1] & Hege Westskog[1]

[1] Center for International Climate and Environmental Research, Oslo, Norway

Correspondence: Hege Westskog, Center for International Climate and Environmental Research, Oslo, Norway. E-mail: hege.westskog@cicero.oslo.no

Abstract

This study explores whether the electricity bill can be used as a medium to deliver information that encourages energy conservation in Norwegian households. Two main categories of information were tested: social comparative feedback and information about the monetary savings that can be made through specific energy-saving measures. The study combines four methods. First, a focus group study gathered advice on the categories and design of information that could influence energy consumption, and addressed general issues on energy consumption. Second, a field experiment monitored actual electricity use in 1000 households over a period of 10 months. In addition, through a survey and a series of in-depth interviews, the experience of the participating households was investigated. The experiment showed no effect on actual consumption. One main barrier was that only half the sample noticed the information. This suggests that the potential for encouraging energy conservation through adding new information to the bill in Norway is limited. The bill is already quite informative and is becoming less relevant as an increasing share of consumers switch to electronic bills and direct debit, which decreases the attention they pay to their bills.

Keywords: information, energy savings, electricity bill

1. Introduction

The question of how to motivate energy conservation in households has been a topic of considerable interest since the 1970s, as the answer is believed to hold great potential for reducing energy consumption and its associated environmental impacts at a small or negative economic cost. Tapping into this potential, however, has proven difficult. By now it is well established that technical and physical improvements in housing are not enough to guarantee reduced energy consumption (Darby, 2006). Consumption in identical homes, even those designed to low-energy standards, can often differ by a factor of two or more depending on the behavior of the inhabitants (Curtis, 1992; Keesee, 2005; Sonderegger, 1978). Further, economic incentives to reduce energy consumption appear to be muted, as there is evidence that foregone energy investment opportunities with a short payback time are ubiquitous (Stern, 2007).

It has been argued that it is difficult to realize the potential for energy conservation because energy use in modern systems is largely an invisible good, which means that it tends to escape human consciousness and reflexivity (Lindén, Carlsson-Kanyama, & Eriksson, 2006; Pedersen, 2000; Winther, 2008). This makes it challenging for users to get a clear idea of how their use changes over time, how it relates to meaningful benchmarks, how it is distributed between different end-uses, and how it would be affected by various energy-saving measures.

This study assesses the potential for using the electricity bill to increase the visibility of the good and thereby motivate electricity conservation. A field experiment is conducted in which different messages seeking to motivate electricity conservation are added to the bills of a sample of electricity consumers, and the level of consumption is measured relative to a control group. The field experiment is supplemented by a focus group study, a survey, and in-depth interviews to broaden the understanding of electricity consumption and of what motivates conservation.

Existing studies on the effect of information on household energy consumption have tested various types of information delivered through a variety of media. The various messages tested in the literature can be divided into antecedent information and feedback. This division is based on a categorization of a broader set of measures by Abrahamse, Steg, Vlek, and Rothengatter (2005). Antecedent information is aimed at influencing determinants prior to the performance of behavior, such as knowledge or attitude. Examples of this would be energy-saving tips and messages underlining the environmental or national energy security importance of conserving energy. Feedback is a way of making the consequences of behavior more salient. Direct feedback is near real-time information (e.g. given in a display), while indirect feedback has been processed in some way before reaching the energy user.

The empirical effectiveness of the different types of messages varies greatly – both within and across categories. Based on a review of existing studies, Abrahamse et al. (2005) conclude that, in general, antecedent information is not very effective on its own. Mass media campaigns about energy problems appear to be successful at increasing knowledge, but little is known about their effects on energy use. The evidence for the effectiveness of feedback is more robust. A review by Fischer (2008) shows that savings generally fall in the range from 1.1% to over 20%, with typical savings in the range between 5% and 12%. However, there are also instances in which no savings are found. Furthermore, the studies finding smaller effects appear more reliable than those indicating large effects, and most studies only look at the short-term effect of information. On aggregate, indirect feedback seems to be somewhat less efficient than direct feedback, with savings in the range of 0-10% (Darby, 2006). At a more disaggregated level, Darby argues that indirect feedback is usually more suitable than direct feedback for demonstrating the effect of changes in space heating, household composition and capital, but less effective for understanding how most appliances use electricity.

A study by Schultz, Nolan, Cialdini, Goldstein, and Griskevicius (2007) on the effectiveness of social comparative feedback has received much attention – such as in the book *Nudge* (Thaler & Sunstein, 2008) – due to the effect obtained from using social comparative feedback. Earlier studies of normative feedback have yielded unexciting results: Out of twelve such studies reviewed by Fischer (2008), none finds a statistically significant difference between the effect of this treatment and other types of information (Note 1). In some of the studies, a boomerang effect is evident. The boomerang effect is defined as a situation where customers whose initial energy use is below the average in the reference group increase their consumption in response to social comparative feedback. What the study by Schultz et al. (2007) suggests is that this effect can be eliminated by adding a message of social approval or disapproval (injunctive normative information). Half of their sample received only social comparisons. The other half also received injunctive normative information: ☺ or ☹, depending on whether they were below or above the average. In both groups, those who initially consumed more than the average reduced their consumption in response to the feedback. However, in the first group, which received only social comparisons, the boomerang effect was evident. In the second group, which also received injunctive normative information, this effect was eliminated; there was no longer a significant increase in the consumption of those who were initially below the average. A problem regarding this result is that the analysis does not control for regression to the mean. A study of social comparative feedback on water consumption (Ferraro & Price, 2011) found no boomerang effect even though no injunctive message was included. Another shortcoming is the small sample (290), and a medium that is difficult to scale up (research assistants leaving messages at participants' doors). Today, however, home energy reports inspired by Schultz et al.'s study are sent to one million households in the USA by the company Opower. Large scale studies have confirmed the effect, with a reduction of about 2% relative to control groups (Allcott, 2011; Ayres, Raseman, Shih, & Research, 2009). These reports include – in addition to neighborhood comparisons - personal historic comparison, targeted energy advice, and injunctive normative feedback, so the studies do not isolate the effect of each piece of information. The present study attempts to do so for the types of information investigated. The study also differs from the Opower studies by including the information with the bill rather than distributing it in a separate report.

Focusing specifically on the electricity bill as a medium for communication, a rigorous quantitative study was carried out by Wilhite and Ling (1995) among households in Oslo in the early 1990s. At the time, Norwegian customers only received information about their actual electricity use once a year. Every three months, they paid one-fourth of the estimated annual consumption based on the previous year. Hence, there was no relationship between the invoiced cost for a given three-month period and actual electricity consumption in that period. One treatment group received the bill for actual use at a greater frequency. This resulted in an 8% reduction in electricity use relative to the control group. Another group received – in addition to more frequent bills – a graph of their current and previous year's consumption. Electricity savings in this group was 12%. However, the difference between this group and the group that only received more frequent billing was not significant, which indicates that the main stimulant was the increased billing frequency. Since 1999, it has been mandatory for Norwegian utility companies to supply quarterly billing based on actual consumption including graphical historical comparative information.

Other studies have investigated the preferences of consumers concerning energy bills. In a study of Swedish households, roughly 75% wanted a graphic presentation of the actual consumption compared with the same month the previous year, about 65% wanted energy-conservation tips incorporated into the bill, and almost 50% wanted comparative statistical information from a comparable household (Sernhed, Pyrko, & Abaravicius, 2003). In a test of social comparative feedback, Wilhite, Hoivik, and Olsen (1999) found that three-quarters of consumers stated that they would be motivated to reduce consumption if they were informed that it was higher than the average. Another study from Sweden (Henryson, Håkansson, & Pyrko, 2000) found that the electricity bill serves two purposes from the perspective of the household: firstly as an invoice, and secondly as a control tool. The latter is achieved through a comparison with previous bills and with those of friends and neighbors. The majority of the households felt that social comparative information would be useful. Consistent findings across studies are that consumers strongly prefer billing for actual use, and that graphic information is easier to process than written information.

The current study contributes to the literature by testing electricity bills systematically as a medium for providing information to customers, both by quantitatively investigating the effects on electricity consumption in a field experiment and by combining this experiment with other methods to understand the effects obtained. Below, the context for the present study is given, explaining why Norway is a particularly relevant case and providing the background for the billing and reporting routines used for the consumers who were part of the study. This is followed by an explanation of the study design and methods used, and a presentation of the results obtained in the field experiment, the follow-up survey, in-depth interviews, and focus group study. Finally, the findings are summarized.

2. Context for the Experiment

In some respects, the Norwegian electricity market is rather atypical. For one thing, 96% of the country's electricity production comes from hydropower (Norwegian Water Resources and Energy Directorat, 2009). The large supply of hydropower has resulted in electricity prices that historically have been low compared to the rest of Europe. However, two aspects make Norway a particularly relevant case. Firstly, Norway has come relatively far in implementing requirements on information that must appear on the electricity bill, which is gradually becoming legally required in other European countries due to EU directives. Secondly, the Norwegian power market has been liberalized, a process several other European countries have also gone through recently, or will be going through. Below, the arguments for these two points are further explained.

The Wilhite and Ling study (1995 – see above) among households in Oslo in the early 1990s, had implications for the billing procedures and the type of information included in the electricity bill Norwegian customers. More frequent billing based on actual consumption as well as historical comparative feedback are now mandatory for the energy suppliers in Norway to provide (Hille, Simonsen, & Aal, 2011). Hence, Norway had already introduced some of the requirements from the EU directive on energy end-use efficiency and energy service when the directive came into force in 2006. This directive introduced strict new requirements for the contents of energy bills, see box 1. Historically comparative feedback is required "where appropriate" while social comparative feedback is required "wherever possible and useful." The proposal for a new directive on energy efficiency goes further, suggesting that these be made unconditional requirements (Commission, 2011).

EU directive on energy end-use efficiency and energy services (Commission, 2006)

Article 13

2. Member states shall ensure that, where appropriate, billing [...] is based on actual energy consumption, and is presented in clear and understandable terms. [...] Billing on the basis of actual use shall be performed frequently enough to enable customers to regulate their own energy consumption.

3. Member states shall ensure that, where appropriate, the following information is made available to final customers in clear and understandable terms [...] in or with their bills [...]:

(a) Current actual prices and actual consumption of energy;

(b) Comparisons of the final customer's current energy consumption with consumption for the same period in the previous year, preferably in graphic form;

(c) Wherever possible and useful, comparisons with an average normalized or benchmarked user of energy in the same user category.

However, Norwegian energy retailers are not required to provide social comparative feedback. (Note 2) Given that the bill is already utilized to deliver feedback on actual consumption, prices, and historical comparisons, is it "useful" to include further information, such as social comparisons? This is one of the elements tested in the field experiment of the present study. Information is considered useful in a strict sense if it leads to observed reductions in energy consumption in this study. However, its usefulness in a looser sense is also judged based on the views of the recipients.

The Norwegian power market was liberalized in 1991, and consumers can choose their power supplier in a competitive market for electricity where access to the network system is given on a transparent and nondiscriminatory basis (see e.g. Bye & Hope, 2005). The power grid is a natural monopoly where the prices in this market are regulated, but owners of the grid have an obligation to deliver electricity to consumers irrespective of which power supplier the consumer chooses. Network companies are required by law to bill household customers at least quarterly (Note 3). For those countries that have also liberalized their electricity markets or are in the process of doing so, knowing more about how information intended to motivate energy savings in a liberalized setting, would be of interest.

2.1 Background for the Field Experiment

The field experiment was conducted in collaboration with the network division of Hafslund. Hafslund is one of the largest utility companies in the Nordic countries. Its network division owns and manages the power grid in Oslo and parts of the adjoining counties of Akershus and Østfold. It services about 550 000 customers, making it the largest network company in Norway. Network customers of Hafslund can choose to receive the bill either by mail or electronically. They can also choose whether they want to pay manually, or through direct debit. Electronic bills are sent directly to the customer's online bank, from which the customer can access the bill by clicking on a link. Payment must be activated by the customer, unless the customer combines electronic billing with direct debit, in which case the amount is automatically deducted. Those who have direct debit without electronic billing receive the bill by mail for their information even though the amount is deducted automatically. When the field experiment was conducted, Hafslund was billing its customers bimonthly, but they planned to change their billing routines to monthly billing (Note 4). This was introduced after the field experiment finished.

All households are requested to report their consumption manually. They are reminded to do so on the first day of every odd month, but can in principle report their meter reading whenever they choose. Meter readings can be submitted by post, phone, sms, or online. If households fail to self-report, their consumption is stipulated by the utility.

3. Method

As the study employs a diverse mix of methods, each will be presented in turn.

3.1 The Focus Group Study

A series of focus group sessions were arranged in Oslo in November and December 2009 to explore consumer preferences for possible designs and content of the information to be tested in the field experiment.

The messages were presented to the focus group participants as examples of ordinary bills from the electricity supplier (Hafslund). The information was either included on the bill or presented as attachments to the bill. The messages discussed in the focus groups included:

- Comparative feedback with information about the electricity consumption in similar homes in their neighborhood. This information was presented as bar graphs comparing the household's electricity consumption to that of the reference group. In addition, one version in which the information was presented as text was included for discussion.

- Comparative feedback with information about the electricity consumption in similar homes in their neighborhood plus an injunctive normative message in the form of a smiley face if the household's consumption was below that of the neighborhood or a grumpy face if the household's consumption was above that of the neighborhood.

- Information on what measures the majority of Norwegian households have already taken to save electricity based on a previous survey (CICERO & Synovate, 2008) and presented as text (Note 5).

In addition to discussing the messages above, the focus group participants were also asked questions about the reasoning behind their energy use. They were asked about which energy savings measures they had already implemented or planned to implement, why they saved energy (if they did), who inspired them in their energy saving decisions, and their views on how a good life was connected to energy use. These questions where included to obtain data for the interpretations of the results from the field experiment. Knowing for instance the arguments for saving electricity, more information that is useful could be found regarding how people reason when it comes to energy savings.

Four focus group sessions in Oslo were arranged, each with eight participants. The recruitment of participants was handled by Synovate (Note 6), via telephone using the phone book or by contacting people who had volunteered to participate in focus groups (Note 7). The participants were divided into age- and gender-specific groups; men aged 30-45, men aged 50-65, women aged 30-45, and women aged 50-65 (Note 8). The focus groups were led by a moderator from Synovate who used the focus group guide developed by the researchers to facilitate the sessions. The focus group sessions and the focus group guide were semi-structured, allowing flexibility in the discussion. The proceedings of the focus groups were documented by professional transcribers. The team of researchers also observed the discussions in real-time through a one-way mirror. The data collected during the focus group sessions were analyzed using the software program NVIVO. This program allows for the organizing and classifying of qualitative data.

3.2 The Field Experiment

3.2.1 Treatments in the Field Experiment

The focus group discussions gave an indication of how to formulate the messages and what messages to include in the field experiment. The following changes in the messages were made based on the focus group discussions:

- The comparative normative feedback messages were only presented as graphs. It was evident from the focus group discussions that a graphical presentation was preferred to receiving the information only as text. Other studies show the same results – graphical presentations are often preferred to plain text (see e.g. Wilhite, 2007; Teisl, Halverson, & Holt, 1997).
- The grumpy face for those consuming electricity above that of their neighborhood was excluded. As one of the focus group participants stated: “You really feel monitored, and that I reject” (Note 9). This sentiment was shared by the majority focus group participants. Based on the negative evaluation of this message as well as the electricity supplier’s unwillingness to give negative feedback on their customer’s behavior, this type of message was excluded from the field experiment.
- Information on what other consumers did to save electricity was excluded. The participants in the focus groups did not find this information interesting: “I think it’s nice to get some concrete advice, but we’ve heard it all before, as you say. And, what most Norwegians do is probably not the most interesting thing. I would rather know what is most effective, because that’s not necessarily what most Norwegians do” (Note 10). Instead, when focus group participants were asked what kind of information they would like to see included in their bills, the most frequent response was information about how much money they could save by carrying out specific measures. “How about giving information like if you turn down the heat two degrees in the whole house, you will save this much money annually. You can cook potatoes on residual heat by turning off the stove when they’re almost done and letting them continue to cook for five minutes, lots of things like that, and if people saw that they could save so much money by doing that, then it would stick better” (Note 11).

Furthermore, the focus group discussions clearly indicated that the information must be presented where it is easily noticed. Attachment to the bill was rarely read which led to a preference of including the information treatments on the actual bill. However, because the front page of the bill is already fully covered by required information, the only option was to use page 2 or 3 of the bill. This could potentially lead to lesser awareness of the information provided, but also made it possible to test the treatments in a realistic setting since this is where the energy supplier would present the information.

The final treatments used in the field experiments were the following:

Treatment 1: Descriptive normative information

This group received a bar graph comparing their own consumption with the median for their postal code. It was accompanied by the following text:

> **New information: Compare your consumption with other detached houses in [postal code]**
>
> Most detached houses in this area normally use less than [median consumption of detached houses within postal code (rounded up to nearest 100 kWh)] in [month 1] and [month 2].*
>
> Your household used x kWh in this period this year.

The difference is y%

How to save electricity: see www.enova.no/hjemme

* Based on the average for 2007-2009.

Treatment 2: Additional injunctive normative information

Following Schultz et al. (2007), an injunctive normative message was added to the second treatment in the form of a smiley face (☺) to households who consumed less than the median.

Treatment 3: Energy-saving tips – gain frame

The monetary savings from 12 commonly mentioned measures were estimated, including both investments and curtailments. On each bill, three measures appeared, chosen for their relevance to the season in which they were distributed. This treatment was introduced in response to the many calls for this type of information by focus group participants (Note 12).

Treatment 4: Energy-saving tips – loss frame

A consistent observation in experimental economics is that people systematically attach greater importance to losses than to gains of similar magnitude (Tversky & Kahneman, 1991). Therefore, variations of the same pieces of information used in treatment 3 were created with a loss framing instead of a gain framing. The monetary figures used were identical, but they were now expressed as the additional expense one would incur through electricity-wasting behavior or by failing to take electricity-saving measures.

3.2.2 Sampling of Households

It was not feasible to obtain a fully random sample of all Norwegian households for this type of study, and some concessions had to be made for practical reasons. Therefore, the results cannot be directly generalized outside the study. However, the collaboration with Norway's largest utility company provided an excellent opportunity to investigate the effect of information in a very realistic and scalable manner.

In order to facilitate social comparisons to be used in treatments 1 and 2, a sample in which the housing stock was relatively homogenous in terms of size, type, and construction year was sought. Several focus group participants stated clearly that it would be meaningless to compare the consumption in a detached house with that in an apartment. The municipality of Asker was chosen. This municipality is dominated by detached houses, and electricity is the dominant source of space and water heating, providing 90% of residential energy use (Statistics-Norway, 2009). Asker has a population of about 55 000, and is located between 15 km and 30 km from the center of Oslo. It has a relatively affluent, well educated, and politically conservative population, while the age structure resembles that of the country as a whole. A central level of analysis in this study is the postal code, since two treatments receive information about their consumption relative to others within the same postal code. There are 14 different postal codes in Asker.

Households were filtered based on a number of criteria. First, the most affluent areas (1390/1391 Vollen and 1397 Nesøya) were excluded in order to somewhat mitigate the issue that the average income in Asker is well above the national average. Next, two postal codes were left out because their low concentration of detached houses made it cost-inefficient to include them (1387 Asker and 1394 Nesbru). Detached houses in the remaining areas were then isolated based on information from the Norwegian Mapping Authority, using Geographical Information System software. Finally, households with a history of not reporting their meter readings were left out due to a lack of reliable data.

A simple random assignment of households to group would have led to high administrative costs, as the reference level consumption for treatments 1 and 2 differ for each postal code. A two-stage cluster sampling was therefore employed. First, three postal codes were randomly drawn. These were 1383 Asker, 1385 Asker, and 1388 Borgen, which are referred to as subsample A. The households in subsample A were randomly assigned to three groups: Group 1, which received the social comparative feedback, Group 2, which received the social comparative feedback plus injunctive normative feedback, and Control Group 1. The remaining households (with postal codes 1384, 1386, 1389, 1392, 1393, 1395, 1396) – which is referred to as subsample B – were randomly assigned to Group 3 which received energy-saving tips with a gain frame, Group 4 which received energy-saving tips with a loss frame, and Control Group 2. Two control groups were chosen in order to ensure that each treatment group was compared with a control group drawn from exactly the same pool, while limiting the number of postal codes implicated by the social comparative information. Hence, Groups 1 and 2 were compared with Control Group 1, while Groups 3 and 4 were compared with Control Group 2. The sample size was chosen based on a standard power analysis. In order to be able to detect a 2.5% reduction in the treatment group relative to the control group, it was estimated that between 65 and 250 households were needed (Note 13). The range of the estimate is quite large, since the sampling had to be done before the all pre-treatment data was collected, and hence based on assumptions. In the treatments where the information was not personalized, a few more households were included as the cost of doing so was relatively small. The number of households in each group is reported in Table 1. Once all pre-treatment data were ready, the means and standard deviations of pre-treatment consumption within the groups were calculated, and the results are also reported in Table 1. An analysis of variance shows no difference between the groups (F=0.14). As will be explained below, the number of households with valid pre-treatment data is smaller than the original sample size reported in the table.

Table 1. Numbers of households in different groups

Group	N	Pre-treatment mean daily consumption (kWh)	Pre-treatment st.d. of daily consumption
1	268	64.50	21.86
2	287	65.20	22.76
3	375	65.11	20.01
4	373	65.78	20.65
Control 1	250	65,77	22.56
Control 2	339	65.97	23.21
Total	1892	100	

3.2.3 Calculation of Postal Code Benchmark Consumption

For each of the three postal codes in subsample A, and for each of the six annual billing periods, the median consumption based on the three years preceding the experiment was calculated. The postal code benchmark consumption levels were used in treatments 1 and 2 to provide comparative feedback.

3.2.4 Running and Analyzing the Experiment

The information was distributed with five consecutive bills, from March to November 2010 and was printed on page 2 or 3 of the bill. The consumption from receipt of the first such bill until the first reading after receipt of the last bill, i.e. January 2011, was recorded.

The dependent variable in the analysis is the change in consumption from the pre-treatment period (March 1st 2009 – January 1st 2010) to the treatment period (March 1st 2010 – January 1st 2011). Only the households that had self-reported the meter readings that were needed to calculate these consumption levels were included, since stipulated meter readings would be a source of noise in the data set. The readings recorded within a time span of 5 days before and 15 days after the date at which they were requested to report were included (Note 14). This implied that 447 households were excluded due to lack of pre-treatment data, and another 367 due to lack of post-treatment data (Note 15). An additional 85 households reporting an annualized consumption below 8000 kWh were excluded, because these were most likely from rental apartments in the basements of buildings registered as detached houses, and because the network company was not required to bill such households bimonthly. Two households were excluded because their consumption from pre-treatment to treatment was either doubled or cut in half, which most likely indicates that the house was only occupied in one of the two periods. Hence, there were 991 usable observations. The attrition rate in the sample of 57% was larger than expected and reduces the statistical power of the analysis relative to what had been planned (Note 16).

3.3 Follow-up Survey and Interviews

After the experiment was completed, the participating households were followed up with a survey as well as in-depth interviews. The purpose was to gain a better understanding of the results obtained in the field experiment. An invitation to participate in the electronic survey was distributed two weeks after the date of the last meter reading to be included in the analysis. To induce participation, respondents were offered the chance to win one out of five "universal" gift cards valued at NOK 1000 (app. €125). A response rate of 17% was obtained. The survey included questions about how they perceived the information that they had received, what measures they had taken to save electricity, whether any of these measures were initiated during the treatment period and because of the information they had received. In addition, respondents were asked which types of information they believed were most effective among those used in the field experiment.

After the survey was completed, a selection of participants in the experiment was contacted by phone and asked to participate in an interview. Each participant was rewarded with a NOK 1000 gift card. Interviewees were recruited with the aim of obtaining a balanced sample in terms of gender, age, and treatment group. Nine interviews were conducted in the period between six and eight weeks after the date of the last meter reading relevant for the experiment. An interview guide developed for the study focused on how the customers perceived the information they received and compared this to the other treatments used in the field experiment. Boards showing examples of the different information designs used were presented to the interviewees. These qualitative data were analyzed using the software tool NVIVO.

4. Results

4.1 The Field Experiment

Figure 1 shows the development over time of electricity consumption in the different groups. There was no apparent systematic difference between the groups during the treatment period. The effect of the treatments is investigated further in Table 2, which reports the mean percentage change in electricity consumption from the pre-treatment period to the treatment period for the different groups. The universal increase in consumption can be ascribed to colder winter temperatures in the treatment period. The variable of interest is the change in consumption from the pre-treatment period to the treatment period. An Analysis of Variance (ANOVA) finds that there are no significant differences between the groups for this variable (F=0.45, t=0.85). Pairwise t-tests between each treatment group and the relevant control group also fail to identify significant differences at the 5% level. Finally, the same non-result is obtained by regressing consumption during the treatment period on pre-treatment consumption, treatment group, and billing form (the latter is included as a control variable) (Note 17): there is no significant effect of belonging to any of the treatment groups (Note 18). Hence, one cannot conclude that there were any non-random differences in the development of electricity use between those who received information and those who did not. Therefore, it cannot be concluded that the information tested was useful in a strict sense by leading to a reduction in electricity consumption. It is also worth noting that the signs of the effects are not as expected based on Schultz et al. (2007), as both groups that received comparative feedback increased their consumption relative to the control group, and this was greater in the injunctive norm group than in the group that received only the social comparison.

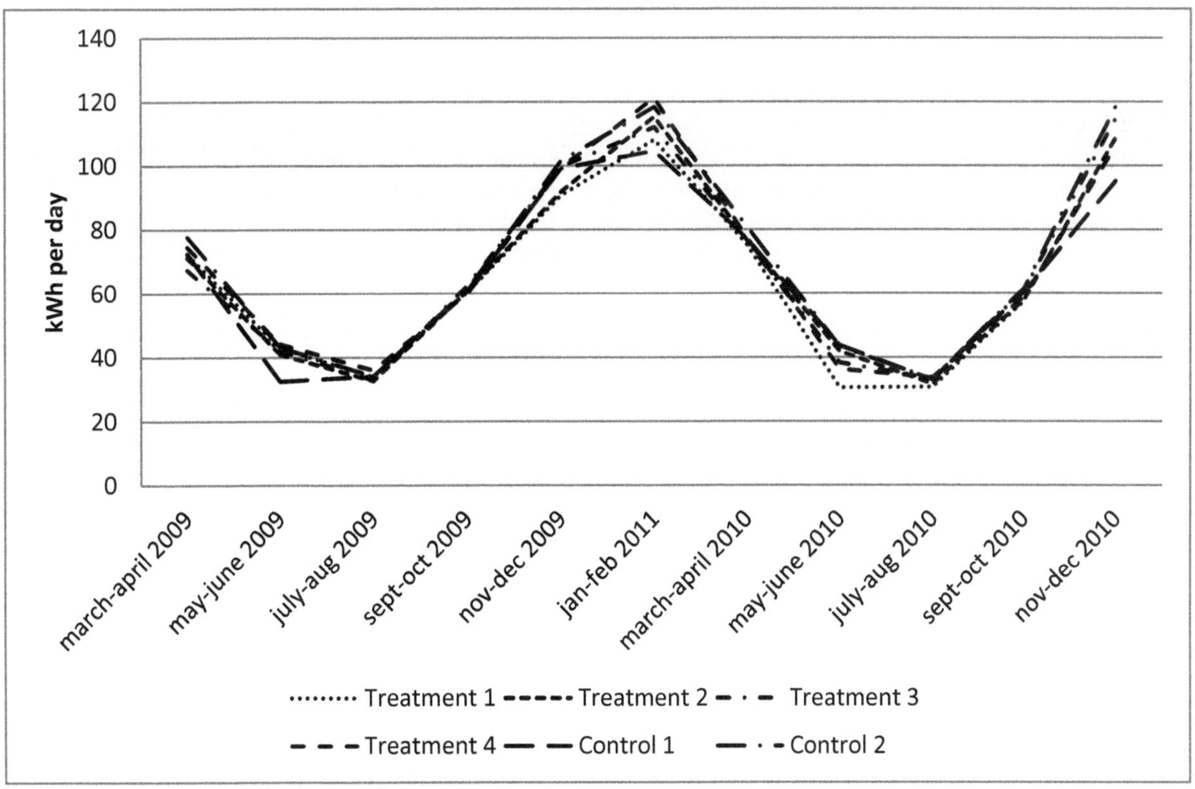

Figure 1. Electricity consumption over time by group

Table 2. Change in consumption during treatment (% of previous year)

Group	Mean	Std. Err.	95% Conf. Interval		No. of usable observations
1	4.47	0.96	2.58	6.35	146
2	6.79	0.96	4.91	8.67	168
3	6.45	0.64	5.19	7.71	219
4	4.21	0.69	2.86	5.57	215
Control 1	4.32	1.00	2.36	6.29	100
Control 2	5.72	0.99	3.78	7.66	143
Total					991

To investigate whether the boomerang effect reported by, amongst others, Schultz et al. (2007) and Ayres et al. (2009) was present in the groups that received social comparative feedback, Table 3 tabulates the mean change in consumption depending on whether the household's consumption in the pre-treatment period was above or below the median. Low-use consumers increased their consumption more than high-use consumers did, but this phenomenon was present in all the groups, not only in Groups 1 and 2, so it appears simply to be a case of regression to the mean.

Table 3. Mean change in electricity consumption (%) by level of pre-treatment consumption

Group	Over median	Under median
1	3.49	5.17
2	4.96	8.54
3	5.49	7.43
4	3.35	5.18
Control 1	3.33	5.15
Control 2	3.36	7.44

4.2 The Survey and the Qualitative Material

The results from the follow-up survey showed that a large proportion of the sample had not noticed the information they were sent, and this proportion varied significantly between the different billing methods, as well as between graphical and written information, as will be explored below (Note 19). Out of those who responded to the survey, 54% stated that they had noticed receiving new information on the bill (Note 20).

However, even among those that had noticed the information, the majority (81%) stated that the information had not induced them to take measures to save electricity. This indicates that the information used did not have the potential for motivating saving of electricity for most people. Hence, the results indicate firstly, that one reason for the lack of effect is that the information failed to reach a large portion of the intended recipients. Secondly, that when noticing the information only a small portion of the respondents became motivated to save electricity. The qualitative material gives us a deeper insight into why this would occur.

First, the qualitative material indicates that people are not careful readers of the information they achieve from their energy suppliers; what they notice is primarily or even exclusively what they are going to pay. After noticing the amount to be paid, some of the respondents said that they might look at the historical comparison. If so, the historical comparison was often used to indicate the reasons for variation in consumption: "Before receiving this information [historical comparison], I compared year-by-year, but now I get it ready-made. It is very useful. For instance, I can see 'This is when our daughter moved out.' But over time, the changes are not so large" (Note 21).

Second, if people are going to notice the information they are achieving, it has to be given to them at the right time. As one informant emphasizes, the information could be valuable for him in a situation when he is about make changes: "For people who move or build a new home. I'm in a situation now where it's expensive to change anything. But if I were to inherit the house of my mother, in such a situation, it would be great to get this kind of information" (Note 22).

Third, how the information – in this case the bill – is distributed to customers also seems to have an important influence on the extent to which its message is transmitted. The survey indicates that those who received the bill electronically or paid by direct debit were the hardest to reach. Whereas 75% of those that received the bill by post had noticed the information, this figure dropped to 57% for those using direct debit (who also received the bill per post, but who may be less likely to pay attention to it since it was debited automatically) and 40% for those who received the bill electronically. The challenge of getting attention to information given in the bill when using electronic billing is also confirmed by one of the focus group participants: "Since I receive the bill electronically, I don't always read it. I have combined electronic billing and direct debit, so it is automatic. I don't even have to approve it. The information that is included here probably escapes me" (Note 23).

One should note, however, that the causal direction between billing form and attention paid to the bill is not unambiguous. The correlation may partly be due to a selection effect, in which those who are most attentive to their electricity consumption choose postal billing. That way, they have the information in an accessible format, which can also be shown to the rest of the household. Hence, the households that received the electronic bills or paid by direct debit may have been less likely than others to see the information even if they had received the bill by post. Nevertheless, the results indicate that the rising number of customers opting for direct debit and electronic billing is not facilitating feedback regarding the customer's consumption for utilities (/utility companies).

On the issue of layout, the focus groups and interviews revealed that graphical information was perceived as easier to understand and increased the likelihood of attention. It was described as being more eye-catching than written information: "It is more eye-catching, the graphs are easier to recall" (Note 24), or as one of the focus group participants stated: "It gives the same information, but visual information is absorbed much quicker and is better retained. Looking at two different numbers requires more brain capacity to deduce and to remember. When you have seen a picture, it gives more information" (Note 25). The analyses of the survey results also found that graphical information was more easily noticed. A distinct difference between those who received graphical and those who received written information in terms of noticing the information was identified (Table 4). An ANOVA shows that there are significant differences between the four groups at a confidence level <1% (F=4.49).

Table 4. Proportion in treatment groups that had noticed the information

Group	Mean	Std. Err.	95% Conf. Interval	
1	0.62	0.07	0.48	0.76
2	0.69	0.06	0.57	0.80
3	0.47	0.06	0.34	0.59
4	0.42	0.06	0.30	0.53

Nevertheless, even if the information is delivered at the ideal time, in the ideal way and with the ideal layout, it may fail to motivate some consumers. The qualitative material indicates that several of the respondents are either not very motivated to save energy or have already done what they perceive as reasonable for them to do in terms of energy saving. When asked how their energy use is connected to the possibility to live a good life, some of the respondents point to this being essential for them: "I think it's the shower. I work out a lot, so if someone had told me that you cannot shower until tomorrow morning - that you have to wait to shower – that would not be negotiable" (Note 26).

Or as another respondent says, she is not motivated to save energy, since she has been involved in savings all her life, now she wants to do what she likes: I've lived with a large family, and we have always had to save. I think that now that we are only two persons, I want to do as I please. If I want 25 degrees, I can. That's how I think today. Of course, I want to protect the environment and such, but I want to do as I please, that's my thinking, because I have been saving all my life" (Note 27).

Others have already done a lot of energy saving measures, and even if they find the information interesting, it is not something that would motivate them to do more.

Interviewer: "Did you take any measures to save energy in 2010 that you have not taken earlier?"

Interviewee: "It was in 2007 that I took major measures. So I was already there. But the example calculation [the information given to Treatment 3] is interesting, I have a fairly small house, but I guess it [the example] is pretty representative. So it shows that you save a fair amount. So, that's very good. I remember well receiving it. I have read it several times to check that it's correct" (Note 28).

This indicates that aside from the problem of reaching people with the information given in the bill, some would not be motivated to implement electricity saving measures regardless of what type of information they received because they do not prioritize saving electricity, they have already done a lot, or the information is not provided at a time where they would be motivated to implement electricity saving measures. On the other hand, the material also indicates that information could play a role in increasing awareness and interest in energy savings, which in the longer term could play a role in changing attitudes towards energy consumption. Out of nine customers interviewed, seven of them stated that they found the information they had received to be interesting. As one of the interviewees said when asked if he thought the information he had received was interesting: "Yes, it was. I would like to be more aware" (Note 29).

Advice from the respondents on how different categories of information could be combined to create more awareness and action was achieved. Several of the interviewees mentioned that they would have liked to get a combination of comparative social feedback and energy-saving tips. The social comparative feedback would motivate households that use more electricity than their neighbors to save and the energy-saving tips would indicate how they could do it and what the most beneficial measures are: "Graphical (in this case social comparative feedback; authors comment) is more interesting as a first message. Then the pieces of advice are useful as message number two" (Note 30, Note 31).

There is also a monetary cost of providing the information. With possibly no effects from information on electricity consumption – only on interest and awareness concerning electricity use – there might be other and more efficient ways to create interest for energy consumption and behavioral change. Some of the results indicate that considerations of energy savings are greatly influenced by external factors. The focus group discussions indicate that consumers' efforts to save on electricity are highly influenced by external motivations such as cost, and that internal motivation for saving electricity are less relevant. As one of the focus group participants observed: "When we arrived here today, the intention was to talk about environmentally friendly solutions, but I keep hearing all the time, myself included, that we are talking about the power of money, money is what we are talking about. We are supposed to save the environment and not use too much electricity, but the issue of money keeps coming up… that is what we care about" (Note 32). This point is also supported by (Sælen, Westskog, & Strumse, 2012) who show that, in Norway, energy is a consumption domain where the correlation between behaviour and internal factors (values and environmental attitudes) is weaker than in other consumption domains such as food and waste reduction. P. C. Stern, Dietz, Abel, Guagnano, and Kalof (1999) argue that information measures work best when internal factors are most important for behaviour, and that incentives are better suited for areas where external factors dominate. Thus, information provision appears to have a smaller potential for leading to behavioural changes within the area of energy than other consumption areas, at least in the case of Norway, which may support an argument for using taxes or other incentive mechanisms to initiate energy savings rather than information.

Furthermore, there is an opportunity cost to providing information, in the sense that if it does not have the desired effect, it may make it more difficult to succeed with future attempts. There is some evidence that information provision may lead to entrenchment of current levels of consumption. In a study on smart energy monitors, Hargreaves, Nye, and Burgess (2013) found that after a relatively short time, such devices are used primarily to control households' discretionary energy use, whilst leaving "normal" usage unquestioned. Hence, providing feedback to households may, as Hargreaves et al. (2013) indicate, result in an increased engagement in energy use, but at the same time this might result in people to a greater extent perceiving their energy use as "normal", and (therefore) not engaging in reducing their energy. It may lead them to be more defensive of their "normal" consumption in the face of future information campaigns. Smart energy monitors might, hence, lead to non-intended effects as Hargreaves (2013) points out. As smart meters and electronic displays are new technologies that are about to be rolled out in several countries (for instance the UK), it is important to understand if and how these adverse effects might be overcome by specific information designs or in combination with other policy instruments. This is a task for further research.

5. Concluding Remarks

In this study, the potential of the electricity bill to provide information that encourages energy conservation to Norwegian customers has been analyzed. A field experiment was used to test the effects on electricity consumption of different types of information provided on the electricity bill. The information used was based on previous studies that had shown promising results and a focus group study on the types of information preferred by customers. The field experiment was followed by a survey and in-depth interviews of customers who had received the information in the field experiment. The field experiment showed no significant effects of the information tested on the level of electricity consumption compared to that of the control groups. The follow-up survey indicated that the information failed to reach about half the customers. It may be that much of the potential for using the bill as a medium to promote electricity conservation has already been realized through the feedback provided in terms of kWh, cost, and historical comparison. The qualitative material supports this finding, by pointing to the difficulties in reaching and motivating people through use of information in the bill. To affect behavior, the information needs to be given to the households at the right time, to be delivered in the right way, and to have the right layout. However, not even optimal information provision is likely to impact all customers, as some informants expressed that they are not interested at all in saving electricity and thus would always find the information irrelevant when it comes to motivating electricity saving. This draws a rather pessimistic picture of the potential of encouraging energy conservation through information on the electricity bills in Norway, at least in the short run. Studies from other countries (Allcott, 2011; Ayres et al., 2009; Darby, 2006), have found that using the right type of information might lead to saving of energy. The present study indicates that this might be overly optimistic in the Norwegian setting due to barriers such as interest in electricity savings, what people already have done, and the difficulty of reaching people with the right type of information when they are about to make changes.

Acknowledgements

The authors would like to thank Tanja Winther, Torgeir Ericson and Einar Strumse for their helpful comments and suggestions. Further, the authors gratefully acknowledge the funding received from the Research Council of Norway.

References

Abrahamse, W., Steg, L., Vlek, C., & Rothengatter, T. (2005). A review of intervention studies aimed at household energy conservation. *Journal of Environmental Psychology, 25*(3), 273-291. http://dx.doi.org/10.1016/j.jenvp.2005.08.002

Allcott, H. (2011). Social norms and energy conservation. *Journal of Public Economics, 95*(9), 1082-1095. http://dx.doi.org/10.1016/j.jpubeco.2011.03.003

Ayres, I., Raseman, S., Shih, A., & Research, N. B. O. E. (2009). *Evidence from two large field experiments that peer comparison feedback can reduce residential energy usage.* National Bureau of Economic Research Cambridge, Mass., USA.

Bye, T., & Hope, E. (2005). Deregulation of Electricity Markets – the Norwegian Experience. Statistics Norway Discussion Papers no. 433. Retrieved August 17, 2012, from http://www.ssb.no/publikasjoner/DP/pdf/dp433.pdf

CICERO, & Synovate. (2008). Den store norske klima- og miljøundersøkelsen. *Synovate 2008.*

Commission, E. (2006). *Directive 2006/32/EC of the European Parliament and of the Council of 5 April 2006 on energy end-use efficiency and energy services.*

Commission, E. (2011). Proposal for a Directive of the European Parliament and the Council on energy efficiency and repealing Directives 2004/8/EC and 2006/32/EC. 2011/0172 (COD). Retrieved February 11, 2011, from http://eurlex.europa.eu/LexUriServ/LexUriServ.do?uri=COM:2011:0370:FIN:EN:PDF

Curtis, T. (1992). Greenhouse Programme: report on the heating advice project carried out on the Packington and Market Estates 1991/92 and 1992/93. *Energy Information Centre, Islington Council.*

Darby, S. (2006). The effectiveness of feedback on energy consumption. *A Review for DEFRA of the Literature on Metering, Billing and direct Displays.*

Ferraro, P. J., & Price, M. K. (2011). Using non-pecuniary strategies to influence behavior: Evidence from a large scale field experiment: National Bureau of Economic Research. http://dx.doi.org/10.3386/w17189

Fischer, C. (2008). Feedback on household electricity consumption: a tool for saving energy? *Energy Efficiency, 1*(1), 79-104. http://dx.doi.org/10.1007/s12053-008-9009-7

Goldstein, N. J., Griskevicius, V., & Cialdini, R. B. (2007). Invoking Social Norms A Social Psychology Perspective on Improving Hotels' Linen-Reuse Programs. *Cornell Hotel and Restaurant Administration Quarterly, 48*(2), 145-150. http://dx.doi.org/10.1177/0010880407299542

Hargreaves, T., Nye, M., & Burgess, J. (2013). Keeping energy visible? Exploring how householders interact with feedback from smart energy monitors in the longer term. *Energy Policy, 52*, 126-134. http://dx.doi.org/10.1016/j.enpol.2012.03.027

Henryson, J., Håkansson, T., & Pyrko, J. (2000). Energy efficiency in buildings through information-Swedish perspective. *Energy Policy, 28*(3), 169-180. http://dx.doi.org/10.1016/S0301-4215(00)00004-5

Hille, J., Simonsen, M., & Aal, C. (2011). Trender og drivere for energibruk i norske husholdninger. *Vest lands for sknings rapport 13/2011.* Retrieved from http://www.nve.no/PageFiles/15077/R-NVE-sluttrapport-endelig%20versjon.pdf

Keesee, M. (2005). Setting a new standard–the zero energy home experience in California. *Sacramento Municipal Utility District.*

Lindén, A., Carlsson-Kanyama, A., & Eriksson, B. (2006). Efficient and inefficient aspects of residential energy behaviour: What are the policy instruments for change? *Energy Policy, 34*(14), 1918-1927. http://dx.doi.org/10.1016/j.enpol.2005.01.015

Norwegian Water Resources and Energy Directorat. (2009). Produksjon. Retrieved from http://www.nve.no/no/Kraftmarked/Sluttbrukermarkedet/Varedeklarasjon1/Produksjon-/

Pedersen, L. (2000). The dynamics of green consumption: a matter of visibility? *Journal of Environmental Policy and Planning, 2*(3), 193-210.

Schultz, P., Nolan, J., Cialdini, R., Goldstein, N., & Griskevicius, V. (2007). The constructive, destructive, and reconstructive power of social norms. *Psychological Science, 18*(5), 429-434. http://dx.doi.org/10.1111/j.1467-9280.2007.01917.x

Sernhed, K., Pyrko, J., & Abaravicius, J. (2003). Bill me this way!–customer preferences regarding electricity bills in Sweden. *Proceedings of the Summer Study of the European Council for an Energy Efficient Economy, 2003*, 1147-1150.

Sonderegger, R. (1978). Movers and stayers: The resident's contribution to variation across houses in energy consumption for space heating. *Saving Energy in the Home: Princeton's Experiments at Twin Rivers*, 207-230.

Statistics-Norway. (2009). Tabell 06926. Retrieved August 20, 2012, from http://statbank.ssb.no/statistikkbanken/Default_FR.asp?Productid=01.03&PXSid=0&nvl=true&PLanguage=0&tilside=selecttable/MenuSelP.asp&SubjectCode=01

Stern, N. (2007). *The economics of climate change: The Stern Review*. Cambridge: Cambridge University Press.

Stern, P. C., Dietz, T., Abel, T., Guagnano, G. A., & Kalof, L. (1999). A value-belief-norm theory of support for social movements: The case of environmentalism. *Human ecology review, 6*(2), 81-98.

Sælen, H., Westskog, H., & Strumse, E. (2012). Values, attitudes, and pro-environmental behaviours – is there a link? Results from a Norwegian survey. *Economics Bulletin, 21*(1), 486-493.

Teisl, M., Halverson, L., & Holt, E. A. (1997). *Information Disclosure for Electricity Sales: Consumer Preferences from Focus Groups--West Coast*: National Council on Competition and the Electric Industry.

Thaler, R. H., & Sunstein, C. R. (2008). *Nudge: Improving decisions about health, wealth, and happiness*: Yale Univ Pr.

Tversky, A., & Kahneman, D. (1991). Loss aversion in riskless choice: A reference-dependent model. *The Quarterly Journal of Economics*, 1039-1061. http://dx.doi.org/10.2307/2937956

Wilhite, H. (2007). An assessment of experiences in the USA with power and emissions disclosure information for energy consumers. *CICERO Report, 2.*

Wilhite, H., Hoivik, A., & Olsen, J. (1999). Advances in the use of consumption feedback information in energy billing: The experiences of a Norwegian utility. *ECEEE 1999 Summer Study Proceedings.*

Wilhite, H., & Ling, R. (1995). Measured energy savings from a more informative energy bill. *Energy & Buildings, 22*(2), 145-155. http://dx.doi.org/10.1016/0378-7788(94)00912-4

Winther, T. (2008). *The impact of Electricity. Development, Desires and Dilemmas*. Oxford: Berhgahn Books.

Notes

Note 1. One reason for the lack of effect historically might be that pre-smart meters and pre-GIS, neighborhood comparisons may not have seemed very credible to customers.

Note 2. The mandatory inclusion of historical comparative feedback on the electricity bill in Norway was a result of inter alia the research done byWilhite and Ling (1995) in the 1990ies. See discussion above.

Note 3. This applies to households with an annual electricity consumption above 8 000 kWh.

Note 4. The stated motivation for introducing monthly payment is to let the customers spread the payments out more evenly. With bimonthly billing the electricity bills could be very high, specifically in periods with high electricity prices.

Note 5. This message was inspired by the study by (Goldstein, Griskevicius, & Cialdini, 2007) on effective messages for improving hotels' linen reuse programs. They showed that by applying a descriptive norm for pro-environmental behaviour, participation by guests in one hotel linen reuse program was improved. Referring to what previous guests did led to that increase. These effects can be explained by Leo Festinger's (1954) theories on social comparisons, which propose that when making decisions under uncertainty, individuals tend to follow the norms of others who seem similar to them.

Note 6. Synovate is a global company specializing in market research. In the present study the authors collaborated with the Norwegian branch of the company.

Note 7. To be eligible to participate in a new focus group, volunteers must not have taken part in a previous focus group in the past six months.

Note 8. Synovate uses these categories as a standard for their recruitment of people to focus groups.

Note 9. Female focus group participant no. 6, Oslo, November 2009, Transcript 88322 Cicero Group 2.

Note 10. Female focus group participant no. 3, Oslo, November 2009, Transcript 88322 Cicero, Group 1.

Note 11. Male focus group participant no. 2, Oslo, December 2009, Transcript 88322 Cicero, Group 4.

Note 12. The calculations were based on a detached home of 120 m^2, four inhabitant, and electricity as the heating source. Assumed annual consumption is 25 000 kWh and the electricity price was set at 0.80 NOK. These assumptions were listed along with the tips.

Note 13. Based on a significance level of 5%, a statistical power of 80%, and standard deviations between 5% and 10 %.

Note 14. If a customer reported multiple readings within this time window, the average value was used.

Note 15. The households who were dropped due to failure to self-report during the experiment did not differ significantly from the rest of the sample in terms of pre-treatment consumption ($t = 1.8$). Hence, this does not seem to have led to a bias in the sample. It is not possible to do the same test for those who had to be dropped because of missing pre-treatment data, since no reliable data on their consumption were available.

Note 16. In addition, the standard deviations of consumption were at the upper limit of what was expected.

Note 17. This is equivalent to an Analysis of Covariance (ANCOVA).

Note 18. The same non-result holds both for a pooled regression where all treatment groups are included and the two control groups merged, as well as for separate regressions for each of the two cluster described in the sampling procedure

Note 19. It is also worth noting that 74% of the respondents were men. This could be taken as an indication that the electricity bill is often addressed to the male member of the household.

Note 20. This probably represents the upper limit of the percentage of all households that were sent the information. People who read their bill carefully may also be more attentive to other information sent by Hafslund, including the invitation to take part in the survey.

Note 21. Male interviewee no. 8, in-depth interviews, Asker, February 2011.

Note 22. Male interviewee no. 3, in-depth interviews, Asker, February 2011.

Note 23. Female focus group participant no. 6, Kirkenes, December 2009, Transcript 88322 CICERO, Group 6.

Note 24. Female focus group participant no. 6, Oslo, November 2009, Transcript 88322 CICERO, Group 2.

Note 25. Male focus group participant no. 3 Oslo, December 2009, Transcript 88322 CICERO, Group 3.

Note 26. Female focus group participant no.1 Oslo, November 2009, Transcript 88322, CICERO, Group 1.

Note 27. Female focus group participant no. 5, Kirkenes, December 2009, Transcript 88322 CICERO, Group 5.

Note 28. Male interviewee no. 6, in-depth interviews, Asker, February 2011.

Note 29. Female interviewee no. 7, in-depth interviews, Asker, February 2011.

Note 30. Female interviewee no. 7, in-depth interviews, Asker, February 2011.

Note 31. The suggestions were noted by a representative of the utility, who took part in the interviews. The authors also presented them when summarizing the results of the study at a seminar with Hafslund. The utility has not yet implemented social comparative feedback or energy saving tips in the format tested in the present study.

Note 32. Male focus group participant no.2, Oslo, December 2009, Transcript 88322 Cicero Group 4.

4

Electricity, Uncertainty and the Good Life

A Comparison of French and Norwegian Household Responses to Policy Appeals for Sustainable Energy

Tanja Winther[1] & Sophie Bouly de Lesdain[2]

[1] Centre for Development and the Environment, University of Oslo, Norway

[2] R&D, Electricité de France, associated with IIAC-LAU (EHESS-CNRS), France

Correspondence: Tanja Winther, Centre for Development and the Environment, University of Oslo, Norway. E-mail: tanja.winther@sum.uio.no

Abstract

Targeted by policies on sustainable energy, ordinary men and women are also engaged in the routines and concerns of everyday life. This paper asks to what extent French and Norwegian households link global warming to their own electricity use. Furthermore, we study people's rationales for selecting renewable sources and reducing electricity consumption at home. Empirical material was collected in France and Norway. We draw on a practice theory framework. The purpose of the paper is to use a comparative analysis for highlighting important aspects of the electricity cultures in each country, thereby accounting for why the French and the Norwegians respond to appeals for sustainable energy in the ways they do.

The results show that people in both countries tend to regard their own electricity use as disconnected from global warming. This is partly due to the high share of non-fossil power production in both countries, but we also suggest that certain denial mechanisms reinforce the idea that electricity consumption is disconnected from environmental challenges. The French respondents relate to the state-owned supplier EDF. They are preoccupied with physical and financial risks and regard electricity as a necessity. Conversely, the Norwegians value electricity in itself as a non-problematic good to which they think they have a claim as Norwegian citizens. Although the cost of electricity has an equal share of the household budget in both countries, the French practice a savings culture to a much greater extent than Norwegians do.

Keywords: Perceptions of electricity, comparison France and Norway, electricity cultures, sustainability

1. Introduction

Stephanie, a mother of two living in Paris, always makes sure to turn off the electric radiator before leaving her flat. She says she wants to reduce the physical danger. Per, a Norwegian who lives with his family in a detached house in the northern part of Norway, says he enjoys having a quick shower in the morning. After showering and shaving, he is done.

Informed by the threat of global warming, current policies are aimed at reducing energy consumption and increasing the share of renewable sources (Note 1). Ordinary men and women are being targeted by such policies on sustainable energy. At the same time they are engaged in the routines and concerns of everyday life. In this comparative paper we ask to what extent French and Norwegian households link global warming to their own electricity use. Furthermore, we study people's degree of interest in renewable electricity as well as their motives for reducing electricity consumption at home.

Our work follows practice theory (Bourdieu, 1977; Sewell, 1992; Warde, 2005; Ortner, 2006) and associated socio-technical approaches for understanding people's energy use (Wilhite, Shove, Lutzenhiser, & Kempton, 2000; Shove, 2004; Henning, 2005; Aune, 2007; Winther, 2008). Changes in consumption patterns require both structural changes and end-users' cooperation, and the field of energy appears particularly resistant to change due to the often habitual, non-reflexive character of such practices (Røpke, 2009; Wilhite, 2008; Maréchal, 2009). When asked in polls, the majority of people in France and Norway express concern about climate change (Ademe, 2012; Commissariat Général au Développement Durable, 2011; CICERO/Synovate, 2010). Europeans are also generally positive towards renewable energy sources (Markard & Holt, 2003; Boardman & Palmer, 2007; Hansla et al., 2008; Litvine & Wüstenhagen, 2011). However, related to the influence of structural factors, people's daily energy practices are often motivated by other concerns which may contradict their "green" dispositions (Røpke, 2009; see also Sælen et al., 2012).

In the present study we seek to understand Stephanie's, Per's and other people's rationales for doing what they do with electricity. The purpose is to use a cross-cultural examination of France and Norway in order to highlight some of the important technical and social aspects that make up the "electricity culture" in each country. Such insight, we argue, is important for understanding how people respond to policy appeals for sustainable energy. In our treatment of electricity, information and uncertainty we will also draw on a cultural concept of risk (Douglas & Wildavsky, 1983; Douglas, 1990; Peretti-Watel, 2010) and the "social organisation of denial" (Zerubavel, 2006).

Within the context of European electricity markets, information measures are considered as an important tool for ensuring sustainable consumption. The electricity bill is the most important means of communication from supplier to customer. The bill conveys information about the level of consumption in a given period of time as well as the amount due to be paid. Within the market logic embedded in the deregulated European electricity sector, customers are expected to reduce consumption or select another supplier when the cost/price becomes high. Furthermore, current EU regulations require that the bill provides information about the electricity's origin (Note 2). The assumption is that environmentally conscious consumers who receive such disclosure information will select renewable products or suppliers that offer electricity with a renewable guarantee (Boardman & Palmer, 2007; Aasen et al., 2012; Winther & Ericson, 2012). Finally, information campaigns are considered as an important means for obtaining energy efficiency (Note 3). What the last two measures have in common is that information is used as a strategy for encouraging individuals to adopt pro-environmental behaviour. According to practice theorist Elisabeth Shove (2010), the underlying thesis informing such policies is rooted in behaviouralism and the view that individual action is both the cause of environmental damage and the key to its solution. Our work includes an examination of people's perceptions of billing information and discusses to what extent such information is relevant for and/or affects electricity use in everyday life.

The outline of the paper is as follows: Section 2 accounts for the methods used for collecting the empirical data in France and Norway. In Section 3 we present the two electricity systems and discuss some of their central characteristics. Section 4 treats people's degree of concern for environmental challenges such as global warming and discusses to what extent they perceive electricity as connected to such challenges. Section 5 focuses on people's perceptions of their suppliers, bills and billing information. Section 6 accounts for people's motives for saving electricity at home. Section 7 provides a summary of contrasts and similarities in French and Norwegian perceptions of electricity. Section 8 concludes.

2. Method

We base the study (Note 4) on in-depth, semi-structured interviews with people in their homes as well as on focus group discussions. The French and Norwegian co-designed interview guides included topics such as people's perceptions of the environment and climate change; their perceptions of electricity and its origin; their thoughts about the possibility of changing supplier; their electricity use at home; how and to what extent they try to reduce consumption; how they manage their electricity bills and other issues. In the Paris region, 20 interviews were conducted in 2011 (Note 5), just before the Fukoshima nuclear accident in Japan. In Norway, 11 interviews were conducted by the Norwegian researchers in 2009 in Kirkenes and Vadsø, two towns in the very northern part of the country. In addition, four focus group discussions involving 32 participants in total were organised in Kirkenes in December 2009.

The sampling method involved selecting individuals from the middle class to obtain a certain degree of diversity of users within this segment in each country. In France, the interviewees, all of whom were EDF customers, were selected to ensure a wide range of ages and generational belonging, the presence or not of children in the household, and diversity in terms of people's marital status, education level, technical type of heating and whether they were homeowners or tenants. In Norway, all the people interviewed owned their own houses and most of them were married with children living at home. Due to the challenge of recruiting people at all, the Norwegian interview sample is likely to contain more individuals who are more interested in energy than is the case among Norwegians in general (Note 6). We balance this by drawing on the material from the focus groups, to which participants were recruited randomly, while ensuring diversity in terms of gender and generation.

3. Two Atypical Electricity Systems

The French and Norwegian electricity systems are unique in a European context, primarily in that power production is mainly based on non-fossil fuels. This implies that production represents a limited amount of greenhouse gas emissions. In France, 74 percent of the production is based on nuclear power (Note 7), whereas in Norway renewable hydropower constitutes as much as 96 percent of the country's electricity production. Both countries exchange power with surrounding countries. Such exchange is particularly important for Norway (Note 8) given that hydropower production depends on variations in rainfall.

Related to their distinct types of production, the two electricity systems have evolved and are organised in radically different ways. The establishment of nuclear power plants in France came in response to the severe energy crisis in the 1970s. The selection of nuclear power required that the state through Electricité de France (EDF) maintained responsibility for production and supply. French households were included in the electricity market in 2007 (Note 9). Currently, eight percent of French customers have chosen a supplier other than EDF (Commission de Régulation de l'Energie, 2011) and some among the alternative suppliers provide 100 percent renewable electricity. A testimony to the continued state control in France is the fact that the government sets EDF's prices after having consulted the regulating authorities. EDF customers may choose between various tariffs, including one that offers different prices during daytime and night-time. There are also social tariffs in France offered to less privileged groups. 21.5 percent of French families say they limit energy consumption for heating purposes because it is too costly, and 15 percent say that they suffer from a low indoor temperature (Devalière, Briant, & Arnault, 2011) (Note 10).

In Norway, the rivers and waterfalls that were tapped as an energy source beginning in the early 20th century are located throughout the country. Thus Norway had, and still has, a range of local producers and suppliers (Note 11), commonly owned by either municipalities, the state or private actors. Together with the political emphasis on economic efficiency at that time, this decentralised structure encouraged the liberalisation of the Norwegian electricity market in 1991, which included households from the start. Today, about one-quarter of Norwegian households have a supplier other than the main one in their area, and the domestic electricity market is considered mature. Prices are highly volatile in Norway and may vary significantly on a daily, monthly or yearly basis, depending on a customer's type of contract. Such variation is market based, and the prices are not set according to the time of day. There are no social tariffs for electricity in Norway. Table 1 summarises some of the important characteristics of the French and Norwegian electricity systems.

Table 1. Characteristics of the French and Norwegian electricity systems (copyright: authors)

	French electricity system	**Norwegian electricity system**
Effect of energy crisis in the 1970s	Severely affected	Not affected
Production (dominant)	Nuclear	Hydropower
Sector structure	Deregulated Centralised, government control	Deregulated Decentralised, many actors, competition
Electricity price	Low, stable	Low on average, but highly volatile
Differentiated prices day/night	Yes (voluntary)	No
Social tariffs	Yes	No
Electricity consumption per capita	Low	High

The two countries have in common that electricity is relatively cheap (Note 12), although consumption levels vary. Per capita, Norwegians consume more than 2.5 times the electricity that the French do (Note 13). It should be noted that fuels other than electricity, such as gas and oil, are also common for cooking and heating in France. Many Norwegians use firewood for heating as a supplement to electricity. In both countries, notably, we find that households spend about four percent of their budget on energy (Commissariat Général au Développement Durable, 2010; our estimations for Norway (Note 14)). Although there is substantial variation in people's ability to pay, the four percent share is relatively low in the European context. We may therefore assume that our middle-class French and Norwegian respondents can afford to pay for electricity. We will now turn to the customers and examine how they perceive the challenge of climate change and a range of issues related to electricity.

4. Concern for Climate Change and the Environment

The majority of the Norwegian respondents related to climate change in an abstract and distant, if also sometimes emotional, way. For example, Linda, a young woman, said she fears how the world will be when her children grow up. Sverre, a 47-year-old man, associated the issue of "environment" (Norwegian: *miljø*) with natural resources. He declared that he believes that "the environmental problem is the most serious threat that the world is facing today". He elaborated by saying that "if everyone is allowed to increase their consumption by two percent, this will not be possible because the resources of earth are limited". When asked specifically about his view on climate change, he said that he is not very concerned about the problem and that he does not know if such changes are caused by human activity or not. More women than men said they were worried, and by and large, our Norwegian respondents seemed modestly concerned about climate change.

In Paris people expressed more concern. When talking about the effects of global warming some would also bring in their personal experiences:

> There are more and more natural disasters, so either they're getting more and more media coverage or something really is out of whack, with increasingly strong hurricanes or that sort of thing. And what's really alarming or weird is that, I don't know, I've been in Paris for 10 years, it's bizarre that winters are increasingly harsh and, consequently, that raises an energy issue because being really well insulated when you are sweating through heat waves in the summer and you've got colder and colder winters, you find yourself with the kind of climate they've got in Canada, and so there's more CO_2, which is making the North Pole melt (Baptiste, 30 years, Paris).

However, some French interviewees also expressed scepticism about the reality of the problem: "We don't know what happened 2,000 or 500 years ago. No one was around to study and see it." (Jacques, 62 years, Paris). In general, though, most respondents in both countries tended to acknowledge the occurrence of human-made global warming. The two groups also generally agreed that it is primarily the responsibility of politicians, rather than individual consumers, to act and provide solutions to the environmental challenges. Some Norwegians regarded the lack of strict environmental policies as a sign that the government itself has doubts about the problem. French respondents tended to stress that the industry and EDF also have a responsibility to provide solutions.

It is important to note that neither the French respondents nor the Norwegians saw a connection between the problem of climate change and individual electricity use. Given the two countries' large share of non-fossil production and people's awareness of this fact, the reason for this disconnection seems obvious at first glance. However, in Norway, the power exchange with other countries implies that the sources of the electricity available consist of a mix of fuel types, ranging from hydroelectric and nuclear power to coal, natural gas and wind-based sources. Although the annual, net share of imported or exported electricity tends to remain in the single digits, the exchange takes place on a daily basis and amounts to considerable quantities of exchanged power. Participants were well aware of this exchange. They spoke about the imported power as being "polluted" (*forurenset*) and contrasted this with the "clean" (*ren*) and "green" (*grønn*) Norwegian electricity. Many also expressed a strong sense of collective ownership of this resource ("our rivers"), which they seem to value as a good in itself (see Westskog & Winther, in progress). However, very few pointed to the environmental aspects of increased imports and nobody referred to polluting imports as a rationale for saving electricity at home. It therefore seems that our respondents also *wanted* to think of Norwegian electricity as being clean, conceptually reproducing the national boundary.

Norwegians' strong conviction that they only consume renewable electricity has turned out to be a barrier to their acceptance of the electricity labelling scheme, in which Guarantees of Origin are traded across national borders. Norwegian producers have responded as anticipated by having Guarantees of Origin issued on the basis on their renewable production. However, most of these guarantees are exported to the European continent. Norwegian households are not willing to pay extra for electricity with a renewable guarantee (Westskog, Winther & Strumse, 2011; Winther & Ericson, 2012). A typical response to such an offer was: "This is ridiculous in Norway. The electricity we get is already green." The resulting paradox is that Norwegians continue to think that they consume renewable electricity, as do customers in Belgium and Germany who have purchased electricity with guarantees issued in Norway.

In contrast to the green and clean electricity perceived by Norwegians, our French respondents were quick to highlight electricity's association with nuclear risk. Such risks were said to be high at the production and waste stages, and low at the consumption stage. People's attitudes towards nuclear production and waste, not in terms of the energy source per se but rather its exploitation by EDF, ranged from full-fledged opposition to acceptance and confidence. Between these two positions, some expressed willingness to compromise. Those with a favourable attitude and those willing to compromise tended to put off questions about risk and waste management. Within this discourse, safety is said to be taken care of in France, and nuclear risk is perceived to be situated far away in other countries associated with less management expertise or even poor management. Today, EDF manages the issue of waste, the refrain goes. Tomorrow is another day, and doing without nuclear energy is not possible: "I don't think we have an alternative to nuclear energy today" (Baptiste, 30 years, Paris). In this case, people's attitude towards risk reflects their attitude towards time, allowing those who favour nuclear energy or who prefer compromise to mentally postpone the waste issue. In contrast, the "anti-nuclear" segment actively focused on future risks: "... the problem is simply being postponed" (Sophie, 37 years, Paris).

In their treatment of European opinions regarding radioactive waste, Boardman and Palmer (2007) observed that it "appears to be a greater concern (...) where there are no nuclear power stations, compared to countries with an active nuclear power program". They concluded that nuclear risk is minimised by those who depend on this kind of source. However, there is another possible interpretation of these stances regarding nuclear energy, which tallies with our interviews: being in favour of nuclear energy, or at least not being against it, does not mean that the risks are ignored, but that they are evaluated in light of the alleged economic risks of phasing out nuclear energy. In a recent study Brouard et al. (2012) found that "only eight percent of people in favour of nuclear energy consider it safe, and a tiny minority believe that nuclear waste does not pose problems". Thus it seems that the current economic crisis has relegated the risk issue related to nuclear energy to a position of secondary importance, supplanted by the issue of unemployment. In a semiotic analysis, Fodor (2012) argues along such lines, observing that the economic crisis replaced environmental issues in the media a few months after the nuclear accident in Fukushima in March 2011 (Note 15). This observation shows that risks and risk management are hierarchically ordered. The concept of risk often also has a flavour of being scientifically objective, which may camouflage its cultural and context-dependent aspects. Peretti-Watel (2010) speaks of a cultural bias in this regard:

> ... on the one hand, our culture gives us a specific perception framework, which determines how we grasp the world around us, how we interpret the information that reaches us, and therefore also how we evaluate risk. On the other hand, our values give the risks around us meaning, filling them with special significance...

The French cultural concern for nuclear risk does not have a parallel in our accounts from Norway. This is not surprising, given that Norway does not have nuclear power plants, though the country did experience some of the effects of the Chernobyl disaster in 1986. In both countries, however, there is a tendency to play down certain facts that in reality implicate consumers in the problem. Most notably, people tend to disregard the links between production and consumption. When the French respondents identified the environmental challenges related to electricity, they primarily did so by limiting the issue to the industry, production and/or the future, disconnected from their own consumption and daily life. In a similar vein, Norwegians tended to disregard the links between imported polluting electricity and their own consumption and insisted that Norwegian electricity is renewable. One may regard such dynamics as reflecting processes of "denial" (Zerubavel, 2006), a concept with close associations to Douglas' cultural model on risk. New information which does not fit with existing ideas is systematically disregarded or collectively denied while culturally anchored ideas remain strong: "nuclear is risky on the production side" (France) and "electricity is renewable per se" (Norway). Compared to, say, tomatoes or coffee, electricity has a particularly complex nature from production to consumption, including its invisibility, instant consumption and integration into multiple systems and markets that stretch beyond national boundaries. We suggest that these multifaceted aspects serve to strengthen processes of denial, thereby reducing the likelihood that customers will absorb new information about electricity and act in response to this.

5. Customers' Perceptions of Suppliers, Bills and Billing Information

Electricity customers' willingness to switch supplier is a central premise for the market to work as intended and for facilitating a shift towards renewable electricity production. In Paris, all the families were EDF customers, which may have made them stress their preferences for the company rather than its competitors. However, EDF has 94 percent of the French household market; thus customer loyalty is generally high. Below we refer to people's accounts in order to understand why the French interviewees hesitate to change supplier and how they perceive the electricity bill. This will be contrasted with the Norwegian material.

When the French were asked if they had ever considered changing supplier, many aspects perceived as barriers came up.

> ...why have I not changed, then? Mostly for reasons of continuity of service. I am concerned that the new suppliers will not be able to maintain service, and if they turn out to be a failure, then what? (Céline, 25 years, Paris)
>
> With respect to changing the contract, I have the impression that it would take time to make comparisons, and I'm afraid it would also cost money (…). As a result, between one and the other, I don't know if I would save very much. And to my mind, I imagine that the competitors use the same electricity as EDF. Does this mean that they purchase electricity from EDF which they would then turn around and charge back to me? In the end, do I really gain anything? I don't know at all. (Solange, 53 years, Paris)

As these quotes reflect, the French arguments for not wanting to change supplier centred on the increased risk of additional costs (money, time, stress) for obtaining what some think would be a highly unreliable energy service. Many also thought that the electricity they would buy from other suppliers would be produced by their current supplier, EDF. Quite a few expressed positive attitudes towards renewable sources, but the perception that "it will always come from EDF" negatively affected their interest in changing to suppliers who offer 100 percent renewable (Note 16). Other studies confirm this observation that not all French are fully aware of how the electricity market works (Note 17). Moreover, our respondents associated the promotion of renewable electricity in France with the world of business, often evaluated in negative terms ("tax exemptions", "unscrupulous installers"). Finally, renewable electricity is more expensive than conventional power from EDF, and this fact probably contributes to French customers' scepticism towards changing supplier.

In striking contrast to the French families' reluctance to leave EDF, Norwegian customers are accustomed to changing supplier. Households are allowed to do this every second week without a fee. Our sample in Northern Norway is untypical in the sense that only six percent of the customers, similar to the French share, had a supplier other than the main one in the area (the national average is 26 percent, as noted). But whereas the French barriers to changing supplier were related to a feeling of risk, people's lack of insight into the market, and EDF's low price, the Norwegian customers under study seemed to have taken a pro-active stance to support their local, municipality-owned supplier. Admittedly, a few showed a concern for what would happen if they changed supplier and it went bankrupt, but they were highly aware of their role as customers in a market. Many said that their reason for not changing supplier is that the prices do not vary significantly. In light of our previous discussion of the labelling scheme for electricity, it is worth noting here that Norwegians have fully accepted the arrangement that electricity can be bought from various suppliers in an open market, and thus also the financial system underlying this market which traces the amounts of electricity supplied and consumed. As seen, however, they do not correspondingly acknowledge the tracing system behind the labelling regime, but instead hold on to the idea that all electricity consumed in Norway is renewable.

People we interviewed in France did not seem interested in changing supplier. Conversely, the Norwegians feel free to leave their local supplier without taking on additional risk. These two distinct customer–supplier relationships are potentially relevant for how people perceive the bills. Both the French and the Norwegians said they normally ignore the billing information except for looking at the amount due and the corresponding date. When asked specifically to comment on the content of the bills, people in both countries said they look for variation compared to previous bills. However, the status of the bill differed in the two contexts. Norwegians said they simply register the amount to be paid to their local supplier without expressing a concern for a lack of choice or mentioning the relatively high share of taxes applied to electricity in Norway. In contrast, the French interviewed associated the bill with a form of tax (Note 18). First of all, the bill was said to be an unavoidable, fixed expense in line with municipality taxes, mortgage payments, and costs for insurance, water and gas. Norwegians did not make similar analogies. Secondly, as in Norway, a substantial part of the French bill is dependent on various governmental taxes, but in Paris this issue notably became a topic during the interviews, which seemed to strengthen the association between bills and taxes. Thirdly, some of our respondents pointed to the fact that bills in France can be used as legal proof of residence; thus bills are documents of authority. In France, there is an important association between electricity bills and the government, whereas Norwegians are more likely to view the bill as reflecting a symmetric relationship between supplier and consumer in an open market.

Interestingly, despite these differences in customer–supplier relationships and the status of the bill, people in both contexts were reluctant to act on the message in additional billing information such as advice on how to save electricity or chose renewable products (Norway). In general, the issue of trust is important when information is assessed, and in both countries trust centred around two main discourses. Firstly, there was a techno-economical discourse, that is, people's perceptions of the supplier's degree of accountability and accuracy in measuring and charging consumption, ensuring security of supply, etc. Here we find that people's level of trust is high in both countries, except for EDF customers' relatively low trust in suppliers other than EDF. Secondly, there is a commercial discourse in which customers evaluate the advice (information) coming from the supplier in relation to their perceptions of the supplier's underlying motives. For both France and Norway we find that customers expect that their supplier is commercially oriented. They therefore question advice about sustainable consumption. As a consequence, the information is disregarded. This indicates that billing information has limited potential for modifying people's practices in a sustainable direction.

One of the central conditions for current EU policies to work as intended is that electricity customers, out of a concern for the environment, are willing to change supplier and that they otherwise act based on information provided on the bills. We have established that both the French and Norwegians, although they have quite different electricity systems, have few incentives for selecting suppliers who offer renewable electricity. Norwegians are generally open towards changing supplier, but not for environmental reasons. In France we found that some people might be motivated to shift supplier out of environmental concern had it not been for the perceived risks involved in leaving EDF, the higher cost of renewable electricity and the general lack of clarity as to how the electricity market works in France. We also know that customers in general tend to disregard billing information and that they lack trust in messages on sustainability coming from suppliers whom they perceive to be commercially motivated. We will now discuss the other important aspect of sustainable electricity in relation to practice, that is, people's motives for saving electricity at home.

6. Electricity Use and Motives for Saving

The French respondents' focus on risk and uncertainty connected to production and the prospect of changing supplier corresponds to how they perceive electricity's uses in everyday life. First of all, several interviewees highlighted the physical danger involved when appliances are used. This makes some of our French respondents adopt a restrictive user-practice: "Anything that can be unplugged is unplugged. I don't like leaving them plugged in as I'm always afraid that there could be a short circuit" (Jacques, 41 years, Paris). In comparison, none of our Norwegian respondents brought up the issue of physical danger connected to electricity use (Note 19). This is a striking finding compared to the highly articulated French concern about avoiding the physical danger of electricity by turning off appliances and thus reducing consumption.

A second reason for saving electricity in France is economically motivated. As noted, the cost of energy in France is among the lowest in Europe. Nonetheless, some of our French respondents said they experience the cost of electricity as a financial burden, and 46 percent of the French have reported that they think their bill is too high (Ademe, 2012). Strategies to reduce the bill include paying on a monthly basis in order to spread the expenses or trying to adjust consumption to the hours of the day when the tariff is lower, such as by washing clothes late in the evening. Most of the French respondents seemed to have a genuine economic motive for trying to reduce consumption, but many felt that they had already reduced consumption to a minimum. They regarded their low bill as proof of their low level of consumption. A similar financial concern was not observed among Norwegians. Here, cost-saving aspects were sometimes mentioned as a rationale for investing in energy saving technology such as heat pumps. But those who had invested in such devices described this more as a smart or clever decision, "why spend more than necessary?", than as an urgent need to save on costs. To the French who regarded their low bill as proof of their low level of consumption, cutting down seemed to require a "mental" effort in return for a limited financial gain: "I'm not going to bother making an effort for just 30 Euros" (Baptiste, 30 years, Paris).

A third aspect connected to savings relates to social norms for electricity use and savings. The French reported that they do many small things on a daily basis to save electricity. Many said they would switch off the lights when leaving a room, turn down the thermostat when they are away, etc. Such acts of curtailment in French homes seemed to be "automatic" gestures, often motivated by a desire to be thrifty. Many expressed a concern for not being wasteful and this stance appears as a cultural value, as a desirable way of living: "I only consume what I need, no non-essentials. That doesn't interest me." (Céline, 35 years, Paris).

The anti-waste attitude probably has its roots in the French energy crisis of the 1970s when extensive television campaigns forwarded the slogan "In France we don't have oil but we have ideas". Measures aimed at careful energy management emerged at that time and still prevail today. For instance, summer time and winter time were introduced (in Europe at large) to ease the impacts of the energy crisis. This situation influenced several generations in the 1970s and was followed by a repudiation of excessive energy consumption by succeeding generations (Note 20). Consequently, today's younger generation also seems motivated by an anti-waste discourse. "Being careful about electricity" was mentioned in line with a whole range of daily gestures, from water use to transport and waste sorting. Together with these related practices, electricity use forms part of a whole, a category for anti-waste behaviour as such. This was also seen when people said they can make up for one particular extravagance by making savings in other areas, which also expresses a feeling of guilt. For example, some said they may have a nice hot bath and make up for it by turning off the tap while brushing their teeth.

On a daily basis, the moral and guilt-inducing dimension of consumption seeps through French media reports on the environment. A lexicon of dependence and addiction is employed in the press, for example by using expressions such as "energy sobriety" (Fodor, 2012). In France, waste is not just an environmental and economic issue, but a moral one as well.

In the case of Norway, people's perceptions of electricity as green and clean produce a very low level of environmental motivation for reducing electricity consumption at home. Also, they did not associate electricity with financial worries. Moreover, our respondents and focus group participants seemed far from adopting a savings culture similar to the one observed in France. On the contrary, to the majority of Norwegians we met, electricity use falls into a category different than other practices associated with environmental behaviour such as waste management and transport. The range and magnitude of steps to save energy reported by Norwegian respondents were far less than what was found in France. Illustrative of this difference is also our observation that the French have greater awareness about the system for labelling appliances than Norwegians (Note 21).

When Norwegians explained their high level of consumption, many mentioned the cold climate (Note 22) as a reason why things are difficult to change, concluding "there is little we can do". A couple of elderly women in a focus group agreed that they have tried to keep consumption at a minimum all their lives and now feel that they have the right to keep it a bit warmer inside. Some participants would blame young people for taking too long showers and changing their clothes every day, while others pointed to the youth as being a particularly conscious group due to the environmental education they have received in school. Initiatives such as using energy-efficient lights and showers were sometimes mentioned, but far from appreciated by all. For example, Birger said that his family used to keep a low-consuming shower, but he does not want to use it anymore: "That is a luxury I allow myself". In line with what many French and Norwegians told us, he explicitly referred to the desire for comfort (Note 23) as a barrier to saving electricity at home. But the Norwegians we met seemed more concerned about "not being fooled" (*ikke bli lurt*), for example when responding to information, than modifying their practices in a sustainable direction. Both the French and the Norwegians expressed a desire for comfort, but in France, the articulated concern for comfort was more balanced with the value of not wasting.

7. A Summary of Contrasts and Similarities in French and Norwegian Perceptions of Electricity

We have examined the ways people talk about electricity, electricity use and the degree to which they try to save electricity at home. The comparative material from France and Norway helped to clarify some of the central aspects of the electricity culture in each country. In France, multiple dimensions of risk are a crucial factor in how people relate to electricity as a technology, to their supplier and bills, and to electricity's services and uses in everyday life. Electricity is perceived as a necessity for living. The French interviewed also expressed a desire to not waste resources; thus we denoted "anti-waste" as a French cultural value and showed that electricity consumption forms part of this category. In comparison, electricity in Norway is valued both as a concept and as a physical good. The idea of having easy access to the clean and safe electricity stemming from "our rivers", as our respondents emphasised, forms part of the good life. So does electricity's various uses. Our material confirms what Wilhite et al. (1996) observed in the 1990s: Norwegians highlight the value of "keeping it warm and cozy" indoors. Norwegians are significantly less preoccupied with trying to reduce electricity consumption compared to the French respondents. Table 2 summarises the findings on French and Norwegian perceptions of electricity and related issues.

Table 2. Perceptions of electricity and related issues among French and Norwegian respondents (copyright: authors)

	France	**Norway**
Concern for climate change	Considerable, but not relevant for household consumption of electricity	Modest, but not relevant for electricity
Responsibility for sustainable energy	Politicians, EDF, industry	Politicians
Overall conceptualisation of electricity	Risky production and waste (nuclear)	Clean, safe and "ours" (hydropower)
Information about electricity's origin	Trusted, but mostly ignored	Campaign not understood, not trusted, ignored
Relationship with supplier	EDF = government, risky to change supplier	Accustomed to changing supplier who is profit seeking
Status of electricity bill	"Tax", proof of residence	Insignificant
Electricity use	May be dangerous, costly	Safe, cheap
Motives for saving electricity	Saving culture/moral dimension Concern for costs (Environmental rationale)	Laissez-faire. Some invest in energy saving equipment. Fear of being fooled.
Perception of electricity in everyday life	A necessity for living the good life	A central part of living the good life

Generally, we follow the literature that highlights the strong, habitual character of people's energy practices at home, and this has relevance for people's responses (or lack of such) to policy appeals for sustainable energy. In addition, the Norwegian perception of electricity as a culturally attractive object and people's explicit focus on comfort and lack of financial or other worries indicate that Norwegian electricity consumption is hard to change through information campaigns. In comparison, many of our French respondents appeared to perceive that they have already done what they can to reduce electricity consumption. They seem nonetheless to be more open to ideas that could help them reduce consumption, especially if the cost and the physical danger were reduced at the same time.

8. Conclusion and Outlook: Whose Responsibility to Provide Change?

This study has shown some important similarities and differences with regard to how the French and Norwegian people under study relate to electricity and sustainable consumption. In this conclusion we discuss our findings in relation to some central concepts in the literature. We seek to highlight some of the difficulties associated with introducing general energy policies in highly distinct contexts. We argue that this endeavour is particularly challenging when the effect of a given policy tool depends on the conscious actions of individual consumers. We end by touching on the issue of responsibility in relation to sustainability and by indicating in what directions French and Norwegian electricity policies are moving at present.

Shove (2010) documents that the concept of individual choice is central within behaviouralist theories and policies. In the case of electricity labelling, the underlying presumption is that environmentally conscious consumers will "be able to choose a supplier on factors other than price, making an informed decision which reflects their environmental values" (Boardman & Palmer 2007). In combination and over time, an open electricity market, disclosure information and environmentally conscious consumers are expected to produce a shift towards less polluting electricity sources. Our empirical results demonstrate that these presumptions are far from realities in France and Norway. First of all, the countries have in common that their power production is primarily based on non-fossil sources, and the fact that production does not leave substantial footprints of CO_2 was reflected when people expressed that they regard electricity consumption as disconnected from the challenge of global warming. However, by drawing on Zerubavel (2006) we found certain processes of denial at play in each country related to people's perceptions of the links, or rather, disconnections between production and consumption, which served to strengthen the conceptual distance between global challenges and people's daily life. One effect of this is that electricity consumers in France and Norway do not feel, or do not want to feel, responsible for solving grand environmental challenges. Another effect of the disconnection is that disclosure information about electricity's origin is perceived as irrelevant by most consumers.

Secondly, the presumption that customers are generally prepared to swap supplier is problematic. The few among the French who might have been interested in pursuing renewable electricity and select suppliers who offer such electricity felt restricted by a range of factors (higher cost, risk of switching supplier). Norwegians are accustomed to changing suppliers, but as noted, they will not do so to obtain electricity with a renewable guarantee. Thirdly, to the extent that French and Norwegian consumers read the information at all, they were sceptical towards receiving advice on sustainable behaviour from perceived profit-maximising suppliers. In sum, the labelling regime introduced throughout Europe is oriented at individual, environmentally conscious customers. Because the premises on which the regime is based do not match realities in our two countries, it has not produced substantially more demand for renewable electricity.

In contrast to behaviouralist approaches focusing on attitudes, behaviour and choice (Note 24), we have tried to base the discussion on an integrated analysis of the socio-technical electricity systems on the one hand (Table 1) and people's perceptions, explanations and habits on the other (Table 2). By employing comparison as a tool we sought to clarify how the various traits in each country make up a specific electricity culture. We argue that insight into such cultures provides a first step to understanding how sustainable consumption could be reached. It is worth listening to our French and Norwegian respondents when they hold policymakers (in France: also EDF and the industry) responsible for providing sustainable solutions. They can be seen as opposing the behaviouralist presumption that environmental damage is a consequence of individual action involving individual responsibility, which Shove (2010) rightly points out is a highly political stance. Our respondents seem to want an alternative policy which does not ask them as individuals to respond to selected pieces of information. As one of our female participants in Norway put it when asked what could make people save energy:

> I think that politicians should set the standard and make some legally binding agreements, and we will also take personal responsibility. But if they do not take responsibility, it is not likely to be of any use. (Focus group participant, Norway)

At present France and Norway seem to be moving in different directions in terms of how to handle future electricity consumption, France by tightening consumption and Norway by expanding production and consumption. The French government has recently proposed a new law introducing progressive tariffs (Assemblée Nationale, 2012). The law aims to place a higher tax per kilowatt hour on consumption over a given level so that those who consume more will pay more than what is the case today. The wording in the proposition includes rather strong judgments, such as "extravagant" and "virtuous" consumers. This might resonate with the anti-waste stance among many of our French respondents but not necessarily with the position of those who think that consumption is unproblematic as long as one pays for it. Norway, by contrast, has recently been included in a market for Tradable Green Certificates (TGCs) with Sweden. As a result, the government has committed itself to substantially expanding the capacity of Norwegian renewable production. Consumption is likely to increase.

Debates are bound to appear in both countries. What limit should define the boundary between normal and excessive consumption in France? What rivers and land should be exploited in Norway for ensuring new renewable production plants? Each of the new policies seems to reinforce important characteristics of the French and Norwegian electricity cultures and may therefore actually succeed by being accepted. For a variety of reasons, many French, like Stephanie whom we introduced in the beginning of this paper, will try to use as little power from EDF as they can. Per and other Norwegians will use even more electricity than before. Forthcoming information campaigns are not likely to change such predicted patterns.

Acknowledgements

We thank our project colleagues at CICERO, Lillehammer University College and EDF, in particular Hege Westskog and Sylvie Douzou, for their kind support. We also thank participants in the panel W050 "Uncertain futures: the cultural dynamics of energy transition" during the EASA Conference in Nanterre, France, 2012, for providing valuable comments. Connie Stultz proofed the English language.

References

Aasen, M., Westskog, H., Wilhite, H., & Lindberg, M. (2010). The EU electricity disclosure from the business perspective-A study from Norway. *Energy Policy, 38*(12), 7921-7928. http://dx.doi.org/10.1016/j.enpol.2010.09.013

Ademe. (2012). *Environnement, énergie: quelles évolutions des opinions et pratiques des Français*. Ademe & vous, Stratégie et études, no 31. Paris, France. Retrieved October 2012, from http://ademe-et-vous.ademe.fr/strategieetudes-n-31-2-fevrier-2012

Ajzen, I. (1991). The theory of planned behavior. *Organizational Behavior and Human Decision Processes, 50*, 179-211. http://dx.doi.org/10.1016/0749-5978(91)90020-T

Assemblée Nationale. (2012). *Proposition de loi instaurant une tarification progressive de l'énergie no 150*. Retrieved October 2012, from http://www.assemblee-nationale.fr/14/dossiers/tarification_progressive_energie.asp

Aune, M. (2007). Energy comes home. *Energy Policy, 35*, 5457-65. http://dx.doi.org/10.1016/j.enpol.2007.05.007

Boardman, B., & Palmer, J. (2007). Electricity disclosure: The troubled birth of a new policy. *Energy Policy, 35*, 4947-58. http://dx.doi.org/10.1016/j.enpol.2007.04.021

Bourdieu, P. (1977). *Outline of a Theory of Practice*. UK: Cambridge University Press.

Brouard S., Gougou, F., Guinaudeau, I., & Persico, S. (2012). *Les Français, le nucléaire et la campagne présidentielle 2012*. TriElec. Paris, France. Retrieved from http://www.cee.sciences-po.fr/en/elections-2012/541-les-francais-le-nucleaire-et-la-campagne-presidentielle.html

Chappells, H., & Shove, E. (2005). Debating the future of comfort: environmental sustainability, energy consumption and the indoor environment. *Building Research & Information, 33*(1), 32-40. http://dx.doi.org/10.1080/0961321042000322762

CICERO/Synovate. (2010). *Den Store Norske Klima- og Miljøundersøkelsen 2010*. Oslo, Norway.

Clancy, J., & Roehr, U. (2003). Gender and Energy: Is there a Northern Perspective? *Energy for Sustainable Development, 7*(3), 44-9. http://dx.doi.org/10.1016/S0973-0826(08)60364-6

Commissariat Général au Développement Durable. (2010). *Le budget 'énergie du logement': les déterminants des écarts entre les ménages*. Le Point sur, no 56. Paris, France.

Commissariat Général au Développement Durable. (2011). *Les perceptions sociales et pratiques environnementales des Français de 1995 à 2011*. La revue du CGDD. Paris, France.

Commission de Régulation de l'Energie. (2011). *L'ouverture des marchés de l'électricité et du gaz naturel pour les clients résidentiels. Baromètre annuel – vague 5*. Energe-Info, LH2. Paris, France.

Commission de Régulation de l'Energie. (2012). *Evolution du mix: les réseaux en première ligne*. Paris, France. Décryptages no 28.

Devalière, I., Briant, P., & Arnault, S. (2011). *La précarité énergétique : avoir froid ou dépenser trop pour se chauffer*. Insee Première no 1351. Paris, France.

Douglas, M. (1990). Risk as a forensic resource, *Daedalus, 119*(4). American Academy of Arts and Sciences.

Douglas, M., & Wildavsky, A. (1983). *Risk and Culture: An Essay on the Selection of Technological and Environmental Dangers*. California: University of California Press.

EC. (2011). *European Commission (2011/0172): Proposal for a Directive of the European Parliament and of the Council on Energy Efficiency*. Retrieved May 2012, from http://ec.europa.eu/energy/efficiency/eed/eed_en.htm

Eurostat. (2012). Communiqué de Presse, 78/2012, 25 May 2012. Retrieved October 2012, from http://epp.eurostat.ec.europa.eu/cache/ITY_PUBLIC/8-25052012-AP/FR/8-25052012-AP-FR.PDF

Fodor, F. (2012). Sustainable Development, Climate Change, Energy Saving: Discursive Developments of an Environmental Ethic. In S. Shmelev & I. Smeleva (Eds.), *Sustainability Analysis* (pp. 179-202). An interdisciplinary Approach. New York: Palgrave Macmilan.

Garabuau-Moussaoui, I. (2011). Energy-related logis of action throughout the ages in France: historical milestones, stages of life and intergenerational transmissions. *Energy Efficiency, 4*(4), 493-509. http://dx.doi.org/10.1007/s12053-011-9124-8

Hansla, A., Gamble, A., Juliusson, A., & Gärling, T. (2008). Psychological determinants of attitude towards and willingness to pay for green electricity. *Energy Policy, 36*, 768-774. http://dx.doi.org/10.1016/j.enpol.2007.10.027

Henning, A. (2005). Equal Couples in Equal Houses: Cultural Perspectives on Swedish Solar and Bio-pellet Heating Design. In S. Guy & S. A. Moore (Eds), *Sustainable Architectures: Cultures and Natures in Europe and North America* (pp. 89-104). New York: Spoon Press. http://dx.doi.org/10.4324/9780203412800_chapter_6

IEA. (2009). Energy statistics for France and Norway. Retrieved September 2012, from http://www.iea.org/stats/electricitydata.asp?COUNTRY_CODE=FR http://www.iea.org/stats/electricitydata.asp?COUNTRY_CODE=NO

Litvine, D., & Wüstenhagen, R. (2011). Helping "light green" consumers walk the talk: Results of a behavioural intervention survey in the Swiss electricity market. *Ecological Economics, 70,* 462-74. http://dx.doi.org/10.1016/j.ecolecon.2010.10.005

Maréchal, K. (2009). An evolutionary perspective on the economics of energy consumption: The crucial role of habits. *Journal of Economic Issues, 43*(1), 69-88. http://dx.doi.org/10.2753/JEI0021-3624430104

Markard, J., & Holt, E. (2003). Disclosure of electricity products - lessons from consumer research as guidance for energy policy. *Energy Policy, 31*(14), 1459-74. http://dx.doi.org/10.1016/S0301-4215(02)00201-X

Ortner, S. (2006). *Anthropology and Social Theory: Culture, Power, and the Acting Subject*. Durham, USA: Duke University Press.

Peretti-Watel, P. (2010). *La société du risque*. Paris, France: La Découverte, collection repères.

Røpke, I. (2009). Theories of practice – new inspiration for ecological economic studies on consumption. *Ecological Economics, 68*, 2490-97. http://dx.doi.org/10.1016/j.ecolecon.2009.05.015

Sælen, H., Westskog, H., & Strumse, E. (2012). Values, attitudes, and pro-environmental behaviours – is there a link? Results from a Norwegian survey. *Economics Bulletin, 32*(1), 486-93.

Sewell, W. H. (1992). A theory of structure: Duality, agency, and transformation. *The American Journal of Sociology, 98*(1), 1-29. http://dx.doi.org/10.1086/229967

Shove, E. (2004). Changing human behaviour and lifestyle: A challenge for sustainable consumption? In L. A. Reisch & I. Røpke (Eds.), *The ecological economics of consumption* (pp. 111-31). Cheltenham, UK: Edward Elgar Publishing. http://dx.doi.org/10.1068/a42282

Shove, E. (2010). Beyond the ABC: climate change policy and theories of social change. *Environment and Planning A, 42*, 1273-85. http://dx.doi.org/10.1080/09613210802079322

Shove, E., Chappells, H., Lutzenhiser, L., & Hackett, B. (2008). Comfort in a lower carbon society. *Building Research & Information,* 36(4), 307-11.

SSB. (2012). Statistics Norway: "Inntektsstatistikk for hushald". Retrieved July 2012, from http://www.ssb.no/emner/05/01/ifhus/

Stern P. (2000). Toward a coherent theory of environmentally significant behaviour. *Journal of Social Issues, 56*, 407-24. http://dx.doi.org/10.1111/0022-4537.00175

Warde, A. (2005). Consumption and Theories of Practice. *Journal of Consumer Culture, 5*(2), 131-153. http://dx.doi.org/10.1177/1469540505053090

Westskog H., Winther, T., & Strumse, E. (2011). Addressing fields of rationality – a policy for reducing household energy consumption? In I. Galarraga et al. (Eds.), *Handbook of Sustainable Energy*, Chapter 22. Edward Elgar Publishing.

Westskog, H., & Winther, T. (in progress). Electricity – commodity or common property? Implications of people's attitudes for the forming of sustainable energy policies.

Wilhite, H. (2008). New thinking on the agentive relationship between end-use technologies and energy-using practices. *Energy Efficiency, 1*, 121-130. http://dx.doi.org/10.1007/s12053-008-9006-x

Wilhite, H., Nakagami, H., Masuda, T., Yamaga Y., & Haneda H. (1996). A cross-cultural analysis of household energy use behaviour in Japan and Norway. *Energy Policy, 24*(9), 795-803. http://dx.doi.org/10.1016/0301-4215(96)00061-4

Wilhite, H., Shove, E., Lutzenhiser, L., & Kempton, W. (2000). The legacy of twenty years of energy demand management: we know more about individual behavior but next to nothing about demand. In E. Jochem, J. Sathaye, & D. Bouille (Eds.), *Society, behaviour and climate change mitigation*. London: Kluwer Academic Publishers.

Winther, T. (2008). *The Impact of Electricity. Development, Desires and Dilemmas*. Oxford: Berghahn Books.

Winter, T., & Ericson, T. (2012). Matching policy and people? Household responses to the promotion of renewable electricity. *Energy Efficiency* (published online 2012).

Zerubavel, E. (2006). *The elephant in the room: Silence and denial in everyday life*. New York: Oxford University Press. http://dx.doi.org/10.1093/acprof:oso/9780195187175.001.0001

Notes

Note 1. For example, the EU has set a goal of reducing energy consumption by 20 percent (compared with projected levels) and achieving a 20 percent share in renewable energy sources by 2020. A desire for energy security also informs this goal.

Note 2. The European Directive 2003/54/EC puts the obligation on national energy authorities to ensure that electricity suppliers provide their customers with information about their fuel mixes on the electricity bill. Within this labelling scheme, suppliers must display the shares of renewables, fossil and nuclear fuels involved in the generation of electricity, as well as the corresponding amounts of CO_2 emissions and nuclear waste. The labelling scheme is based on an underlying market for issuing and trading Guarantees of Origin.

Note 3. EU's proposed Energy Efficiency Directive states that information campaigns are obligatory (EC, 2011).

Note 4. The study forms part of a research project led by the Center for International Climate Research (CICERO) in Oslo, called "Do customer information programs influence energy consumption?" funded by the Research Council of Norway from 2009 to 2011 (project no. 190769/S60).

Note 5. These interviews complete research studies we carried out on the environment and energy management. The French part of this study was conducted under the framework of the European Energy Efficiency Research Centre (ECLEER), in partnership with Celsa-Paris-Sorbonne – our thanks to Claire Burlat.

Note 6. Based on a list of customers obtained through the local supplier, the Norwegian interviewees were recruited by the researcher by telephone. At first, people were very hesitant to agree to an interview, but this changed slightly after the researcher had been interviewed on the local radio NRK Finnmark and could refer to this interview when approaching people on the phone. It seemed that most of those who agreed to an interview had a particular interest in energy, electricity and new energy technology, so they are not fully representative of the population.

Note 7. The remaining production in France includes 11 percent thermal power and 15 percent renewable sources, of which 12 percent is hydro (Commission de Régulation de l'Energie, 2012).

Note 8. Norway exchanges power with Sweden, Russia, Finland, Denmark, the Netherlands and Germany.

Note 9. The European Directive 2003/54/EC required EU member states to open up their national energy markets for competition.

Note 10. The French term "précarité énergétique" implies that a household spends more than 10 percent of their budget on energy, cf. the notion of "energy poverty" (Clancy & Roehr, 2003).

Note 11. As of April 2012, 89 suppliers offering electricity to Norwegian households were listed on the home page of the Norwegian Competition Authority (www.konkurransetilsynet.no).

Note 12. The total price of electricity (energy, transport and taxes) to households in France and Norway is about 11 cents per kWh, that is, 66 percent of the average in the EU/EFTA region (for France: Eurostat, 2012; for Norway: SSB, 2010).

Note 13. The average, annual consumption of electricity per capita in the residential sector is 2701 kWh in France and 7272 in Norway (IEA, 2009). We assume that the population of France is 63 million and that of Norway is 5 million.

Note 14. In this estimation we use 55 000 EUR as the median income per household after tax (SSB, 2010), 22 000 kWh as the average level of annual consumption within our research area and 9 cents as the electricity cost in our research area in 2009.

Note 15. As Fodor (2012) notes, in 2007 the French media perceived climate change as a fact requiring immediate action. Two years later "climate sceptics" called it into question and experts debated the issue. In 2011, just after Fukoshima, media reports were more consensual: climate change was a recognised fact, and Fukushima had demonstrated the need for innovation. However, only a few months after the nuclear accident, the European financial crisis was a reality and became the dominant, pressing concern in French public debate at the expense of environmental risks.

Note 16. On its bills, EDF states the fuel mix of the electricity it sells to customers. Our French respondents had limited knowledge about other suppliers and their offers of 100 percent renewable electricity. There are websites in France (e.g. at Commission de Régulation de l'Energie) where customers can compare companies and make a choice of supplier depending on price, the electricity's origin, etc. None of our respondents referred to such websites/lists.

Note 17. 42 percent of French households confirm that they know they can choose electricity supplier, and 57 percent consider themselves to be well informed about this option (Commission de Régulation de l'Energie, 2011). Among those who have changed supplier, as many as 33 percent believe that they continue to receive electricity from EDF (ibid.).

Note 18. In French, the word for income or municipality tax is *impot*, whereas value added taxes (VAT) are referred to as *tax* (TVA). In the accounts we refer to, the word *impot* was not used.

Note 19. In a pre-study recently conducted for another Norwegian project we focused on people's preferences for shifting electricity prices according to the hour of the day (not offered today). Interestingly, in four out of nine conversations, the risk of fire was brought up as an issue. This may indicate that some Norwegians already tacitly take precautions by not using appliances at night or when away from their homes, but avoid bringing the issue up in conversations about the present system. This issue needs further investigation.

Note 20. See Garabuau-Moussaoui (2011) for a treatment of the significance of generational belonging for people's energy savings habits in France.

Note 21. In France, people stressed more often and in more detail the advantage of choosing A up to A+++ labelled appliances (which are technically more energy efficient compared to appliances marked B-E). When Norwegians were asked about the label of their various appliances, they often did not know.

Note 22. In this region there is midnight sun from May until late July when the average temperature is 12 degrees Celsius (www.eKlima.no). From November to late January the sun is always below the horizon and the mid temperature in January is minus 12 degrees Celsius.

Note 23. Perceptions of comfort are culturally determined (Wilhite et al., 1996, Chappells & Shove, 2005). Our material confirms that people often use physiological arguments, for example relating to body and health, when accounting for the notion of comfort (Chappells & Shove, 2005).

Note 24. Shove (2010) refers to Ajzen's (1991) notion "ABC" (attitudes – behaviour – choice) and insists that Stern (2000), although denoting C as "contextual factors", has maintained conceptual focus on individual choice.

5

The Use of the Leopold Matrix in Carrying Out the EIA for Wind Farms in Serbia

Bosko Josimovic[1], Jasna Petric[1] & Sasa Milijic[1]

[1] Centre for Spatial Development and Environment, Institute of Architecture and Urban and Spatial Planning of Serbia, Belgrade, Serbia

Correspondence: Bosko Josimovic, Centre for Spatial Development and Environment, Institute of Architecture and Urban and Spatial Planning of Serbia, Belgrade, Serbia. E-mail: bosko@iaus.ac.rs

Abstract

Although the Energy Sector Development Strategy of the Republic of Serbia until 2015 envisages an increasing share of renewable energy (RE) in the total energy balance of the country and although there are plenty of wind energy initiatives resulting in the elaboration of planning and project documentation, no wind farms have yet been built in Serbia. One of the significant problems in the realization of these projects is the unjustifiable resistance of a part of local authorities and population, particularly related to the possible environmental impacts of the planned projects. In this context, the paper addresses the possibility of using the Leopold matrix in carrying out the Environmental Impact Assessment (EIA) for "Kladovo" Wind Farm in Serbia (case study). The Leopold matrix is a framework method for assessing the environmental impact of a project. The novelty in this method, presented herein, refers to the evaluation of planned project activities relative to a group of criteria related to: significance (spatial dispersion), probability and duration of impact. The obtained EIA results have enabled the precise identification of possible environmental impacts of "Kladovo" Wind Farm project, as well as removal of dilemmas and problems related to the public resistance to the realization of the project through a transparent relationship with stakeholders.

Keywords: Environmental Impact Assessment (EIA), Leopold matrix, wind farm

1. Introduction

In 2005, the National Assembly of the Republic of Serbia adopted the Energy Sector Development Strategy of the Republic of Serbia ("Official Gazette of the Republic of Serbia", No. 44/05) defining the elements for determining the energy sector strategy of Serbia. The Strategy also defines the development and use of new sources of renewable energy, as well as the most efficient technologies and devices. According to the abovementioned, the most important goals of Serbian energy policy are harmonized with the EU practice and regulations aiming at increasing energy efficiency and intensifying the use of new sources of renewable energy as a major precondition for achieving the sustainable socio-economic development of the country, as well as for establishing an ecological energy balance. However, little has been done so far in wind energy implementation in Serbia. No wind farms have yet been built, while activities carried out with the aim to build them are still in the stage of the elaboration of the planning/technical documentation and obtaining of building permits (Josimovic & Pucar, 2010).

Besides the slow state administration, the significant problem in the realization of the wind farms in Serbia is also reflected in the resistance of a part of local authorities and population because of environmental impacts of wind farms. In this context, possible impacts of land-use change and higher intensity noise that may jeopardize the health and welfare or safety of the people who live very close to wind farms are emphasized as the specific problem. In addition, a part of stakeholders express their concern for possible dramatic wind farm impacts on ornithofauna and chiropteran fauna in a wider area.

Analyzing the abovementioned problems and their causes, it has been concluded that carrying out an Environmental Impact Assessment (EIA) presents the key instrument by which it is possible to eliminate these problems, as well as that a special attention within the EIA needs to be paid to:

- Avoid rough estimates, predictions and assessments based only on textual comments;

- The assessment of impacts of the planned project that will be based not only on identification of majority of impacts but also on consideration of different aspects of possible impacts.

In this context, the subject research has focused on further development of one of the current methods used in carrying out an EIA, which has been practiced in Serbia for more than 20 years (Crnčević, Marić, & Josimović, 2011). Here, it is of particular importance to chose a method which is based on the application of MCDM (Multiple Criteria Decision Making) in the field of environmental protection, because the use of the MCDM enables an analysis of different aspects of project impacts necessary for making decisions which optimize effects of certain activities relative to the capacity of space and environmental capacity. Methods based on the MCDM have a wide application in environmental impact assessment of proposed projects. From the aspect of the environmental impact assessment of wind-energy projects, the ARAS (Additive Ratio Assessment) method is also worth mentioning, which was elaborated in detail by Chatterjee and Bose (2013). From the aspect of quality and applicability, the AHP (Analytical Hierarchy Process) method is of particular importance. It has been widely used after the study of "A new Environmental Performance Index using analytic hierarchy process: A case of ASEAN countries" (Ismail & Abdullah, 2012).

The Leopold matrix has been chosen as a suitable method. There are several major reasons why precisely this approach to EIA has been chosen as appropriate for further development:

- The Leopold matrix presents a framework approach to impact assessment of a project because of which it is possible to creatively work on its further development;
- It is widely applicable in carrying out an EIA for different types of projects. Given that it has been used for many decades now, its major principles are known to scientific and professional circles, thereby implying that the elaboration of the obtained results will be understandable to many scientists and experts;
- The way of presenting the results can be appropriate for overcoming the existing problems in the realization of wind farm projects in Serbia because descriptive and rough assessment of possible impacts is avoided in the practice of carrying out an EIA;
- Leopold matrix fits in the framework MCDM method.

In this paper, we refer to the Leopold matrix as one of the pioneering and widely used approaches to impact assessment, which was the first comprehensive listing of environmental and socio-economic factors. This matrix is two-dimensional, where the stages of almost any type of construction project (actions) are assessed in relation to the existing environmental characteristics and conditions that may be affected during the execution of those actions.

2. About the Methodology of the Leopold Matrix

The Leopold matrix was conceived by geologist Luna B. Leopold and his colleagues in 1971, as a response to the US Environmental Policy Act of 1969, which didn't give clear instructions to the Federal Government agencies for preparing an impact report or for examining the environmental effects of the projects that an agency plans. The Leopold matrix addressed this challenge by 'providing a system for the analysis and numerical weighting of probable impacts'. As pointed out by the creators of the Leopold matrix, there is a clear advantage of using the matrix 'as a checklist or reminder' of the large scope of actions and impacts on the environment that can relate to the proposed actions (Leopold, Clarke, Hanshaw, & Balsley, 1971).

According to the Leopold matrix method, EIA should consist of three basic elements: a) a listing of the effects on the environment that the proposed development may induce, including the estimate of the *magnitude* of each of the effects (Note 1); b) an evaluation of the *importance* of each of listed effects (e.g., regional vs. local) (Note 2); and c) a summary evaluation, which is a combination of *magnitude* and *importance* estimates.

The format of Leopold matrix evolves around 8,800 possible interactions since the number of actions listed on the horizontal axis in the sample matrix is 100, and the number of environmental factors listed on the vertical axis of this matrix is 88. As the authors of this method pointed out, there are, however, only a few of the interactions that would deserve comprehensive consideration due to their expressed magnitude and importance, varying between 25 and 50 interactions on average for a typical project (Leopold, Clarke, Hanshaw, & Balsley, 1971).

In order to achieve higher efficiency of the matrix, a starting point is to check each significant action listed in a horizontal axis. The experience often proves that 'only about a dozen actions will be important' (Leopold, Clarke, Hanshaw, & Balsley, 1971). Each checked action is evaluated in terms of *magnitude* of effect on environmental factors that are listed in the vertical axis. In the matrix, across each box where significant interaction is expected, a slash (/) is placed diagonally from upper right to lower left angle of the block. In the text which accompanies the Leopold matrix, the evaluator has to indicate whether the assessment is for short-term or long-term impacts. Then, in the upper left-hand angle of each box with a slash, the evaluator should place a number from 1 to 10, which indicates the *magnitude* of the possible impact, where 10 represents the greatest magnitude of impact and 1 the least (no zeros should be assigned). The scale of *importance* (placed in the lower right-hand angle of each box with a slash) may also range from 1 to 10, with the same principle applied – the higher the value, the higher the *importance*. Assignment of numerical value for *importance* is based on subjective judgment of a multi-disciplinary team working in the EIA. In addition to assigning the numerical values to each marked box, plus (+) or minus (-) sign can be used to show whether an impact is beneficial or adverse.

The EIA is a discussion of the individual boxes marked with greater numbers for *magnitude* and *importance*. This is commonly applied to a simplified or reduced matrix, which consists only of those actions and environmental characteristics that have been identified as interacting. The rating scheme requires quantification that expresses evaluator's judgment (line of reasoning) in terms of probable impacts. This scheme then enables the reviewer to systematically follow the evaluator's standpoints, to assist in identifying points of agreement and disagreement.

Although most matrices are developed for specific application, Leopold matrix is quite general. That is a plus point for its wider application, as well as the fact that the method can accommodate both quantitative and qualitative data. On the other hand, discrimination between these two types of data is not here provided neither it is possible to explicitly relate the magnitudes of predictions to the 'with action' or 'without action' future states Bhatnagar (2009).

There is an opportunity in the Leopold matrix to double count impacts, therefore the method is not mutually exclusive. One of the strongest points of the Leopold matrix is that it draws a clear line and safeguards evaluator's opinion from the value judgments and political interference by policy makers. This is achieved by disaggregated presentation of detailed assessment results, leaving policy makers with full responsibility for the evaluation synthesis and for drawing its policy implications, and such point has been accepted as an evaluation standard for decades, being also maintained in the EU Impact Assessment Guidelines (SEC(2005)791) (Radej, 2010).

In terms of communication, even though Leopold matrix does not indicate the main issues or the groups of people who are most likely to be affected by the impact, it is still quite effective in communicating the results by providing a visual display on a single diagram (Decree on the Ecological Network, 2010).

3. CASE STUDY: The Use of Leopold Matrix in Carrying out the EIA for the "Kladovo" Wind Farm, Serbia

3.1 Description of Location

The area where the "Kladovo" Wind Farm is planned is located in the Ključ (Key) region, a large key-like river meander that the Danube forms right after it leaves the Iron Gate Gorge. It is surrounded by the Danube from the north, east and partly from the south, and the M-25 main road (E-771) from the west (Figure 1). The project covers an area of approximately 509 ha.

The location comprises only agricultural land with intensively cultivated monoculture fields, mainly without overgrown boundaries between them. There is a small number of neglected fields, as well as steppes, forest vegetation and wetland habitats.

The "Mala Vrbica" fish pond is located north and east of the immediate surrounding of the location. Because of its rich ornithofauna, the "Mala Vrbica" has a status of internationally and nationally Important Bird Area (IBA), protected by law as an ecologically significant area (Decree on the Ecological Network, 2010). The eastern boundary of the Đerdap National Park is less then 15 km west from the research area. This area also has the IBA status because of its rich ornithofauna. The area is also characterized by a rich chiropteran fauna. Nevertheless, due to the distance of this protected natural value and the fact that it is separated by the zone of changed urban and agricultural landscapes, the expected impact on the state of ornithofauna in this location is small, while neglectable on the chiropteran fauna.

The monitoring of ornithofauna and chiropteran fauna was carried out in the wider area planned for the "Kladovo" Wind Farm in the period between March 2011 to February 2012. The presence of representatives of 100 bird species was recorded, while the total of 173 bird species were recorded in a wider area of Ključ, including IBA of Mala Vrbica and a part of the Danube river course. Representatives of many of the recorded types were present in extremely small number.

The activity of 9 identified species of bats at the location planned for the “Kladovo” Wind Farm was recorded using ultrasound detectors. Despite the considerable number of recorded species, it should be pointed out that, during the monitoring, most of them were recorded in a relatively very small number, i.e. only several times in the total of 4,273 over-flights/contacts. Based on the abovementioned, it can be concluded that, trophically, as well as cryprically, resources for bats in the wind farm location were scarce during the monitoring period.

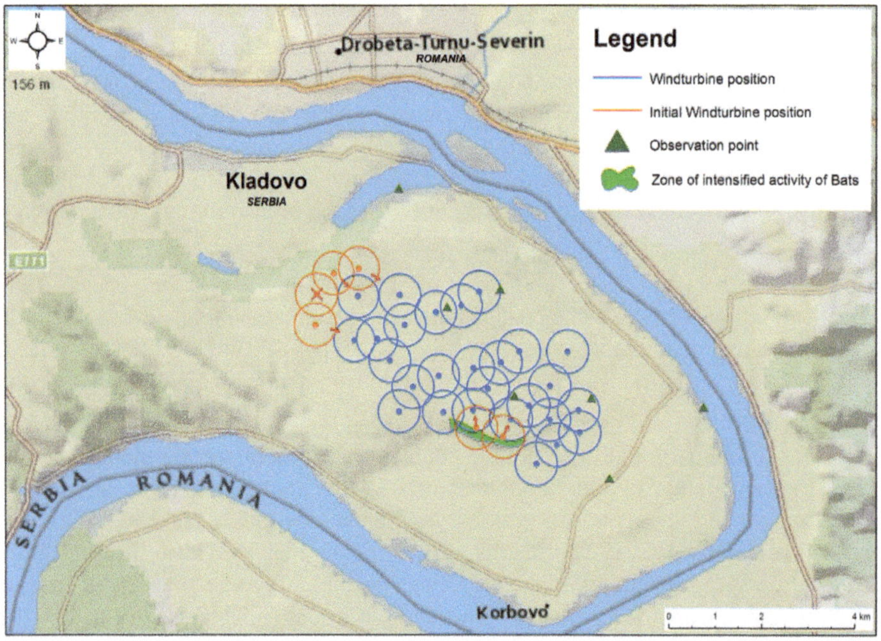

Figure 1. The location for the “Kladovo” Wind Farm relative to its wider surrounding (Source: Google Earth)

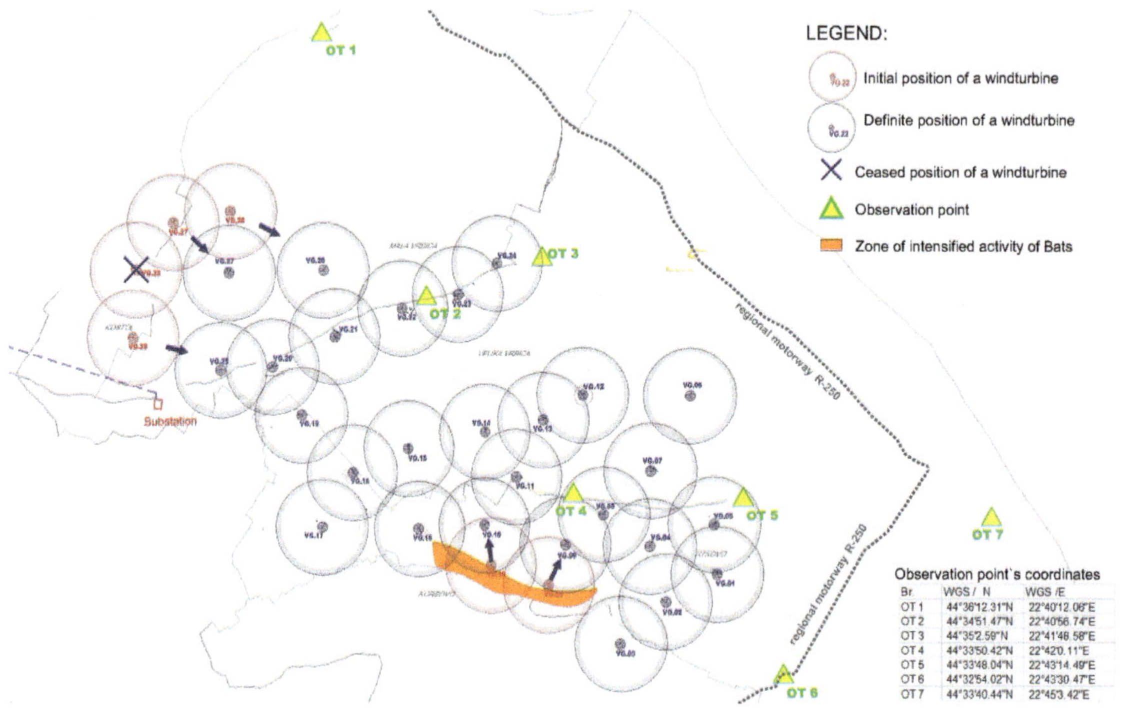

Br.	WGS / N	WGS /E
OT 1	44°36'12.31"N	22°40'12.06"E
OT 2	44°34'51.47"N	22°40'56.74"E
OT 3	44°35'2.59"N	22°41'48.58"E
OT 4	44°33'50.42"N	22°42'0.11"E
OT 5	44°33'48.04"N	22°43'14.49"E
OT 6	44°32'54.02"N	22°43'30.47"E
OT 7	44°33'40.44"N	22°45'3.42"E

Map 1. Final spatial disposition of wind turbines in the “Kladovo” wind farm adjusted to the results of one year monitoring of ornithofauna and chiropteran fauna

On Map 1, positions of wind turbine columns, later dislocated in order to protect ornithofauna and chiropteran fauna, are denoted by red color. In addition, one position at the west end of the location has been abandoned as unfavorable relative to possible impacts on certain protected bird species. The project envisages 27 wind turbines of 2 MW each, i.e. the total installed capacity of 54 MW.

3.2 Environmental Impact Assessment Using the Leopold Matrix

For the analysis of possible impacts of certain activities and procedures during the construction of "Kladovo" Wind Farm and its exploitation on the environmental elements, 9 possible factors, which are actually activities on the realization of the "Kladovo" Wind Farm project, have been singled out from a wider list of potential factors of impacts that can be expected for such type of interventions in nature. Although it is possible to partially determine aggregate, i.e. average assessment of impact factors for each of these components, we consider that it is sufficiently appropriate and functional to present them as a whole. The fact is that some of them are synergistic ones, mutually reinforcing their effects, so that this matching of information should be maintained in the analysis. A synthetic presentation of endangering factors is given through mean values and not through aggregate assessment, which will be later scaled.

Impact factors have been evaluated separately for each environmental component relevant for the scope of this study, and scored on a scale from 0 to 5 for <u>impact magnitude</u>, according to the following scale:

0 – no observable effect;

1 – low effect;

2 – tolerable effect;

3 – medium high effect;

4 – high effect;

5 – very high effect (devastation).

In addition to the standard form of the Leopold matrix, the following criteria have also been used:

<u>Impact significance</u> with designations from L to M, according to the following scale:

L – limited impact on location;

O – impact of importance for municipality;

R – impact of regional character;

N – impact of national character;

M – impact of cross-border character.

<u>Impact probability</u> with designations from M to I, according to the following scale:

M – impact is possible (probability of less than 50%);

V – impact is probable (probability of over 50%);

I – impact is certain (100% probability).

<u>Impact duration</u> with designation P (occasional/temporary) and D (long-term/permanent).

Furthermore, physical, biological and socio-cultural environmental characteristics of the subject location have been separated and, within them, 16 environmental components have been defined.

Effects of impact factors have been assessed for environmental components of the subject location (!). The results of the analysis are shown in Tables 1, 2, 3 and 4 for all environmental components and impact factors in the form of Leopold matrix, and then elaborated in an appropriate way.

3.3 Elaboration of the Assessed Effects of Impact Factors on Environmental Components

3.3.1 Physical Components

The subject project will not affect groundwater regime and quality. Wind turbines do not consume water, thus there is no waste water. Furthermore, the project will not affect microclimatic characters and parameters, neither will it stimulate the process of erosion.

As for possible impact on land, it can occur primarily as a result of the construction of turbine column foundations, manipulation of construction equipment at the location and inadequate waste material treatment during the wind farm construction. The total average value of expected impact magnitudes is within low effects

(1.33), dominantly of local character, with permanent or occasional effects.

Certain negative effects of the plan can occur in the phase of wind farm construction as a result of the realization of certain planning concepts, primarily construction of roads for the needs of wind farm complex, as well as wind farm administration complex. These impacts include air pollution caused by construction equipment and vehicles, as well as by created dust. The total average value of magnitudes of expected impacts of construction phase of the subject project on air quality is within low effects (0.77) of local character.

Besides total power capacity and size of wind turbines, the spatial aspect is particularly important when considering the noise intensity. The wind turbine noise decreases with increasing distance from wind turbines. In order to model the noise intensity levels, interpretation of the Regulation on the Methodology for Determining the Acoustic Zone was first interpreted (2010). The complex of the planned wind farm should be considered as the zone VI – industrial zone, while the nearest residential buildings should be considered as zone II – small and rural settlements. The abovementioned conclusion was taken into consideration in assessing the noise intensity levels from and their spatial dispersion in the area covered by the project. The modeling of wind turbine noise levels for the planned wind farm was performed using the Danish "calculator for noise levels from wind turbines" (Figure 2).

Table 1. Matrix of magnitude of the impact of factors on environmental components for WF, "Kladovo"

		PROJECT ACTIVITIES										
	Envisaged impact factors	Placement of wind turbines	Foundation of columns	The use of building materials	Substation construction	Transmission line construction	Construction of internal roads	Operation of construction equipment	Waste material Treatment	Project exploitation	Sum of IF values by types and bio. comp..	Average values
PHYSICAL COMPONENTS	Water	0	0	0	0	0	0	0	0	0	0	0.00
	Microclimate	0	0	0	0	0	0	0	0	0	0	0.00
	Land	1	2	1	1	1	1	2	2	1	12	1.33
	Erosion	0	0	0	0	0	0	0	0	0	0	0.00
	Air	0	0	0	1	0	2	3	1	0	7	0.77
	Noise	1	1	1	2	1	2	3	0	2	14	1.55
BIOLOGICAL COMPONENTS	Diversity of flora	0	1	0	0	1	0	1	1	0	4	0.44
	Diversity of fauna	2	1	1	1	1	0	2	2	2	12	1.33
	Ornithofauna	2	1	1	1	1	0	2	2	2	12	1.33
	Chiropteran fauna	2	1	1	1	1	0	2	2	2	12	1.33
	Barriers/corridors	2	1	1	1	1	0	1	1	2	10	1.11
SOCIO-CULTURAL COMPONENTS	Landscape	2	2	1	2	1	1	1	3	2	15	1.66
	Land use	1	2	1	1	1	1	1	2	1	11	1.22
	Economy	0	0	0	0	0	0	0	0	0	0	0.00
	Cultural heritage	0	2	0	0	0	0	0	0	0	2	0.22
	Accidents	2	0	0	1	0	0	0	2	2	7	0.77
Cumulative values of IF according to environmental factors		15	14	8	12	9	8	18	18	16		
Average		0.93	0.87	0.50	0.75	0.56	0.50	1.12	1.12	1.00	**IF = 0.82**	

Table 2. Matrix of significance of the impact of factors on environmental components for WF, "Kladovo"

	Envisaged impact factors	PROJECT ACTIVITIES: Placement of wind turbines	Foundation of columns	The use of building materials	Substation construction	Transmission line construction	Construction of internal roads	Operation of construction equipment	Waste material Treatment	Project exploitation
PHYSICAL COMPONENTS	Water									
	Microclimate									
	Land	L	L	L	L	O	L	L	L	L
	Erosion									
	Air				L		L	L	L	
	Noise	L	L	L	L	L	L	L		L
BIOLOGICAL COMPONENTS	Diversity of flora		L			L		L	L	
	Diversity of fauna	L	L	L	L	L		L	L	M
	Ornithofauna	L	L	L	L	L		L	L	M
	Chiropteran fauna	L	L	L	L	L		L	L	M
	Barriers/corridors	L	L	L	L	L		L	L	L
SOCIO-CULTURAL COMPONENTS	Landscape	L	L	L	L	L	L	L	L	L
	Land use	L	L	L	L	L	L	L	L	L
	Economy									
	Cultural heritage		L							
	Accidents	L			L				L	L

Table 3. Matrix of probability of the impact of factors on environmental components for WF, "Kladovo"

	Envisaged impact factors	PROJECT ACTIVITIES: Placement of wind turbines	Foundation of columns	The use of building materials	Substation construction	Transmission line construction	Construction of internal roads	Operation of construction equipment	Waste material Treatment	Project exploitation
PHYSICAL COMPONENTS	Water									
	Microclimate									
	Land	M	I	M	I	V	V	M	M	M
	Erosion									
	Air				V		M	I	M	
	Noise	V	V	M	V	V	V	I		V
BIOLOGICAL COMPONENTS	Diversity of flora		M			M		M	M	
	Diversity of fauna	M	M	M	M	M		M	M	M
	Ornithofauna	M	M	M	M	M		M	M	M
	Chiropteran fauna	M	M	M	M	M		M	M	M
	Barriers/corridors	M	M	M	M	M		M	M	M
SOCIO-CULTURAL COMPONENTS	Landscape	I	M	M	I	I	V	V	M	I
	Land use	I	I	V	I	I	I	I	M	I
	Economy									
	Cultural heritage		M							
	Accidents	M			M				M	M

Table 4. Matrix of duration of the impact of factors on environmental components for WF, "Kladovo"

		PROJECT ACTIVITIES								
	Envisaged impact factors	Placement of wind turbines	Foundation of columns	The use of building materials	Substation construction	Transmission line construction	Construction of internal roads	Operation of construction equipment	Waste material Treatment	Project exploitation
PHYSICAL COMPONENTS	Water									
	Microclimate									
	Land	P	D	P	D	P	P	P	P	D
	Erosion									
	Air				P		P	P	P	
	Noise	P	P	P	P	P	P	P		D
BIOLOGICAL COMPONENTS	Diversity of flora		D			P		P	P	
	Diversity of fauna	D	P	P	D	M		P	P	D
	Ornithofauna	D	P	P	D	M		P	P	D
	Chiropteran fauna	D	P	P	D	M		P	P	D
	Barriers/corridors	D	P	P	D	M		P	P	D
SOCIO-CULTURAL COMPONENTS	Landscape	D	D	P	D	D	D	P	P	D
	Land use	D	D	P	D	P	D	P	P	D
	Economy									
	Cultural heritage		D							
	Accidents	P			P				P	P

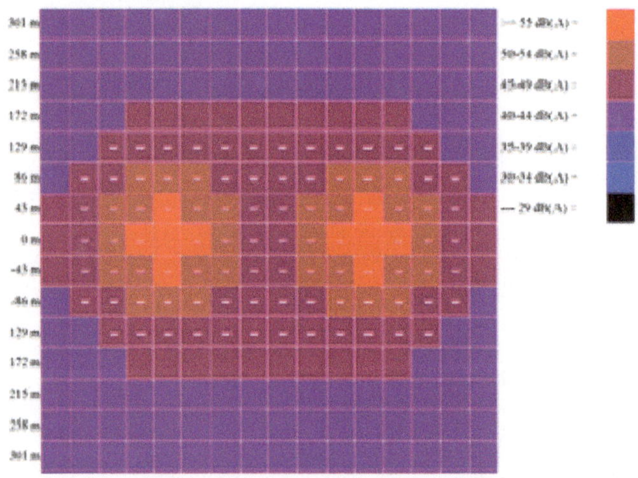

Figure 2. Spatial dispersion of wind turbine noise levels (Source: Danish Wind Industry Associoation)

Considering that in determining the disposition of wind turbine columns, the aspect of possible impact of noise has primarily been taken into account. Based on the model shown in Figure 2, it was found that wind turbine noise levels were lower than legally prescribed noise levels. The total average value of magnitudes of expected impacts of the project on noise levels is within low effects (1.55).

3.3.2 Biological Components

Based on analyzed data on the presence, directions and heights at which different types of birds fly over the area, as well as on the basis of the disposition of wind turbines, it can be concluded that the greatest impact of the construction and operation of the "Kladovo" Wind Farm will be on the most numerous and most frequent species of birds, such as common buzzard (*Buteo buteo*), as well as common raven (*Corvus corax*), although their number is incomparably smaller than the number of common buzzards. Other target species of birds flying at critical heights were smaller in number, or the critical heights were mostly out of the coverage of the parts of research location in which installation of wind turbines is planned. The shown positions of wind turbines have been carefully chosen, considering that they should be at least 1,000 m far from the Important Bird Area, as well as from the bed and right bank of the Danube river for which it was assumed to be an important migration corridor. In the location, no significant negative impacts have been assessed. Numerosity of birds belonging to each identified species is relatively small, because of which there are no significant effects on them. On the other hand, it is also necessary to emphasize the assessment of possible positive impacts of the wind farm construction and operation, as well as the construction of accompanying infrastructure, on certain species of birds. The total average of magnitude of expected impacts of the subject project on ornithofauna is within low effects (1.33).

As for bats, during the construction phase of the project, due to works on infrastructure, a part of woody and bushy plants in the location and its immediate surrounding, where an insignificant part of these types of plants is found, will be removed. Removal of woody plants in these zones will lead to the loss of these shelters, but this will have a negligible effect on local and migratory population, because almost the whole fund of their recorded and potential shelters is outside the zone of this impact. For some of these species, there are no, or no significant, hunting territories during all phases of their life cycle. Removal of woody plants in the zone of forest fragments near the "Đalu mare" hunting ground, which is the hunting territory of at least some significance in the location itself, would lead to the loss of the function of shelters. This would have a low effect on local and migratory population of these species, because the greatest part of their hunting territories is outside the wind farm location. However, after making suggestions to the investor on the importance of this forest fragment for chiropteran fauna found in the location, the disposition of wind turbines has been changed, thus the total risk of losing the hunting territory in the location is negligible. The total average of magnitude of expected impacts of the subject project on chiropteran fauna is within low effects (1.33).

Positioning and operation of wind turbines or other infrastructure in important flight corridors, or their immediate surroundings, would lead to the creation of barriers to normal daily transit activities of bats. Given that such activities of these or other species have not been recorded in the location, it can be concluded that the risk of losing flight corridors, if any, is negligible. Important migratory corridors of these species have not been recorded in the location. The total average of magnitude of expected impacts of the subject project on the creation of barriers/corridors is within low effects (1.11).

3.3.3 Socio–Cultural Components

The total average impact factor for these components falls into the category of low effects. The reason for this is because the location is isolated. The most important average value of impact factor is associated with general appearance of the landscape. Continual landscape transformation, as well as the complexity of landscape functions implies difficulties in meeting the obligations set out by documents at international and national levels (Maksin & Milijić, 2010). Visual effects on its surroundings are actually an subjective impression which, besides observer's perception, can also depend on the type of landscape and its specific visual characteristics. Analyzing the subject location, it has been concluded that wind turbines will dominate the surrounding, so that the landscape will be only partially changed by the construction of the planned wind farm. It has also been concluded that the planned disposition of wind turbines, although originally based on wind source assessment, will not disturb the landscape characteristics, but will create a specific visual identity. On the other hand, the exposure/visibility of the location is small considering the distance from the nearest buildings and road infrastructure, thus diminishing the importance of potential impacts of the planned wind farm on landscape and ambience values of a wider space. The total average of magnitude of expected impacts of the subject project on the landscape is within low effects (1.66).

Taking into account the terrain configuration, disposition of wind turbines and existing buildings in the surrounding of the planned location, as well as the sun trajectory, it can be concluded that the planned wind turbines will not cast shadows on neighboring properties.

Physically, wind turbines occupy only several percents of the land covered by the wind farm, while the rest of the land between turbine foundations and around internal roads can be used for other purposes, i.e. for original purposes without any limitations. In this case, the area can be further used for agricultural production. Therefore, there are no significant impacts in terms of land-use change. The total average of magnitude of expected impacts of the subject project on land use is within low effects (1.22).

Although there are no registered protected archaeological sites, neither there are other elements of cultural heritage in the space covered by the subject project, it is still possible to come across undiscovered archeological remains during excavations for wind farm foundations. In such situation, it would be necessary to undertake adequate measures in order to preserve the remains, or to stop the works and inform the Cultural Heritage Preservation Institute. The total average of magnitude of expected impacts of the subject project on cultural heritage is within theoretical possibility (0.22).

The risk of wind turbine accidents is very small. In the subject location, there is no immediate danger for the population even in case of serious damage. There are no residential settlements and roads in the surrounding area, thereby no any danger of wind turbine related accidents. In order to avoid possible accidental situations, it is also necessary to define within impact assessment the adequate protection measures and measures to reduce accident risks. The total average of magnitude of expected impacts of the subject project on the possibility of implying the accidental situations is within low effects (0.77) of local character.

3.4 Effects of Cumulative Impact Factors on Environmental Components

In Table 1, the activities in the phase of wind farm construction (operation of construction equipment and waste material treatment, installation of wind turbines and construction of foundations and substation), as well as possible impacts during the period of exploitation when possible impacts on ornithofauna and chiropteran fauna and increase in noise levels close to sound source can occur, are emphasized as factors having the greatest cumulative impact on environmental components.

The biological components are the most vulnerable primarily because of possible impacts on flying fauna. However, their effects range low. They are actually effects of low intensity only in the boundary area and mainly with limited time duration. The average assessment of effects of cumulative impact factors on environmental components is 0.82. This impact factor value is within low effects.

Nevertheless, if negative effects of these and others factors are considered as a whole, then it should be pointed out that qualitative and quantitative losses in living fauna will still be negligible and, practically, spatially and time limited only to the location itself. The survival of any of species of important, vulnerable or rare ecosystems, or other natural values, will not be called into question.

4. Discussion and Conclusions

Taking into account the problems in the realization of the wind farm project in Serbia that are caused by the resistance of a part of stakeholders resulting from the fear of negative effects on the environment, the idea of the research was to supplement the current methodology used in EIA. It has been decided to carry out an EIA because it presents an instrument for environmental protection which is, according to the Serbian legislation, the most transparent to stakeholders since they are involved in all phases of the elaboration of this document. Thus, an opportunity has been created for presenting to the public the extent and significance of possible negative impacts on space and environment in a simple and unambiguous way, as well as for overcoming the existing problems through discussions.

The initial assumption in the research was that the realization of the planned wind farm project implies the use of renewable energy, the wind energy, by using clean technology for its exploitation so that significant adverse impacts to the environment are not to be expected, as well as that the resistance of a part of stakeholders has resulted from the fear of the unknown because no wind farm in Serbia has been realized that could prove or disapprove the existing assumptions.

A special attention in carrying out EIA for the "Kladovo" Wind Farm has been dedicated to the analysis of the state of the environment of the location planned for the construction of the "Kladovo" Wind Farm and its wider surrounding. By their nature, wind farms can imply certain adverse impacts on the environment, both in the phase of construction (impacts of temporary character) and in the phase of its exploitation. Possible dominant negative effects have been related to the flying fauna, i.e. ornithofauna and chiropteran fauna. In this context, within the analysis of the state of the environment, monitoring of ornithofauna and chiropteran fauna was carried out in the period between March 2011 and February 2012 for the needs of undertaking the preventive measures for the protection of flying fauna. Based on observations, the final spatial disposition of wind turbines has been determined, i.e. certain wind turbine columns have been relocated to locations in which the possibility of jeopardizing the flying fauna has been minimized (preventive measures for the protection of flying fauna), while the results of observations have been incorporated into EIA and presented to the public in the earliest stage of EIA.

After the analysis of the state of the environment, as well as the analysis of project documentation, a multi-criteria evaluation of possible impacts of the "Kladovo" WF on environmental components using the Leopold matrix has been carried out. For the needs of the evaluation, 9 possible factors resulting from planned project activities have been singled out from the list of potential impact factors (jeopardizing factors) which could be expected for such type of interventions in nature. Impact factors have been evaluated for each environmental component relevant for EIA, as well as in relation to the impact magnitude. Furthermore, physical, biological and socio-cultural characteristics of the environment in the subject location have been separated, and the total of 16 environmental components defined within them.

All the above mentioned is the standard procedure in using the Leopold matrix which is a procedure for quantitative expert evaluation of planned activities related to the magnitude (intensity) of impact. However, by applying such approach, the spatial dispersion of identified impacts, probability of identified impacts actually occurring and, finally, the duration of identified impacts, still remains unclear. In this context, the idea was to additionally build the additional aspects into the Leopold matrix in form of a group of criteria that would be used in additional evolution of the planned project activities, assuming that the presentation and elaboration of obtained results would give to the public a complete picture of real impacts that could be expected in the realization of the "Kladovo" Wind Farm project. In this context, the following three additional criteria have been derived for: significance (spatial dispersion), probability and duration of impact, with the total of ten additional criteria. The evaluation of planned activities has been carried out for each group of criteria using the Leopold matrix. In addition to tabular presentation, the (positive and negative) results have also been elaborated in detail both for each factor and cumulatively in relation to environmental components.

The presented results that have been obtained using multicriteria analysis and evaluation have been surprisingly well received in the procedure of discussion and public insight in EIA for the "Kladovo" Wind Farm. The reason lies in the fact that adverse reaction of a part of the public has changed when it has been proven that certain identified negative effects will be micro-localized and that they will be only of a temporary character, as well as that they will occur only in a short period of time, which is, for example, the case with possible pollution during the wind farm construction. Or, when it has been shown that increased levels of noise caused by the rotation of wind turbine blades will be dominant only near the sound source, as well as that spatial dispersion of noise levels will exclude the possibility of exposure to higher levels of noise of the people living in the closest residential buildings considering their distance from the wind farm.

It can be concluded that the applied approach has yielded good results in overcoming the problems in the realization of the "Kladovo" Wind Farm project and that it is also possible to successfully use this approach for other similar projects in the field of implementation of renewable energy (RE) sources. In addition, the use of multicriteria analysis and evaluation provide to decison makers the possibility of considering possible effects of the project on the environment, as well as make the decision-making regarding the choise of optimal solutions a lot more simple. This is particularly important for Serbia because this will enable easier realization/implementation of renewable energy projects from the aspect of environmental conservation.

Future research directions for the EIA method, as well as for its development, could be also reflected in highlighting the positive implications of projects in the field of implementation of renewable energy (RE) sources, actually that the multi-criteria evaluation should not be based only on the identification and dominant presentation of negative effects of a planned project, but also on a balanced presentation of both the positive and negative effects, in this way bringing greater objectivity to the process of impact assessment. In this process, a special attention is to be dedicated to the use of: GIS tools; AHP and similar mathematical methods and models in the process of environmental impact assessment in order to minimize subjectivity in this process, as well as in the decision-making process.

Acknowledgments

This paper has resulted from the research conducted within the scientific projects "Spatial, ecological, energy, and social aspects of settlement development and climate change - their mutual impacts", TR 36035; "Sustinable spatial development of Danubian Serbia", TR36016; and "The role and implementation of the national spatial plan and regional development documents in renewal of strategic research, thinking and governance in Serbia", III47014, all financed by the Ministry of Education, Science and Technological Development of the Republic of Serbia in the period 2011-2014.

References

Bhatnagar, K. C. (2009). Environmental management - New Directions for the 21st Century, New Delhi: Global India Publications Pvt Ltd.

Chatterjee, N. C., & Bose, G. K. (2013). Selection of vendors for wind farm under fuzzy MCDM environment.

Crnčević, T., Marić, I., & Josimović, B. (2011). Strategic Environmental Assessment and Climate Change in the Republic of Serbia – Support to the Development and Adjustment Process, SPATIUM. No 26, 14-19.

Decree on the Ecological Network (“Official Gazette of the Republic of Serbia”, No. 102/10)

Energy Sector Development Strategy of the Republic of Serbia until 2015 (“Official Gazette of the Republic of Serbia”, No. 44/05)

Ismail, W. K. W., Abdullah, L., Zhang, J., & Zhang, W. (2012). A new Environmental Performance Index using analytic hierarchy process: a case of ASEAN countries.

Josimović, B., & Pucar, M. (2010). The strategic environmental impact assessment of electric wind energy plants: Case study ‘Bavanište’ (Serbia). *Renewable Energy, 35*(7), 1509-1519. http://dx.doi.org/10.1016/j.renene.2009.12.005

Leopold, L. B., Clarke, F. E., Hanshaw, B. B., & Balsley, J. R. (1971). A Procedure for Evaluating Environmental Impact in Geological Survey Circular 645, USGS, Washington DC.

Maksin, M., & Milijić, S. (2010). Strategic planning for sustainable spatial, landscape and tourism development in Serbia, SPATIUM. No 23, 30-37

Radej, B. (2011). Synthesis in policy impact assessment. *Evaluation, 17*(2), 133-150.

Regulation on the Methodology for Determining the Acoustic Zone ("Official Gazette of RS", No. 72/10)

6

Enhancing Furnace Thermal Efficiency by Adjusting Fuel Temperature

Chien-Li Lee[1] & Chih-Ju G. Jou[2]

[1] Research and Development Center for Water Resource and Conservation, National Kaohsiung First University of Science and Technology, Taiwan

[2] Department of Safety, Health and Environmental Engineering, National Kaohsiung First University of Science and Technology, Taiwan

Correspondence: Chih-Ju G. Jou, Juoyue Road, Nantsu District, Kaohsiung 811, Taiwan.
E-mail: george@ccms.nkfust.edu.tw

Abstract

Increasing fuel temperature leads to lower fuel density so that speed of the fuel leaving the nozzle increases causing the fuel to reach the burning point sooner. Hence, a higher temperature can be maintained in the furnace to reduce heat loss. Increasing the fuel temperature also causes higher temperature of the hot gas surrounding the burner nozzle, faster flowing velocity for the hot gas flow in the furnace, more even thermal distribution in the furnace, as well as faster thermal flow rising velocity and higher temperature in the furnace radiation zone. Hence, the fuel consumption and carbon dioxide emission are significantly reduced. In this study, results obtained with a full-scale furnace show that when the fuel temperature is raised from 95 °C to 130 °C, the hot gas temperature in the convective zone of the furnace rises by 21 °C; the furnace pressure increases by 2.5 mm H_2O so that oxygen concentration in the excess air drops by 1.2 vol %. Under these conditions, 7.5 x 10^5 m^3 of fuel cost can be saved, and 858 ton per year of CO_2 emission can be reduced.

Keywords: fuel temperature, carbon dioxide, thermal furnace

1. Introduction

Increasing fuel demands as a result of booming global economy (Yang, Wang, Zhou, & Liu, 2012) leads to gradual depletion of petroleum fuel in addition to causing global warming problem (Pitt, Randolph, Jean, & Chang, 2012). Raising the thermal efficiency for furnaces burning petroleum fuel (Zhang, Wang & Zhang, 2012), improving the incineration efficiency (Flamme, 2001), and recovering energy from waste heat and waste gas become the most pertinent strategies before an alternative energy source can be realistically developed (Kimemia & Annegarn, 2012).

Many advanced industrial nations have carried out research to develop new or improved technologies for improving the thermal efficiency of industrial heat exchanger, boiler and furnace (Lunghi & Burzacca, 2004). Some of these technologies include adjusting the furnace vacuum, recovering hot exhaust gases, regulating the residual oxygen concentration in excess air, and developing alternative energy. For example, Lee and Jou used the heat recovered from the flue gas to raise temperature of the pre-heated air for reducing the thermal loss and raising the furnace efficiency to save fuel consumption (Lee & Jou, 2011). Lee and Jou reduced the oxygen concentration in the flue excess air for raising the thermal efficiency of boiler and furnace to reduce fuel consumption and carbon dioxide emission (Lee & Jou, 2012); Lee, Jou, Tsai and Wang (2007); Jou, Lee, Tsai and Wang (2008); Jou et al. (2010) applied the recovered tail gas from petrochemical processes as an alternative fuel to replace natural gas and diesel fuel to power a full-size furnace as well as a steam-gas cogeneration high-pressure boiler and furnace, the results show that using the recovered tail gas has the potential of saving fuel cost and reducing carbon dioxide emission.

The fuel composition influences the flame appearance significantly (Ko, Hou, & Lin, 2003). For example, Choudhuri and Gollahalli (Choudhuria & Gollahalli, 2003) added various portions of natural gas and propane fuel to hydrogen gas for studying variations of the resulting flame appearance. The results show that at fixed Reynolds number, the hydrogen flame length increases but the flame width decreases with more natural gas or propane in the fuel. Additionally, the hydrogen-propane mixture has longer flame length than the hydrogen gas fuel. The hydrogen gas flame is bright orange in color from the flame center to the far burning region with blue and slight reddish color near the front of burner tip. In contrast, the hydrogen-propane flame has the same bright organic color from the flame center to the far burning region; but it is slightly shorter with reddish color near the front of the burner tip. The change in flame appearance is caused by higher C/H ratio in propane that leads to the formation of ashy smoke to result in more flame formation and oxidation.

There is little information published in literature concerning the influence of fuel oil temperature on the overall furnace thermal efficiency. In this study, the PSA (Pressure swing adsorption) method is used to recover hydrogen gas from the tail gas emitted from the hydrogen producing process. The recovered hydrogen gas is reused as a fuel additive to power a full-scale distillation unit furnace for studying on-site the influence of fuel temperature on the furnace cost-effectiveness.

2. Experimental Equipment and Methods

2.1 Equipment

The experimental equipment is a set of heating furnace used in the distillation operation (Figure 1) (Petro-Chen Development Co. Inc., 1993). The furnace set consists of two identical furnaces with dimensions of 11.4 m in length, 5.0 m in width and 10.0 m in height. It is equipped with 14 burners with a total of 28 burners that have a total rated heat releasing capacity of 68.15 mm kcal/hr (LHV). Both furnaces are capable of burning fuel oil as well as mixture of gas and No. 6 fuel that contains (wt%) 86.03% C, 12.5% H, 0.5% S, 0.35%N, 0.625% O, and 0.05% ash is used. The average hydrocarbon components include (in volume) 64.2 mole% H_2, 11.1 mole% CH_4, 8.4 mole% C_2H_6, 0.4 mole% C_2H_4, 6.0 mole% C_3H_8, 0.2 mole% C_3H_6, 1.5 mole% n-C_4^+, 1.8 mole% i-C_4^+, 0.7 mole% i-C_5^+, 0.3 n-C_5^+, and 0.7 mole% n-C_6^+.

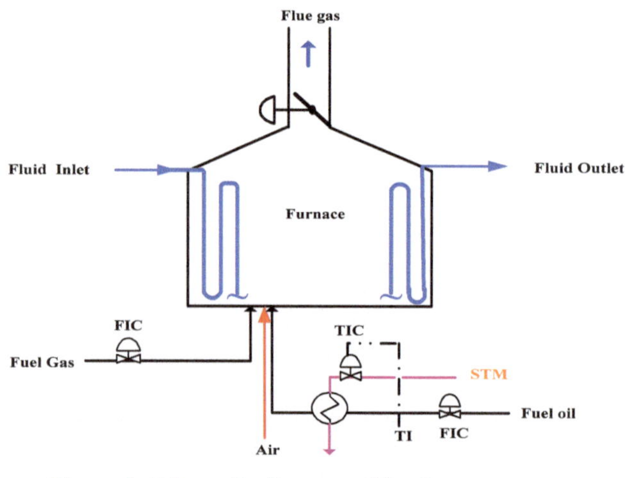

Figure 1. Schematic diagram of the furnace system

2.2 Methods and Analyses

This study was conducted by using the ACC (Automatic Combustion Control) system to operate the furnace. The temperature of the pre-heated gas prior to entering the furnace was used as a control parameter; it was maintained constant at 330 °C. During the experiment, the exit temperature of fuel gas and pre-heated gas mixture was controlled by adjusting the fuel input in response to variations of the calculated exit temperature. Fuel oil (FO) and fuel gas (FG) have different heat values; 1.3 m^3 of FG generate the same heat as 1 m^3 of FO; and the volumetric flow rate of fuel is calculated by multiplying the percentage of opening for control value by the maximum value volumetric flow rate and then by the ratio of fuel heat value.

The furnace was loaded 100% during the study period with air baffle and flue damper maintained at the same degrees of opening. The fuel temperature that was raised by the low pressure steam through a heat exchanger was gradually increased from 95 °C to 100 °C, 105 °C, 110 °C, 115 °C, 120 °C, 125 °C and ultimate to 130 °C by controlling the temperature control valve for investigating the influence of fuel temperature on the overall furnace thermal efficiency. When the system operation reaches steady-state after each adjustment of fuel temperature, duplicate operational data were recorded every minute for 4 hours and repeat the secondary. The O_2 concentration was continuously monitored on-line with CEMS (Continuing Emission Monitor System, E.S.A.,

MIR 900); it has 5.9% relative accuracy and 0.67 confidence coefficient with ±2 °C temperature error. The monitoring system was calibrated every six months according to the standard procedures in NIEA A442.70C. Temperature was measured using a thermocouple (K-Type, error is 0.3% of full scale); pressure was measured using Bourdon Tube with 0.3% error of the full scale (Ashcroft Co.). The control value is a gate value (Fisher Co.) with an error of 0.3% when it opens fully.

3. Results and Discussion

3.1 Relationship between the Convection Region Temperature and Furnace Vacuum

Raising the fuel temperature causes the fuel density to decrease. If the fuel flow rate is maintained unchanged, the fuel flow velocity must be increased. The higher speed for fuel exiting the burner nozzle shortens the time for the fuel to reach the burning point so that the flame volume is reduced. This causes higher temperature and longer residence time for the hot gas flowing in the furnace so that the heat can be evenly distributed in the furnace radiation zone. The rapid rise of the thermal flow from the furnace radiation zone to the convection zone raises temperature of the furnace convection zone. Results shown in Figure 2 indicate the tendency of rising temperature in the furnace convection zone. If the fuel temperature is raised from 95 °C to 130 °C, the furnace convection zone temperature increases from 674 °C to 695 °C. On the other hand, if openings for fresh air baffle and flue gas damper unchanged when raising the fuel temperature is raised from 95 °C to 130 °C, Figure 3 shows that the furnace pressure increases from -6.8 to -4.3 mm H_2O.

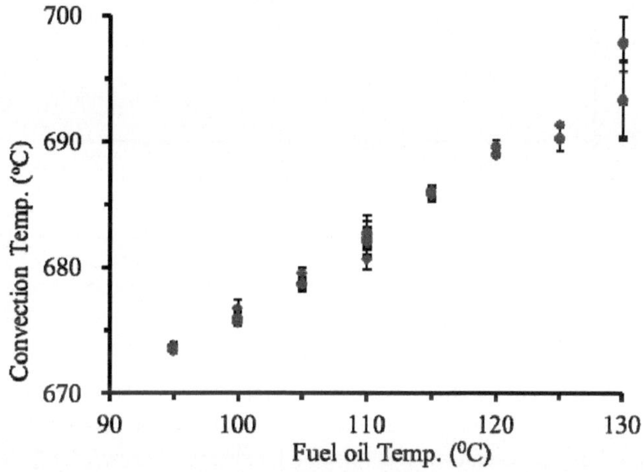

Figure 2. Relationship between fuel temperature and furnace convection zone temperature

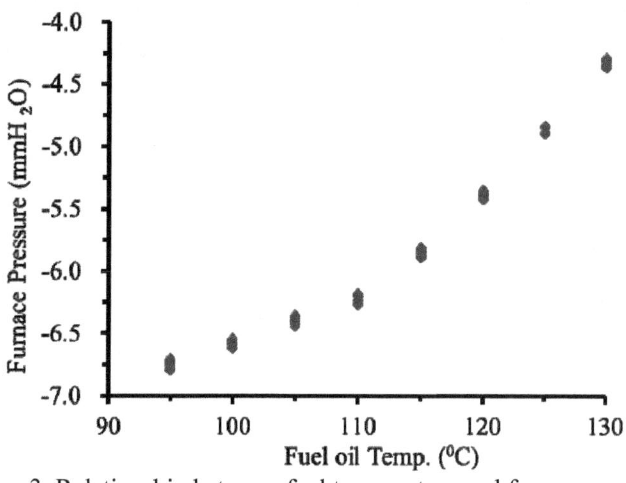

Figure 3. Relationship between fuel temperature and furnace pressure

3.2 Influence of Fuel Temperature on Oxygen Concentration in Flue Gas

As shown in Figure 3, when the fuel temperature is raised from 95 °C to 130 °C, the furnace internal pressure increases from -6.8 to -4.3 mm H_2O. The higher furnace pressure prolongs the residence time for the heat energy staying in the furnace radiation zone. Additionally, results shown in Figure 2 reveal that higher furnace internal pressure causes the furnace convection zone temperature to increase from 674 °C to 695 °C. Hence, when mixture of air and fuel is used, and the furnace is subject to higher temperature, the oxygen concentration in the uprising hot gas flow is lowered causing a decreasing tendency of O_2 concentration in the flue gas (Figure 4). The oxygen concentration in the flue gas drops from 3.2 vol % to 2.9 vol % when the fuel temperature is raised from 95 to 105 °C; it further drops to 2.6 vol % and 2.0 vol % if the fuel temperature is increased to 115 °C and 130 °C, respectively.

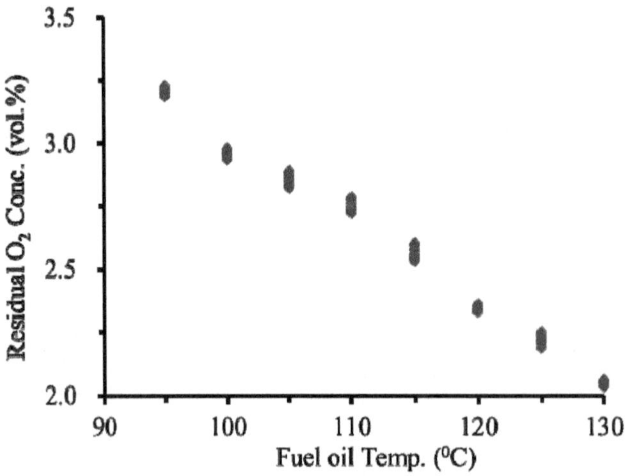

Figure 4. Relationship between fuel temperature and oxygen concentration in the flue gas

3.3 Influence of Fuel Temperature on Fuel Flow Rate

Lower oxygen concentration in the excess air indicates higher furnace burning chamber temperature. The higher temperature causes more efficient transfer of thermal energy across the furnace metal wall to raise the conductive thermal flow across the metal furnace wall and hot fluid pipe leading to a higher furnace heat absorbance efficiency. Additionally, when the hot gas stream in the furnace has higher temperature, the time for the fuel to reach the burning point is shortened thus the furnace energy consumption can be reduced. Results shown in Figure 5 indicate that the quantity of fuel consusmed in the furnace has a decreasing tendency with increasing fuel temperature. When the fuel temperature is maintained at 95 °C, the fuel gas flow rate is 1183 m^3/h. Increasing the fuel temperature to 105 °C, 115 °C, and 130 °C causes the gaseous fuel flow rate to drop to 1161 m^3/h, 1145 m^3/h and 1098 m^3/h, respectively. Therefore, maintaining the fuel temperature at 130 °C will save 7.5 $\times 10^5$ m^3 of fuel consumption and reduce 858 tons per year of CO_2 emission.

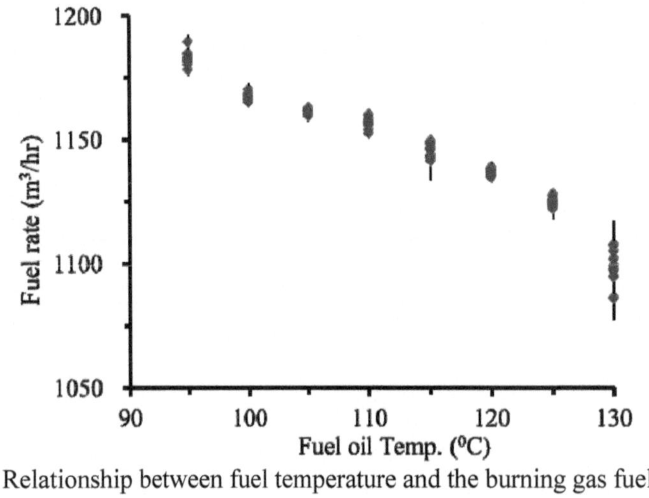

Figure 5. Relationship between fuel temperature and the burning gas fuel flow rate

4. Conclusion

The heat generated in a furnace is transferred via radiation from the fuel burning chamber to the thermal fluid confined in the thermal flow pipe so that the heat is carried away to achieve the objective of hating. Higher fuel temperatures causes the fuel density to drop so that the fuel emitting speed leaving the burner increases leading to shorter time for the fuel to reach the burning point, higher temperature in the burning chamber and less thermal loss. In the study, when the fuel temperature is gradually raised from 95 °C to 130 °C, on-site results obtained with a currently operated full-scale furnace show that the raised fuel temperature causes the temperature of the furnace convection zone to increase from 674 °C to 695 °C, furnace pressure to rise from -6.8 mm H_2O to -4.3 mm H_2O, O_2 concentration in the flue gas to drop from 3.2 vol.% to 2.0 vol.%. Because higher temperature in the burning chamber causes lower oxygen concentration in the excess air, more heat is obviously absorbed and carried away by the thermal flow from the furnace. Additionally, when the hot gas flow in the furnace is under higher temperature, the time for the fuel to reach the burning point becomes shorter so that less fuel is consumed. If the fuel temperature is increased from 95 °C to 130 °C, the fuel consumption is reduced by 85 m^3/h with 7.5×10^5 m^3/yr of fuel saved and 858 tons per year of CO_2 reduced.

References

Choudhuria, A. R., & Gollahalli, S. R. (2003). Characteristics of hydrogen–hydrocarbon composite fuel turbulent jet flames. *International Journal of Hydrogen Energy, 28*, 445-454. http://dx.doi.org/10.1016/S0360-3199(02)00063-0

Flamme, M. (2001). Low NOx combustion technologies for high temperature application. *Energy conversion and management, 42*, 1919-1935. http://dx.doi.org/10.1016/S0196-8904(01)00051-6

Hainsworth, D., Pourkashanian, M., Richardson, P., Rupp, J. L., & Williams, A. (1995). The Influence of Carbon Dioxide on Smoke Formation and Stability in Methane-Oxgyen-Carbon Dioxide Flames. *Fuel, 75*, 393-396. http://dx.doi.org/10.1016/0016-2361(95)00265-0

Jou, C. J. G., Lee, C. L., Tsai, C. H., & Wang, H. P. (2008). Reduction of energy cost and CO_2 emission for the boilers in a full-scale refinery plant by adding waste hydrogen-rich fuel gas. *Energy and Fuels, 22*, 564-569. http://dx.doi:10.1021/ef070180z

Jou, C. J. G., Lee, C. L., Tsai, C. H., Wang, H. P., & Lin, M. L. (2008). Enhancing the performance of a high-pressure cogeneration boiler with waste hydrogen-rich fuel. *International Journal of Hydrogen Energy, 33*, 5806-5810. http://dx.doi.org/10.1016/j.ijhydene.2008.05.114

Jou, C. J. G., Wu, C. R., & Lee, C. L. (2010). Reduction of Energy Cost and CO_2 Emission for the Furnace Using Energy Recovered from Waste Tail-Gas. *Energy, 35*, 1232-1236. http://dx.doi.org/10.1016/j.energy.2009.11.002

Kimemia, D. K., & Annegarn, H. J. (2012). Productive Uses of Basic Energy and Fuel Transitions in Urban South Africa. *Energy and Environment Research, 2*, 103-112. http://dx.doi.org/10.5539/eer.v2n2p103

Ko, Y. C., Hou, S. S., & Lin, T. H. (2003). Incineration of oil gas at various compositions and supply rates. *Proceedings of 14th International Symposium on Transport Phenomena*, pp. 445-449.

Lee, C. L., & Jou, C. J. G. (2011). Saving Fuel Consumption and Reducing Pollution Emissions for Industrial Furnace. *Fuel Processing Technology, 92*, 2335-2340. http://trove.nla.gov.au/version/172679177

Lee, C. L., & Jou, C. J. G. (2012). Improving Furnace and Boiler Cost-Effectiveness and CO_2 Emission by Adjusting Excess Air. *Environ Progress, 31*, 157-163. http:// dx.doi.org/10.1002/ep.10543

Lee, C. L., Jou, C. J. G., Tsai, C. H., & Wang, H. P. (2007). Improvement of a medium-pressure-boiler performance through adjustments on the hydrogen-rich fuel for refinery plant. *Fuel, 86*, 625-631. http://dx.doi:10.1016/j.fuel.2006.09.001

Lunghi, P., & Burzacca, R. (2004). Energy recovery from industrial waste of a confectionery plant by means of BIGFC plant. *Energy, 29*, 2601-2617. http://dx.doi.org/10.1016/j.energy.2004.05.016

Pitt, D., Randolph, J., Jean, D. S., & Chang, M. (2012). Estimating Potential Community-wide Energy and Greenhouse Gas Emissions Savings from Residential Energy Retrofits. *Energy and Environment Research, 2*, 44-61. http://dx.doi.org/10.5539/eer.v2n1p44

Yang, G., Wang, H., Zhou, J., & Liu, X. (2012). Analyzing and Predicting the Economic Growth, Energy Consumption and CO2 Emissions in Shanghai. *Energy and Environment Research, 2*, 83-91. http://dx.doi.org/10.5539/eer.v2n2p83

Zhang, S., Wang, S., & Zhang, R. (2012). Analysis on the Potential of Greenhouse Gas Emission Reduction in Henan's Electricity Sector. *Energy and Environment Research, 2*, 195-204. http://dx.doi.org/10.5539/eer.v2n1p195

7

A Study of Energy Conservation Policies at (Primary) Eco-Schools in Istanbul

Kutlu Sevinç Kayıhan[1] & Seda Tönük[2]

[1] Department of Architecture, Faculty of Architecture, Gebze Institute of Technology, Kocaeli, Turkey

[2] Department of Architecture, Faculty of Architecture, Yıldız Technical University, İstanbul, Turkey

Correspondence: Kutlu Sevinç Kayıhan, Faculty of Architecture, Gebze Institute of Technology, Çayırova, Gebze, Kocaeli, 101, 41400, Turkey. E-mail: sevinc@gyte.edu.tr

Abstract

Approximately 50% of the energy produced on earth is used by buildings. As for all building types, ensuring the energy conservation in primary school buildings is important in many ways. In addition to environmental and economic benefits, the systematic application of sustainable energy consumption in primary school buildings also makes a positive contribution to the education of future generations. The "Eco-Schools International Programme" aims to introduce environmental management systems into primary schools within the scope of ISO 14001/EMAS.

This study examines primary schools in Istanbul that are part of the Eco-Schools International Programme. A questionnaire-based survey was administered on the theme of energy management within primary schools located in Istanbul city, and the findings of a field observation were examined. Questionnaire findings were categorized under three main themes: *energy usage and conservation policies at the eco-schools' buildings' envelope*, *energy usage and conservation policies in the indoor environments of eco-schools*; and *types of energy sources used at eco-schools*. According to the main results of the study; solar control elements on the facades, methods aimed at energy conservation from windows, energy-effective woodwork applications and independent control of energy systems (which are used in a small proportion of eco-schools) should be discussed and integrated in all eco-school buildings. It is seen that natural gas is used by 100% of survey respondents and that renewable energy resources are used only at low levels for testing purposes and/or to make small contributions to energy gain. In this context, expanded use of renewable energies at all the eco-schools should be considered.

Keywords: energy conservation, eco-schools programme, Istanbul eco-schools

1. Introduction

Rapidly increasing world population, technological developments and increase in standards of living have resulted in significant increases in demand for resources, particularly in energy consumption. The processes of energy generation and consumption are the most important factor in the depletion of global natural resources, the increase in CO_2 emissions to the atmosphere and, indirectly, global climate change. The necessity of energy for both natural and industrial systems is defined by Kibert, Sendzimir and Guy as follows: "Both natural and industrial systems require energy to reproduce and maintain their functions. Natural systems, for the most part, use solar flux or stored solar energy in the form of biomass for their functioning, whereas industrial systems use a wide variety of energy sources. In the present era, industrial systems operate largely by using stored solar energy in the form of fossil fuels, but these are being consumed at a pace of the order of 10,000 times their regeneration rate" (Kibert, Sendzimir, & Guy, 2002). Pitts makes the following explanation for the increased energy requirement and changes triggered by the industrialization process: "One of the most significant changes brought about by the industrialization process has been the almost complete reliance on non-renewable fossil-fuel based energy sources to provide the power, not just to drive the industrial processes but also to support the infrastructure, including construction, buildings and especially transport." (Pitts, 2004).

The oil crises of 1973 and 1979 demonstrated to the industrialized countries their dependency on energy in order to sustain the living standards they had achieved. International Energy Agency (IEA) projections point out that, as long as the current energy policies and energy supply preferences continue to be pursued, global primary energy demand will increase by 40% in the period 2007-2030 (www.iea.org). While the world population has doubled since 1950, the energy demand has increased by 6 times. Global population is estimated to be 6.5 billion today and to increase to 7.2 billion in 2015 and 8.9 billion in 2050 (http://www.energy.itu.edu.tr/iTUOnerileri.pdf).

Building-related energy consumption represents a huge share of total energy consumption: "Buildings consume more than half the energy used worldwide (Gissen, 2002, p. 19). Due to their embodied and operational energy consumption, buildings have significant environmental effects. According to Graham, buildings affect the environment in the following ways (Graham, 2003):

- The physical disruption caused by mining energy resources,
- Pollution produced by energy conversion processes,
- Greenhouse gas emissions,
- Dealing with wastes.

Sassi suggests a three-step approach to minimize the environmental effects caused by the energy consumption of buildings: "First, building fabric design alternatives selected to provide the same performance with reduced energy requirements. Second, if a zero-energy design solution is not possible, active systems should be selected that use energy in an efficient way. Third, the resulting reduced energy requirements should be provided by alternative, low CO_2-emitting energy sources" (Sassi, 2006).

Energy conservation and management policies can be achieved by introducing the subject into laws and regulations, making comprehensive studies on this issue and raising public awareness. CHPS (Collaborative for High Performance Schools), one of the largest-scale studies on sustainable school buildings, emphasizes the environmental, economic and educational advantages of the subject as follows: "Energy-efficient schools cost less to operate, which means that more money can be used for books, computers, teacher salaries and other items essential to the educational goals of schools. According to the criteria specified under the "energy" title of CHPS, an additional 2-10 points can be awarded if the school building design contributes to reducing consumption by 10% or 15-35% of the National "Title 24-2001 California energy efficiency standards (CHPS, v. II.-III., 2009).

It is known that developed countries make various attempts to raise environmental awareness among primary education-aged children, who constitute the future of the society; School buildings serve as a useful tool in realizing such awareness-raising attempts. In this scope, many organizations have been established in order to minimize the negative environmental effects of school buildings; to take specific measures in various fields such as energy, water and waste; and to raise environmental awareness among children. The "Eco-Schools Programme", successfully implemented in different parts of the world, can be such an organization. Many Turkish schools are participants in the Eco-Schools Programme.

The present study assesses the energy usage and conservation methods at primary schools which are part of the Eco-Schools Programme and located in Istanbul. Istanbul is the largest city in Turkey with an area of around 5750 km^2 and a population of around 13.2 M (2012). The scope of the study includes detection and assessment of the following policies pursued by the eco-schools participating in the study:

- Energy usage and conservation policies at the eco-schools' buildings' envelope (building façade, windows and entrances of the school building),
- Energy usage and conservation policies in indoor environments (doors, lighting components, independent control of energy, motion/daylight sensors),
- Types of energy sources used at eco-schools.

Before discussing energy conservation in the eco-schools in Istanbul, the study briefly summarizes energy consumption in Turkey: During the period 1990-2008, the annual increase in primary energy demand was 4.3%, on average. During the last decade, Turkey had the most rapid increase in energy demand among OECD countries. Since 2000, Turkey has also seen the second-largest increase in demand for electricity and natural gas, after China. Projections of the Ministry of Energy and Natural Resources of the Republic of Turkey show that this tendency will continue in the mid-term (http://www.enerji.gov.tr/index.php?dil=tr&sf=webpages&b=enerji&bn=215&hn=12&nm=384&id=384).

When compared to global reserves, the primary energy resources in Turkey are low in terms of quantity and quality. The failure to meet the required quantity and quality and the high cost of energy generation from these resources, results in an energy deficit in Turkey. Turkey as an energy-importing country, smart transfer of energy technologies should be adopted to increase energy efficiency. In order to reduce and/or eliminate the energy deficit primarily appropriate energy-related policies should be determined and put into implementation (www.dtm.gov.tr/dtmadmin/upload/ead/tanitimkoordinasyondb/6sa7.doc).

2. Eco-Schools International Programme

Eco-Schools is a programme for environmental management and certification, designed to implement sustainable development education in schools by encouraging children and youths to take an active role in how their school can be run for the benefit of the environment (http://www.eco-schools.org). The Eco-Schools International Programme was developed as a response to some of the needs identified at the UN Conference on Environment and Development in 1992. The Programme started in 1994, with the support of the European Commission (DG XI and DG XXII), in 4 countries: Denmark, Germany, Greece and United Kingdom. In 1999, the organization received the "Worldaware Award for Global Education" from the North-South Centre of Council of Europe and NCDO. In 2003, Eco-Schools was identified by the United Nations Environment Programme (UNEP) as a model initiative for Education for Sustainable Development (http://www.eco-schools.org).

The Eco-Schools Programme employs an holistic, participatory approach, combining learning and action, thus providing an effective method for improving the environments of schools, raising awareness and producing behavioral change in young people, school staff, families, local authorities and others with a significant influence within local communities. Eco-Schools is one of the programmes of the FEE - Foundation for Environmental Education (www.fee-international.org) and, as such, it is implemented through FEE Member organizations (one per country). Currently, the Programme is being implemented in 47 countries around the world, involving 32,156 schools (9.898 of which have already been awarded with various Eco-Schools certificates), 9,125,460 students, 628,005 teachers and 5,013 local authorities. The Eco-Schools Programme is based on ISO14001:2004. The Eco-Schools methodology encompasses seven steps that any school can adopt (http://www.eco-schools.org).

In Step 2, work begins with a review or assessment of the environmental impact of the school. The results of the school's Environmental Review are essential to build the subsequent Action Plan, assisting the school to decide whether change is necessary, urgent or not required. The Review also helps schools to set realistic targets and measure their success. Schools may firstly quantify their impacts for the theme they are working on or cover all the areas in which the school may have environmental impacts, namely: waste management, waste minimization, school soils, biodiversity, energy, water, transport, health and well-being, and sustaining the world (http://www.eco-schools.org/page.php?id=53).

The action plan targeted creating in Step 3 is the core of the Eco-Schools work and is developed using the results of the environmental review (http://www.eco-schools.org/page.php?id=54).

Information and advice presented under the Eco-Schools theme of "energy" are included below:

The energy issues within the Eco-Schools programme suggest ways in which all members of the school can work together to increase awareness of energy matters and to improve energy efficiency within the school. Eco-Schools are required to carry out an energy audit as part of their Environmental Review and then set targets for reducing unnecessary energy use through their Action Plan. The "Environment and Innovation" project encourages Eco-Schools to come up with innovative and creative solutions to environmental problems. Winning entries receive a grant to help them put their ideas into action and to work with the local community to tackle issues.

The project's new, two-year cycle on "Climate Change: Let's Save Energy!" is active in six countries: Denmark, Finland, Italy, Norway, Turkey and Portugal (www.eco-schools.org/page.php?id=30).

From October 2008 as part of the European Energy Performance of Buildings Directive, all state schools with a floor area over 1000 m2 are required to display a certificate that rates their energy use on a scale of A-G. The new Display Energy Certificate (DEC) looks like the A-G rating you would expect to see when purchasing a new electrical appliance. It is part of the Green Flag criteria to include the school's DEC with the application to the Eco-Schools programme. This is so that the Eco-Schools project administrators can collect data to establish any trends (http://www.eco-schools.org.uk/nine-topics/energy.aspx).

3. Material and Method

The city selected for the present study, Istanbul, is the largest city by population and, economically, the most important city in Turkey. Istanbul has the 34th largest economy in the world and is the most crowded city in Europe (Wikipedia, 2010a). In this scope, data from the studies conducted in Istanbul will reflect the wider Turkish case at a significant level.

Directly proportional to economic development level and population, the number of educational institutions in Istanbul is higher than in other Turkish cities. Moreover, a great majority of the schools involved in Eco-Schools programs are in Istanbul. As of mid-2009, Istanbul had 55 schools participating in the Eco-Schools programme. Questionnaires were distributed to these 55 schools and 37 (67%) schools returned completed questionnaires. The distribution of participating schools located in the European and Asian sides of Istanbul is shown in Figure 1.

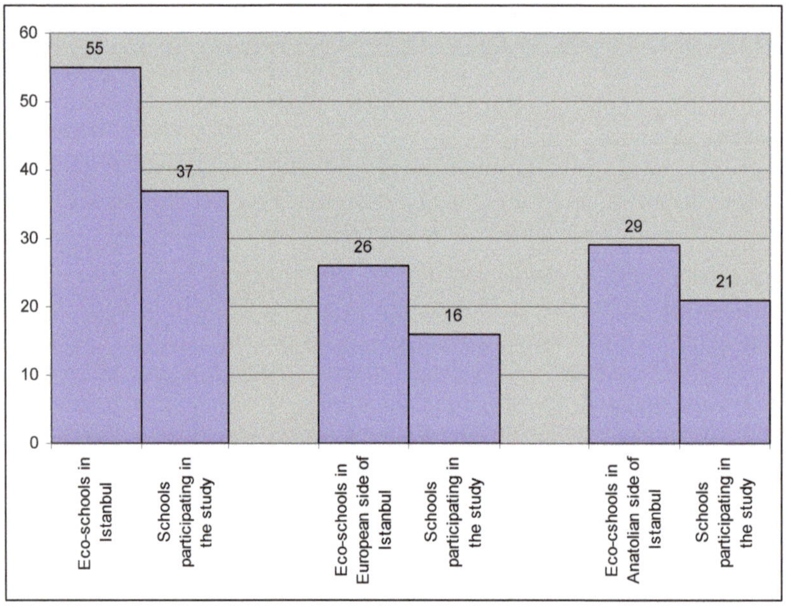

Figure 1. Distribution of Participating Eco-Schools in Istanbul

The present study used a structured review method. This is a quantitative research method that is used in survey studies. A previous study called *"Formal Environmental Review"* (Formal Environmental Review, 2010), produced by the Eco-Schools International Programme, was used as the basis when preparing the questionnaire. However, the questionnaire was revised according to the characteristics of the study site, in accordance with suggestions from the Eco-Schools Programme. The final questionnaire format consisted of 11 questions on the theme of energy conservation. The questionnaire generally consisted of closed questions, but an open-ended question was included to encourage additional comments from respondents. Questionnaire (Note 1) forms were emailed to Eco-Schools coordinators and administrators in private and state primary schools in Istanbul that participate in the Eco-Schools Programme. The contact details of coordinators and administrators was provided by the National Coordinator of Eco-Schools (Turkish Foundation for Environmental Education, 2009) of the Turkish Foundation for Environmental Education, which is the representative of the Eco-Schools International Programme in Turkey. The database used in the present study consisted of questionnaires completed and returned by 37 schools (see Figure 1). The frequency analysis functions within the SPSS statistical program were used to analyze questionnaire responses.

4. Case Study

Communication was established with Eco-Schools coordinators and administrators in these schools via e-mail, and they were sent the questionnaire. The main themes of the questionnaire study involve policies and interventions towards energy-efficiency related to the eco-schools' building facade, entrances and doors, selection of energy-efficient building equipment and components, energy sources and usage at eco-schools.

4.1 Energy Usage and Conservation Policies Related to the Building Envelope of Eco-Schools

According to Buchanan, in addition to efficient building insulation, it is of great importance to implement three basic strategies in ensuring energy conservation: "The whole form and organization of buildings should be shaped to be far less dependent on fossil-fuel energy; any mechanical plant (if indeed there is any) should be as efficient as possible, as should be the whole system of which it is part; and the building and its environmental systems should harvest and be fuelled by constantly replenished ambient energies (replenishable sources)" (Buchanan, 2005). In this scope, building form, the organization of the interior space, building orientation and building envelope play a significant role in ensuring energy conservation in buildings. In sustainable school design, these four design decisions should be tested separately in terms of energy conservation and energy saving (Tönük, 2001). However, the questionnaire administered to the eco-schools examined only the properties related to the building envelope.

Building envelope, in general terms, acts as a barrier between the indoor and outdoor environments. According to Tuluca, the basic function of the building envelope is as follows; "Increasingly, the building envelope is viewed not simply as protection from, but as connection to the environment. This connection is achieved by integration with the lighting, HVAC, and control systems. Envelope subsystems integrate several envelope functions (e.g. thermal protection, structural support, or water and weather protection) into a single, cost-effective component" (Tuluca, 1999).

4.1.1 Building Facades

Today, building facades are designed by taking into consideration factors such as reduction of energy consumption, prevention of environmental pollution and use of recyclable materials. Climatic facade designs are of great importance for the optimization of the energy loads caused by heating-cooling and ventilation systems of buildings. Such facades and the solar control elements used in facades aim to meet the comfort requirements of the building users by controlling external factors in order to ensure optimum energy consumption. Solar control elements can be placed either between the two layers of the double-skin building facades or on facade surfaces as an additional building element. According to Çetiner and Aygün, the energy cost of buildings with a double-skin facade is significantly lower than that of buildings with a single-skin facade (Çetiner & Aygün, 2005). Today, building facades may be equipped with overhangs; vertical fins; light shelves; low shading coefficient glass; interior space elements which prevent glare, such as venetian blinds; and adjustable louvers for solar control purposes.

Table 1. Energy efficiency methods used at building facades of eco-schools in Istanbul

Questions	Answers	Yes	No
Are there solar control elements on the school building's facades where required?	Frequency	3	34
	%	8.1	91.9
Does the school building have a double-skin facade?	Frequency	21	16
	%	56.8	43.2

Solar control elements in the building facades are used by only 8.1% of the eco-schools included in the present study (see Table 1), which emphasizes a deficiency in the "design with climate" approach adopted in the architectural designs of the study schools. In this scope, depending on the natural illumination conditions of the school building facades, it is suggested to equip facades with solar control elements. However, in order to prevent the retrofitting of solar control elements from damaging the structural system of the building, static calculations of the building and facades should be reviewed. It was found that 56.8% of the study schools use double-skin facade systems in the school building design.

4.1.2 Windows

The window design of primary education buildings is important in terms of both energy conservation and interior comfort. Window elements can significantly affect the heating and cooling requirements of a building. The study titled "The Efficient Windows Collaborative Tools for Schools" lists design and technical criteria related to the windows of the school buildings, as follows: "If the building location, window orientation, placement and size allow for sufficient daylighting, an integrated daylight control system should be considered early in the design process. Once orientation, daylighting, window area, and shading conditions are known, the window type must be chosen. When the window type is chosen, the effect of the window design on HVAC demand should be taken into account. Energy-efficient window design often allows for smaller and less costly HVAC systems, thus freeing funds that can be allocated to the efficient window technologies" (http://www.efficientwindows.org/ToolsForSchools.pdf).

Table 2. Energy efficiency methods used at the windows of eco-schools in Istanbul

Questions	Answers	Yes	No
Does your school building employ preventative methods to reduce heat losses from windows?	Frequency	17	20
	%	45.9	54.1
Are double-layer and energy-efficient window frame applications used?	Frequency	7	30
	%	18.9	81.1

Basic methods of reducing heat loss from windows:

- Draught excluding strips: Draughts are caused by cold air forcing its way through gaps around windows or doors. By blocking the gap, you will stop the draught, but it is very important not to block out the airflow designed to pass under the building via the air vents at ground level.
- Solar reflecting film: Such types of films attached to the window glass block solar UV radiation by 99.99% and reduce heat transfer by 60%.
- Responsive class window monitors: These elements, which are integrated particularly into classroom windows, adjust artificial illumination according to the natural illumination and turn on/off artificial illumination elements.

Table 3. Preventive method/methods towards reducing heat losses occurring from windows at eco-schools in Istanbul

State the preventive method/methods used at your school to reduce heat losses from windows			
Elementary Schools	Draught excluding strips	Solar reflecting film	Response. class window monitors
Irmak	√	-	-
Fenerbahçe	-	-	-
Kültür	-	-	-
Marmara	-	-	-
MEF	-	√	√
Önde	-	-	-
Sevgi Çiçeği Anafen	-	-	-
Sezin	-	-	-
Umutfono	-	-	-
Uğur	√	-	-
TED	√	√	-
Yenibosna Doğu S.	√	-	-
HEV Kemerköy	-	-	-
Anabilim	-	√	√
Altınyıldız	-	-	-
Eyüboğlu	√	-	-
İstek Acıbadem	-	-	-
İstek Acıbadem N.	√		√
Atanur Oğuz	√	-	-
İstek Belde	√	-	-
Mimar Sinan	√	-	-
ALEV	-	-	-
Atacan	-	√	-
Ataşehir Adıgüzel	-	√	-
Avrupa	-	-	-
Bahçeşehir Florya	-	-	√
Beykoz Doğa	√	-	√
Kartal Doğa	√	-	-
Üsküdar Doğa	√	-	-
Cent Nursery	√	-	-
Cihangir	√	√	-
Çamlıca Kalem	-	-	-
Kalemlik Nursery	-	-	-
Natuk Birkan	-	-	-
Derya Öncü	-	-	-
Elit Gençler	√	-	-
Enka	√	-	-
%	43.2	16.2	13.5

According to the results shown in Tables 2 and 3, 45.9% of schools used measures to reduce heat losses from windows, which suggests that this could be given greater emphasis in these schools. In this scope, it is suggested to implement easy-to-apply measures in order to decrease energy losses.

4.1.3 Building Entrances

Article 1.1.6. of Turkish Standard (TS) 9518 offers the following definition for the entrances and entrance gates of primary school buildings; "The main entrance and student entrance of the school building should be arranged separately; minimum width of the entrance gates should be 2.00 m and entrances should be equipped with a windbreak..." (TS 9518, 2000). In addition to the above-listed recommendations made in relation to school doors, a "self-closing mechanism" applied on gates can be used as another measure to prevent energy loss. The eco-schools questionnaire study also includes a question on this mechanism.

Table 4. Preventive method/methods towards reducing heat losses occurring from building entrances of eco-schools in Istanbul

Questions	Answers	Yes	No
Is there a self-closing mechanism on your school building's entrance gates?	Frequency	6	31
	%	16.2	83.8

4.2 Energy Usage and Conservation Policies for Indoor Environments of the Eco-Schools

There are various "sustainable school buildings guidelines" developed to direct sustainable energy usage and energy conservation policies to be adopted by schools. Most of these design handbooks include information and suggestions related to implementing energy management in the building design, maintenance and operation processes. The energy title of the BREEAM Education Guideline, published in England, emphasizes the reduction of CO_2 emissions as a top priority; "To recognise and encourage buildings that are designed to minimise the CO2 emissions associated with their operational energy consumption" (BREEAM Education, 2008). LEED for Schools, implemented in the USA, explains the target under the title of "optimize energy performance" as follows: "To achieve increasing levels of energy performance beyond the prerequisite standard, to reduce environmental and economic impacts associated with excessive energy use" (LEED for Schools, 2009). The Massachusetts High Performance Green Schools Guidelines state: "Model the school using LEED energy savings calculations protocol (version 2.1 of LEED) to show that it will achieve 20% less energy cost than a Massachusetts energy code minimum building (780 CMR Chapter 13), regulated loads only" (http://www.chps.net/content/034/MA-CHPS_2006.pdf). CHPS, which issues various guidelines on sustainable school buildings, promotes 16-46% reductions in the energy consumption of school buildings (http://www.chps.net/content/032/CA_CHPS_Criteria_2009.pdf).

Findings related to indoor energy usage in eco-schools are listed below. The questionnaire is intended for interior doors, energy and illumination equipment and components.

4.2.1 Interior Doors

According to Article 1.2.12.2 of TS 12014, "Classroom doors should be placed on the wall between the front desk and the board and they should open outward. Classroom doors opening to the same corridor should not oppose each other, to ensure easy evacuation in case of panic." (TS 12014, 1996).

As with the building entrance gates, a self-closing mechanism -to be placed on the interior doors- retains indoor heat and ensures energy conservation. As stated by Stitt, "Doors represent a compromise between conflicting needs: it will not be possible to move from room to room, to foster or prevent air exchange, and to interpose a temporary wall across an opening, for privacy" (Stitt, 1999).

Table 5. Preventive method/methods towards reducing heat losses occurring from interior doors of eco-schools in Istanbul

Questions	Answers	Yes	No
Is there a self-closing mechanism on your school's interior doors?	Frequency	29	8
	%	78.4	21.6

According to the findings, presented in Tables 4 and 5, while a self-closing mechanism is applied on the exterior doors of the school buildings at a low rate of 16.2%, it is applied on interior doors at a high rate of 78.4%. It is suggested that this system should also be applied on the exterior doors of school buildings in order to ensure energy conservation.

4.2.2 Energy and Lighting Equipments and Components

The energy consumed for artificial illumination represents an important proportion of the total energy consumed by buildings. According to Bougdah, "Globally, electric lighting consumes approximately 20 percent of the world's generated electricity. In buildings, electric lighting typically accounts for around 50 percent of total electricity use in offices, 25 percent in hospitals and 15 percent in schools and homes. Energy efficiency in electric lighting involves the correct choice of lamp, luminaire and control strategy (so lighting is supplied only when it is needed) (Bougdah, 2010). Determining appropriate illumination systems for school buildings is an important step towards energy saving; however, the most important saving in this scope is natural illumination of the school building. According to the information included in High Performance School Buildings, "Daylighting can also save school money. Properly designed systems can substantially reduce the need for electric lighting, which can account for 35% to 50% of a school's electrical energy consumption. As an added benefit, waste heat from the lighting system is reduced, lowering demands on the school's cooling equipment" (Illinois Resource Guide for Healthy, High Performing School Buildings, 2006). Moreover, natural illumination has been proven by many studies to have positive effects on the sustainability of student performance as well as on energy saving (Heschong, 2009; Kuller & Lindsten, 1992).

Classrooms that face different frontages are affected differentially by climatic conditions and, in turn, have different energy needs. In this context, energy systems need to be managed independently for each classroom, which will bring important advantages in terms of energy saving.

Table 6. Usage of energy efficient building equipment and components at the eco-schools in Istanbul

Questions	Answers	Yes	No
Are low-energy consumption bulbs or fluorescent lights used at your school?	Frequency	14	23
	%	37.8	62.2
Is there independent control of energy systems in your school's classrooms?	Frequency	5	32
	%	13.5	86.5

The Massachusetts High Performance Green Schools Guidelines provides for a maximum of 1.0 Watts/ft^2 as the average illumination equipment energy density for the whole school. In addition, it recommends that technologies such as "occupancy sensors, timed lighting schedules, and/or timed switches, which shut interior lights off when spaces are unoccupied for 15 minutes or more" should be used in school buildings. Light switches shall be installed such that more than one level of artificial illumination is possible. Each regularly occupied perimeter and non-perimeter space enclosed by ceiling-height partitions shall have a manual control to allow the occupant to uniformly reduce the connected lighting load by at least 50% (MA-CHPS_2006).

Table 7. Usage of motion/daylight sensors at the eco-schools in İstanbul

Question	Answers	Yes	No	Sometimes
Are lights and other electrical components at your school fitted with motion/daylight sensors?	Frequency	7	15	15
	%	18.9	40.5	40.5

The limited use, in 13.5% of schools, of energy control systems that can be managed independently for each classroom; and sunlight-sensitive sensor systems (which lead artificial illumination elements) in 18.9% of schools suggests that the study eco-schools do not attach due importance to these issues. It is suggested that schools ensure the installation of such systems in the school buildings, both to reduce costs and to increase awareness of electricity use in their schools.

4.3 Types of Energy Sources and Usage at Eco-Schools

The Organization of Eco-Schools summarizes the energy-related steps to be taken by a candidate eco-school as follows: "School energy conservation efforts usually begin with a comprehensive energy audit, frequently conducted by the students (of any age), to determine where the school is wasting the most energy--and money. Many schools have saved energy and money by following the students' recommendations and changing these simple practices. Some schools also seek to conserve energy through the use of strategically planted shade trees that cool air-conditioning units (in warm climates) and reduce the amount of heat absorbed by surrounding asphalt. Other schools have gone to the next step and decided to generate energy on-site. Solar panels and wind turbines are the most common choices at schools, and can be installed in a variety of sizes and combinations. Small systems can do things such as power a pump for a schoolyard pond or provide the energy for night lighting in a schoolyard. Larger systems can be connected to the school's main power supply, or the neighbourhood power grid, and offset-or entirely replace-the school's energy needs. A few schools have tried other renewable energy systems that require further modifications of their buildings' structures or other specialized equipment. Some have installed geothermal energy systems that circulate incoming air through underground ducts to pre-heat or pre-cool it before it enters the rest of the ventilation system. Other schools use passive solar energy systems that rely on the interaction between the climate and the building's design for much its heating, cooling, and lighting (http://www.ecoschools.com/Energy/Energy_wSidebar.html).

CHPS California Guidelines also provide for the use of on-site renewable energy resources to meet a part of the energy consumption of the school building and specifies this rate to be 1%-90%. (CHPS, 2009). Similarly, the LEED for Schools Guidelines emphasize the use of on-site renewable energy self-supply rather than fossile fuels (LEED for Schools, 2009). "To encourage the development and use of grid-source, renewable energy technologies on a net zero pollution basis. Engage in at least a 2-year renewable energy contract to provide at least 35% of the building's electricity from renewable sources, as defined by the Centre for Resource Solutions' Green Energy product certification requirements" (LEED for Schools, 2009).

Aware and economic use of energy, as well as utilization of clean energy resources and energy generation from natural resources, constitutes an important part of the themes of "sustainable energy management" and "sustainable energy policies". The Energy Guidelines (2009) state: "Energy sources must be monitored for verification of actual energy usage for operation of the facility. Electrical energy, natural gas or propane systems should be monitored 24-hours a day, 365-days a year. The building automation system should be used to constantly monitor, archive data and totalize the electrical energy usage" (Energy Guidelines, 2009).

Table 8. Distribution of the energy sources used at eco-schools in Istanbul

	Which of the following energy sources are used at your school?						
Elementary Schools	Wind generator	Solar water PV heating panels	Petroleum-based energies	Ground source heat pump	Coal	Natural gas	Other
Irmak						√	
Fenerbahçe						√	
Kültür						√	
Marmara						√	
MEF						√	
Önde						√	√
Sevgi Çiçeği Anafen						√	
Sezin						√	
Umutfono						√	
Uğur						√	
TED						√	
Yenibosna Doğu S.						√	

Which of the following energy sources are used at your school?							
Elementary Schools	Wind generator	Solar water PV heating panels	Petroleum-based energies	Ground source heat pump	Coal	Natural gas	Other
HEV Kemerköy						√	
Anabilim			√			√	
Altınyıldız						√	√
Eyüboğlu						√	
İstek Acıbadem						√	
İstek Acıbadem N.						√	
Atanur Oğuz						√	
İstek Belde						√	
Mimar Sinan						√	
ALEV						√	
Atacan		√					
Ataşehir Adıgüzel						√	
Avrupa							
Bahçeşehir Florya						√	
Beykoz Doğa						√	√
Kartal Doğa						√	
Üsküdar Doğa						√	
Cent Nursery						√	
Cihangir			√			√	
Çamlıca Kalem						√	
Kalemlik Nurs.						√	
Natuk Birkan		√		√		√	
Derya Öncü						√	
Elit Gençler						√	
Enka						√	
%	0	5.4	5.4	2.7	0	100	8.1

According to the findings listed in Table 8, all schools use natural gas as the main energy source for heating. Other renewable energy resources are used at very low levels, to support natural gas systems. This situation results from extensive use of natural gas in Istanbul. Active use of renewable energy systems such as ground source heat pump, wind and solar energies remains at significantly low rates both within Istanbul and across Turkey. However, solar and wind technologies are widely and actively used on the Aegean and Mediterranean coasts, particularly within tourist facilities. The development of a renewable energy resources industry in Turkey and public tendency to use such resources will be decisive in the adoption of such systems in Turkey.

In addition to the energy conservation strategies listed in the tables above, the additional information given under the title of "other" (Note 2) shows that several of the study eco-schools have adopted a range of additional energy-conservation technologies, as follows:

In the indoor environments of Beykoz Doğa College, transformers are used to reduce 220 volt to 12 volt to use 4×18 energy saving fluorescence lights in interior illumination and time-adjusted illumination elements for outdoor illumination. Photoelectric bulbs are preferred in barns, animal shelters and similar outdoor spaces. In Hisar Education Foundation, only the electrical equipment in the toilets is reported to be of sensor type. In Atanur Oğuz Primary School, time-adjusted solar sensors are used to ensure energy control in the classrooms. In Hisar Educational Foundation Kemerköy Schools, special attention is paid to the use of low-consumption ampoules. A Saving Committee -established by the school organization- record all the consumptions such as energy, water, paper, paper cups, natural gas, etc., in order to find consumption solutions. In Umutfono Primary School, an "eco-team" composed of students and personnel raise awareness among other students about the conscious use of energy resources. In Önde College, waste hot water obtained from the thermal plant is used for heating and natural gas is used only in the cafeteria. In Natuk Birkan Primary School, solar energy is used in the sports tents, geothermal energy in the lobby and swimming pool and natural gas in the other places.

5. Results and Discussion

Schools in countries with different development levels can participate in the Eco-Schools Programme. It is therefore impossible to achieve an equivalent level of success in terms of energy conservation in all Eco-Schools. However, the documentation of the activities carried out by the member schools as part of the program will enable local and global comparisons of these schools. Such data will reveal the strengths and weaknesses of these schools and will enable project administrators to make future projections on these issues.

The questionnaire investigated the strengths and weaknesses of the energy conservation policies adopted by Eco-Schools in Istanbul (55 total, 37 survey respondents).

The study findings show that there are some deficiencies in the awareness and economic use of energy in the indoor and outdoor environments of the eco-schools located in Istanbul that participated in the study. Based on the findings of the present study, the following recommendations can be made for the progression of eco-schools in Istanbul:

- Taking into consideration the findings listed in Table 1:
  - Solar control elements should be applied on the facades, according to how much sunlight they receive,
  - The use of double-skin facades should be discussed when constructing new buildings and in comprehensive building renovations,
- Taking into consideration the findings listed in Table 2:
  - Preventive measures should be considered in all eco-schools to reduce heat loss from windows and double-skin and energy-effective woodwork applications (which are in limited use in the eco-schools) should be generalized to all eco-schools,
- Taking into consideration the findings listed in Table 3:
  - Methods aimed at energy conservation from windows (which are used in a small proportion of eco-schools) should be discussed and should be integrated in all eco-school buildings,
- Taking into consideration the findings listed in Table 4:
  - Self-closing mechanisms (which are in limited use on the entrance gates of some eco-schools) should be integrated to the entrance gates of all eco-schools,
- Taking into consideration the findings listed in Table 6:
  - Low-energy consumption bulbs and fluorescence lights, which are used widely in the indoor environments of the eco-schools, should be generalized to the indoor environments of all eco-schools,
  - Independent control of energy systems, installed in a small proportion of the eco-schools- should be integrated to all eco-school environments,
- Taking into consideration the findings listed in Table 7:
  - The use of sensors mounted on illumination elements and other electrical elements should be generalized in the eco-schools,
- Taking into consideration the findings listed in Table 8:
  - It is seen that natural gas is used by 100% of survey respondents and that renewable energy resources are used only at low levels for testing purposes and/or to make small contributions to energy gain.

The initial investment costs of advanced technologies -used for renewable energy gain purpose- remain at significantly high levels. Although this presents less of a problem for developed countries, these high costs are a real problem for developing countries. However, it is concluded that the study schools, which are labeled as "eco-schools", should pioneer the use of such technologies and should create special budgets for the adoption of renewable energy resources.

Considering the general results, it is recommended:

- To gradually apply the same or similar applications in all primary schools in Istanbul and Turkey.
- To add educative seminaries and/or compulsory lessons on the topics of "environment", "environmental awareness", "energy conservation and management" to the curriculum in addition to basic education subjects.
- To seek active support from professionals in the fields of "education" and "environment and renewable energy systems" (universities, those working in applied fields, non-governmental organizations, provincial governments etc.).

Acknowledgements

The authors are grateful to Istanbul Ministry of National Education Istanbul Provincial Directorate for National Education, Turkish Foundation for Environmental Education (TÜRÇEV) National Coordinator of Eco-Schools, coordinators and the administrators of eco-schools participating in the study for their valuable contributions and support.

References

Assessment Tool. (2009). *Criteria for New Construction Major Modernizations*. Retrieved June 21, 2013, from http://www.massschoolbuildings.org/sites/default/files/edit-contentfile/Guidelines_Forms/Guidelines_Policies/MA-CHPS_Green_School_Guidelines_2009.pdf

Bougdah, H., & Sharples, S. (2010). *Environment, Technology and Sustainability* (pp. 53-179). UK: Taylor & Francis.

BREEAM (BRE Environmental Assessment Method) Education. (2008). Scheme Document SD 5051, BRE Global Ltd., UK, p. 117. Retrieved January 17, 2013, from http://www.breeam.org/page.jsp?id=20

Buchanan, P. (2005). *Ten Shades of Green Architecture and the Natural World* (p. 30). New York: The Architectural League of New York.

Çetiner, I., & Aygün, M. (2005). Investigating the Energy Efficiency of Double-Skin Glass Facades for an Office Building in İstanbul. In J. Yang, P. S. Brandon, & A. C. Sidwell (Eds.), *Smart & Sustainable Built Environments*. UK: Blackwell Publishing.

CHPS (Collaborative for High Performance Schools). (2006). *Massachusetts High Performance Green Schools Guidelines: Criteria* (p. 42). Massachusetts: The Collaborative for High Performance Schools Inc.

CHPS (Collaborative for High Performance Schools). (2009). Best Practices Manual, Volume II, USA, p. 12.

CHPS (Collaborative for High Performance Schools). (2009). Best Practices Manual, Volume III., USA, p. 22.

CHPS. (2009). *California, Assessment Tool*. Retrieved January 17, 2013, from http://www.chps.net/content/032/CA_CHPS_Criteria_2009.pdf

CHPS. (2006). *Massachusetts High Performance Green Schools Guidelines*. Retrieved January 17, 2013, from http://www.chps.net/content/034/MA-CHPS_2006.pdf

Eco-School Design. (n.d.). Retrieved January 17, 2013, from http://www.ecoschools.com/Energy/Energy_wSidebar.html

Eco-Schools England, Nine Topics, Energy. (n.d.). Retrieved January 17, 2013, from http://www.eco-schools.org.uk/nine-topics/energy.aspx

Eco-Schools England, Nine Topics, Energy. (n.d.). Retrieved January 17, 2013, from www.eco-schools.org.uk/nine-topics/energy.aspx

Foundation for Environmental Education (FEE International), Eco-Schools. (n.d.). Retrieved January 17, 2013, from http://www.eco-schools.org

Foundation for Environmental Education (FEE International), Eco-Schools, Methodology, Step 2. (n.d.). Retrieved January 17, 2013, from http://www.eco-schools.org/page.php?id=53

Foundation for Environmental Education (FEE International), Eco-Schools, Methodology, Step 3. (n.d.). Retrieved January 17, 2013, from http://www.eco-schools.org/page.php?id=54

Foundation for Environmental Education (FEE International), Eco-Schools. (n.d.). Retrieved January 17, 2013, from www.eco-schools.org/page.php?id=30

Gissen, D. (2002). Big & Green: Toward Sustainable Architecture in the 21th Century, Princeton Architectural Press, New York, p. 19.

Graham P. (2003). *Building Ecology, Blackwell Science, Oxford* (p. 56). Retrieved January 17, 2013, from http://www.en-wikipedia.org/wiki/Airlock

International Energy Agency. (n.d.). Retrieved January 17, 2013, from www.iea.org

İstanbul Technical University. (2007). *Türkiye'de Enerji ve Geleceği: İTÜ Görüşü (Energy and the Future of Turkey. ITU Opinion), İstanbul.* Retrieved January 17, 2013, from http://www.energy.itu.edu.tr/iTUOnerileri.pdf

Kibert, C. J., Sendzimir, J., & Guy, B. (2002). *Construction Ecology: Nature as the Basis for Green Buildings* (p. 18). London: Spon Press.

Kösetorunu, A. (1996). Türkiye'de Enerji Sektörünün Geleceği (The Future of Energy Sector in Turkey), Undersecretariat of the Prime Ministry for Foreign Trade. Retrieved January 17, 2013, from www.dtm.gov.tr/dtmadmin/upload/ead/tanitimkoordinasyondb/6sa7.doc

North Carolina Department of Public Instruction, Public Schools of North Carolina, State Board of Education. (2009). *Energy Guidelines For K-12 Public Schools, USA*, pp. 13-14. Retrieved January 17, 2013, from http://www.schoolclearinghouse.org/pubs/ENERGY.pdf

Nottinghamshire County Council. (2010). *Environmental Review for Eco-Schools*. Retrieved January 17, 2013, from www.nottinghamshire.gov.uk/environmentalreviewform2009.doc

Pitts, A. (2004). *Planning and Design Strategies for Sustainability and Profit: Pragmatic Sustainable Design on Building and Urban Scales* (pp. 10-11). Boston: Elsevier Architectural Press.

Regents of the University of Minnesota, Twin Cities Campus, College of Design, Center for Sustainable Building Research, Efficient Windows Collaborative. (n.d.). Retrieved January 17, 2013, from http://www.efficientwindows.org/ToolsForSchools.pdf

Republic of Turkey, Ministry of Energy and Natural Resources, Energy. (n.d.). Retrieved January 17, 2013, from http://www.enerji.gov.tr/index.php?dil=tr&sf=webpages&b=enerji&bn=215&hn=12&nm=384&id=384

Sassi, P. (2006). *Strategies for Sustainable Architecture* (p. 204). New York: Taylor & Francis.

Seal of The State of Illinois. (2006). *Illinois Resource Guide for Healthy, High Performing School Buildings, USA* (p. 22). Retrieved January 17, 2013, from http://www.cdb.state.il.us/schools/HealthySchoolsGuide.pdf

Stitt, F. A. (1999). *Ecological Design Handbook* (p. 112). London: McGraw-Hill.

Tönük, S. (2001). *Bina Tasarımında Ekoloji (Ecology for Building Design)* (pp. 54-59). Istanbul: Yıldız Technical University Print Publications Center.

Tuluca, A. (1999). *The Building Envelope: Walls, Roofs and Foundations* (p. 185). In F. A. Stitt (ed.), *Ecological Design Handbook*. London: McGraw-Hill.

Turkish Standard (TS) 12014 (ICS 13.100.13.120). (1996). *Environment Health - Schools, Turkish Standards Institution (TSE), Ankara* (p. 4).

Turkish Standard (TS) 9518 (ICS 03.080.30.03.180). (2000). *Primary Schools – Physical Settlement – General Rules, Turkish Standards Institution (TSE), Ankara* (p. 4).

Türkiye Foundation for Environmental Education. (2009). Retrieved January 17, 2013, from www.turcev.org.tr

USGBC (US Green Building Council). (2010). *LEED 2009 for Schools New Constructions and Major Renovations, USA* (pp. 35, 38, 45). Retrieved January 17, 2013, from http://www.usgbc.org/ShowFile.aspx?DocumentID=7248

Notes

Note 1. The results of the questionnaire study, which is conducted under the approval and permission of Istanbul Provincial Directorate for National Education (İİMEM), have to be published and these publications must also be delivered to İİMEM.

Note 2. The necessary permission to publish school names in academic studies and publications conducted and to be conducted on eco-schools in Istanbul was received from the National Coordinator of the Turkish Foundation for Environmental Education (TURÇEV) Eco-Schools Program via e-mail in April, 2012.

8

Lipase Catalysed Ethanolysis of Jatropha Oil for Biodiesel Production

Elizabeth Funmilayo Aransiola[1]

[1] Department of Chemical Engineering, Obafemi Awolowo University, Ile-Ife, Nigeria

Correspondence: Elizabeth Funmilayo Aransiola, Department of Chemical Engineering, Obafemi Awolowo University, Ile-Ife, Nigeria. E-mail: aransiola4@yahoo.com & aransiolaef@gmail.com

Abstract

This study investigated ethanolysis of both crude and pre-treated Jatropha oil using immobilized lipase from *Pseudomonas cepacia*. The methods employed were stepwise addition of ethanol and n-hexane as solvent. The ethanol was added three times in a stepwise way so as to prevent enzyme inhibition. The effects of reaction temperature and oil to ethanol ratio were studied. Maximum yields of fatty acid ethyl ester (FAEE) of 66.7 wt%, 72.1 wt% and 58.8 wt% were obtained at oil to ethanol ratio of 1:4 for crude *Jatropha curcas* oil at reaction temperatures of 30, 40, and 50 °C respectively. Each run of the experiment took an overall reaction period of 48 h. These results showed the best optimized reaction temperature to be 40 °C. A maximum FAEE yield of 96.1 wt% was obtained when pre-treated *Jatropha curcas* oil was transesterified at the best optimized conditions. The loss noticed after 13 cycles in the activity of the enzyme was quite negligible. This proved it to be a stable one. The physical properties of the biodiesel obtained from the pre-treated Jatropha oil met the ASTM standard of D-6751.

Keywords: biodiesel, ethanolysis, Jatropha oil, lipase enzyme

1. Introduction

The energy trend of our present generation is fast moving towards a renewable one due to high diminishing rate of fossil fuels, its environmental problems and increasing prices of crude oil. Biodiesel, a form of biofuel is being intensively researched to as a solution to the problems generated by fossil fuel.

Recently, much attention is being paid on it as green, non toxic, environmentally safe and renewable fuel (Yucel, 2011). The major feedstocks for biodiesel production are edible, non-edible, used vegetable oils and animal fats. Edible oils' utilization for production of biodiesel is being disannulled as it is highly competitive with consumption; hence attention is being shifted to non-edible and waste or used vegetable oils (E. F. Aransiola, Betiku, Ikhuomoregbe, & Ojumu, 2012; E. F. Aransiola, Ojumu, Oyekola, & Ikhuomoregbe, 2012; Jain & Sharma, 2010; Tsai, Lin, & Yeh, 2007; R. Wang et al., 2011). In this study, a non-edible oil known as *Jatropha curcas* oil was used.

Jatropha curcas L. also known as physic nut is from Euphorbiaceous family. The seeds are very rich in oil (30 to 50% by weight of the seed) (Pramanik, 2003). It is a plant that grows on prolonged period of abnormally low rainfall and also on unused land. These characteristics make it uncompetitive with food crops.The toxicity of the seeds and the oil is as a result of the phorbol esters present in them (Kumari, Mahapatra, Garlapati, & Banerjee, 2009). *Jatropha curcas* oil has 21.6%, 45.4% and 33% of saturated, monounsaturated and polyunsaturated fatty acids (Akbar, Yaakob, Kamarudin, Ismail, & Salimon, 2009; E. F. Aransiola et al., 2012). The most suitable vegetable oil for producing biodiesel must be readily available, its plant should be easy to cultivate and its composition must include a high percentage of monounsaturated fatty acids when compared relatively to other fatty acids (R. Wang et al., 2011). Based on these criteria, *Jatropha curcas* oil is potentially seen as a useful renewable feedstock for biodiesel production.

The common methods used in producing this biodiesel (fatty acid alkyl esters) are acid and alkali methods of catalysis. These methods have some problems associated with it: downstream processing costs, by-products recovery and environmental problems. Al-Zuhair et al. (2007) also reviewed other drawbacks associated with the chemical catalyst as being energy intensive, unwanted side products generated from the use of alkali catalyst especially in the presence of free fatty acids in the oil, hence the need to search for better and alternative methods of production. Enzyme method of catalysis can act as a substitute to the chemical ones due to these characteristics attached to it; its reusability, specificity, ability to accept new substrates, thermo stability, mild reaction conditions in terms of low temperature and its environmental friendliness (Akoh, Chang, Lee, & Shaw, 2007; Al-Zuhair et al., 2007; Modi, Reddy, Rao, & Prasad, 2007; Noureddini, Gao, & Philkana, 2005; Ognjanovic, Bezbradica, & Knezevic-Jugovic, 2009). These characteristics have factored the reactions with the use of enzyme to be termed green (Akoh et al., 2007).

Enzymes are used for transesterification reaction, though not commercially developed. As reviewed by Akoh et al. (2007), recent awareness is on the use of lipase as a biocatalyst for fatty acid methyl ester production. This is because it is economically feasible in terms of energy consumption, and also due to its high efficiency and selectivity.

Some researchers had revealed enzyme to be a promising catalyst and in order to establish its use industrially, the reaction conditions (temperature, oil/alcohol molar ratio, amount of enzyme used, reaction time) need to be optimized (Akoh et al., 2007; Modi et al., 2007; Moreira, Perez, Zanin, & de Castro, 2007; Noureddini et al., 2005; Shah & Gupta, 2007; Shah, Sharma, & Gupta, 2004; Wang, Liu, Zhao, Ding, & Xu, 2011; Yucel, 2011).

When synthesizing biodiesel using lipase enzyme, the high cost and inactivation by methanol are considered its major limitations. The problem of high cost can be solved to a greater extent by employing the process of immobilization and genetically engineering them. The issue of inhibition by methanol makes the type of alcohol for enzyme transesterification an important factor for consideration. Another type of alcohol that can be considered is ethanol; it is readily available and cheaper. A lot of studies had been conducted by using these two types of alcohol. Triolein's transesterification with methanol and ethanol using lipase from P.cepacia was investigated by Salis et al (2005), and a conversion of 38% and 90% were obtained respectively. Noureddini et al. (2005) also reported 67% conversion when soybean oil was transesterified with methanol using lipase from *P.cepacia.* Abigor et al. (2000) investigated the conversion of palm kernel oil using ethanol and methanol and *P. cepacia* as a lipase and conversions of 72% and 15% were attained respectively. Although lipase deactivation was reported due to contact between alcohol and the enzyme (Shimada, Watanabe, Sugihara, & Tominaga, 2002), this was minimized by stepwise addition of the alcohol in 2 or 3 steps (Shimada et al., 1999; Watanabe et al., 2000). From all the findings above, ethanol is seen as a better alcohol for enzyme catalyzed reaction, as it can also be synthesized from non toxic renewable biomass categorized as wastes, such as molasses, cassava bagasse, sawdust, banana peels. Comparing ethanol with methanol from economical point of view, ethanol is a cheaper substrate for biodiesel production as it can be produced from cheaper and readily available sources. The biodiesel produced when ethanol is used as alcohol and enzyme used as catalyst in conjunction with non edible oil makes it a total (100%) green product. This study is therefore aimed at producing biodiesel by ethanolysis of non edible oil; *Jatropha curcas* oil using immobilized lipase enzyme from *Pseudomonas cepacia.*

2. Materials and Methods

2.1 Materials

The acquirement of the Jatropha seeds was done from a town in Nigeria called Ikere-Ekiti. Analytical grade chemicals used were Ethanol, Acetone, Lipase immobilized in Sol-Gel-Ak from Pseudomonas cepacia 40U/g, Oleic acid, 1–Oleoyl rac–glycerol, 1, 2 Dioleyol–rac–glycerol, Triolein, Oleic acid ethyl oleate, Acetonitrile and n-Hexane. These were purchased from Sigma Aldrich in Germany.

2.2 Experimental Set up

2.2.1 Pretreatment of the Crude Oil

Akintayo (2004) method was employed in extracting the oil from the Jatropha seed. The free fatty acid was tested and discovered very high. The method used for the pretreatment was that of Berchmans and Hirata (2008). The pretreatment was carried out via acid esterification process in order to decrease the FFA content. Methanol and sulphuric acid were mixed together and warmed for 15 minutes. The quantity of methanol and sulphuric acid added slowly to the oil were 60% w/w and 1% w/w of oil respectively. These were allowed to agitate for one hour at 60 °C-65 °C. Afterwards, the mixture got settled and the top layer having the majority of the FFA was separated leaving the bottom layer which is the pretreated oil. The FFA of the pretreated oil was further tested.

2.2.2 Enzymatic Catalysis of *Jatropha curcas* Oil

Batch transesterification of crude *Jatropha curcas* oil was performed in an Erlenmeyer flask of 250 ml containing 11.2 gram of the oil. These were carried out in an incubator shaker at different temperatures and reaction periods of 30 °C, 40 °C and 50 °C and 6 h, 12 h, 24 h and 48 h respectively. The reaction mixtures were done at oil to ethanol molar ratios of 1:3, 1:4 and 1:6 using the 3-stepwise method of reaction. One third molar of the required ethanol was added every one third hour of the reaction time. To the reaction mixture, 4 wt% of the immobilized enzyme was added and also 0.6 ml of hexane in enhancing its solubility. The addition of hexane and the enzyme had been done before adding the first one third molar of the alcohol added. The agitation rate was kept constant at 200 rpm.

On completion of the various experimental runs, the product was centrifuged in order to separate the upper layer which is the ethyl ester. This ester with other glycerides was analyzed for the different parameters required. The enzyme was washed with hexane and reused for the next experiment. Based on the optimal conditions obtained when the ethyl ester from transesterification of crude *Jatropha curcas* oil was analyzed, this was used for the pre-treated *Jatropha curcas* oil.

The experiment was scaled up in order to be able to get enough biodiesel samples for characterization according to the American Standard Testing Method.

2.2.3 The Immobilized Enzyme's Performance

In order to determine the ability or performance of an immobilized enzyme, its stability must be significantly considered. (Kumari et al., 2009). Factors like too much alcohol and glycerol do have a negative effect on the activity of the enzyme. This is due to hydrophilic and insolubility nature of glycerol in the oil giving room for adsorption on the surface of the enzyme. The reduced life span of the enzyme is also associated with excess alcohol (Dizge & Keskinler, 2008).

At the end of every experimental run, the immobilized enzyme was rinsed with n-hexane so as to remove the glycerol attached to it. The stepwise adding of ethanol was adopted in order to avoid enzyme inhibition. The relative activity of the enzyme after 13 cycles was seen relative to the yield of the FAEE obtained.

2.3 Analytical Methods

2.3.1 Percentage Oil Yield and Its Physico-Chemical Analysis

The analysis on the oil; its percentage oil yield and physicochemical properties (saponification and acid values) had been carried out in some of our previous studies (E. F. Aransiola et al., 2012; E. Aransiola, et al., 2012).

2.3.2 Ester and Glyceride Analysis by HPLC Method

LC – 20AB HPLC connected to an ultraviolet detector consisting of a controller, auto sampler and binary pump was used in measuring the concentration of the ester produced, as well as the other glycerides present. The standards used were 1–oleoyl rac–glycerol, 1, 2 dioleyol–rac–glycerol, triolein and ethyl oleate. The mobile phases are Acetone and Acetonitrile at ratio of 49 to 51 at a flow rate of 0.7 Ml.min-1. The column used was 4.6 mm ID x 15 cm CLC-ODS (T) of 5 μm particles and 100 Å pore size. The preparation of the samples and analysis were done according to Aransiola (2012), Dubé et al. (2004) and Darnoko and Cheryan (2000).

2.3.3 Biodiesel Characterization

The methods employed in characterizing the biodiesel produced was according to our previous study Aransiola et al. (2010).

Table 1. Jatropha Biodiesel Characteristics and Biodiesel Standards

Properties	Jatropha Biodiesel*	Biodiesel Standard ASTM 6751	Test Method
Flash point (°C)	Not determined	130 (min)	ASTMD – 93
Moisture content	Nil	0.050 max	ASTMD – 2709
Kinematic viscosity (mm^2/s)	5.6	1.9 t0 6.0	ASTMD – 445
Cloud Point (°C)	3	-	ASTMD – 2500
Specific gravity at 15/15°C	0.88	0.860 to 0.900	-
Pour point (°C)	-3	-15 to 10	ASTMD – 97

*These are mean values of data measured in duplicates.

3. Results and Discussions

3.1 The Physico-Chemical Properties of the Jatropha curcas Oil

The physico-chemical properties of the oil are shown in Table 2. This is quite comparable to the results revealed by Akintayo (2004).

The percentage oil content obtained was approximately 52% comparable to the ones obtained by Akintayo (2004), Martín et al. (2010) and Akbar et al. (2009) respectively. This high percentage of the oil and its high content of monounsaturated value are the indicative measures that it is quite suitable as a substrate for biodiesel production.

The percentage free fatty acid values of the crude and pre-treated oil were 18.1% and 0.6% respectively.

Table 2. The Physico-chemical properties of *Jatropha curcas* oil

Parameters	Values
Acid value (mgKOH.g^{-1})	35.8
Percentage oil yield (%)	52.2
Density at room temperature (kg.m^{-3})	895
Viscosity at room temperature (cSt)	41.4
Saponification value (mgKOH.g^{-1})	193
Free fatty acid of crude oil (%)	18.1
Free fatty acid of pre-treated oil (%)	0.6

These are mean values of data measured in duplicates.

3.2 Enzyme Catalysis of Jatropha curcas Oil to Biodiesel

The alcohol-to-oil ratio and reaction temperature effects were studied for enzyme catalysis of crude and pre-treated *Jatropha curcas* oil for biodiesel production. Stirring speed of all experiments was fixed at 200 rpm.

3.2.1 Effect of Oil to Alcohol Molar Ratio

Figures 1 depict the oil-to-alcohol ratio effect on the biodiesel at 30 °C (Figure 1A), 40 °C (Figure 1B), and 50 °C (Figure 1C). In all these temperatures, ester concentration increases from oil to alcohol molar ratio of 1:3 to 1:4 but decreases at a ratio of 1:6. This may be due to the excess ethanol inhibiting the enzyme and thus reducing its percentage conversion to ester. The three stepwise reactions are carried out so as not to allow the excess alcohol to inhibit the enzyme For the oil to ethanol of 1:6, two times of what was supposed to be introduced per one–third of the stoichiometric ratio was used for the reaction which may really be the cause of the low fatty acid ethyl ester concentration produced. The best results of 66.7 wt%, 72.1 wt% and 58.8 wt% were obtained for temperatures of 30, 40, and 50 °C respectively at oil to ethanol ratio of 1:4. Typical compositions profiles of TG, DG, MG and Fatty Acid Ethyl ester for the enzyme catalysis of the crude Jatropha oil at the temperatures of 30, 40, and 50 °C for oil to ethanol ratio of 1:4 are shown in Figures 2A-C. The yields are not so high and this was attributed to the presence the phospholids and free fatty acid in the oil. This led to the use of pre-treated *Jatropha curcas* oil which was transesterified by using the lipase enzyme at the best condition obtained for the above results. Surprisingly, it gave a very good yield of 96.1 wt%. This is shown in Figure 3.

3.2.2 Effect of Reaction Temperature

The use of enzymes is very sensitive to temperature effects. This temperature effect against the best oil to alcohol ratio (1:4) for crude Jatropha oil is shown in Figure 4. In this figure, temperature of 40 °C gave a best result of 72.1 wt%. The pre-treated oil gave a yield of 96.1 wt% at the same reaction temperature (Figure 3). This agrees with the study conducted by Shimada et al (2002) and Dizge and Keskinler (2008) who confirmed that lipase enzyme operates best between temperatures of 30 °C and 40 °C.

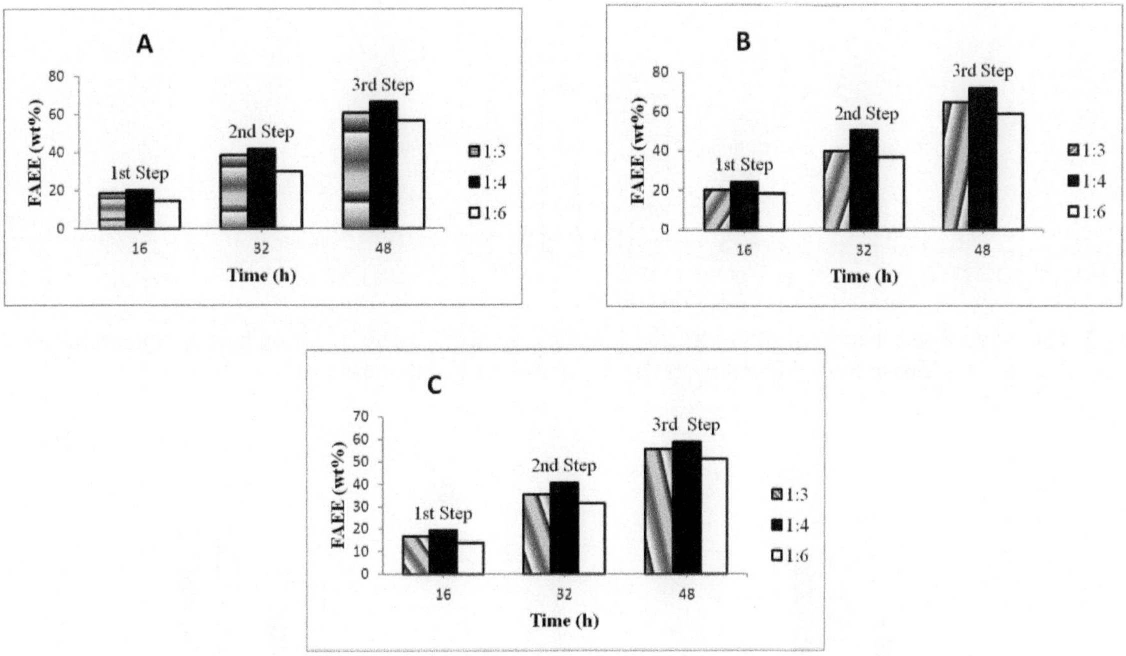

Figure 1. Results of oil-to-ethanol molar ratio on FAEE concentration for crude *Jatropha curcas* oil using immobilized Lipase from *Pseudomonas cepacia* for 48 h. **A.** at 30 °C; **B.** at 40 °C and **C.** at 50 °C

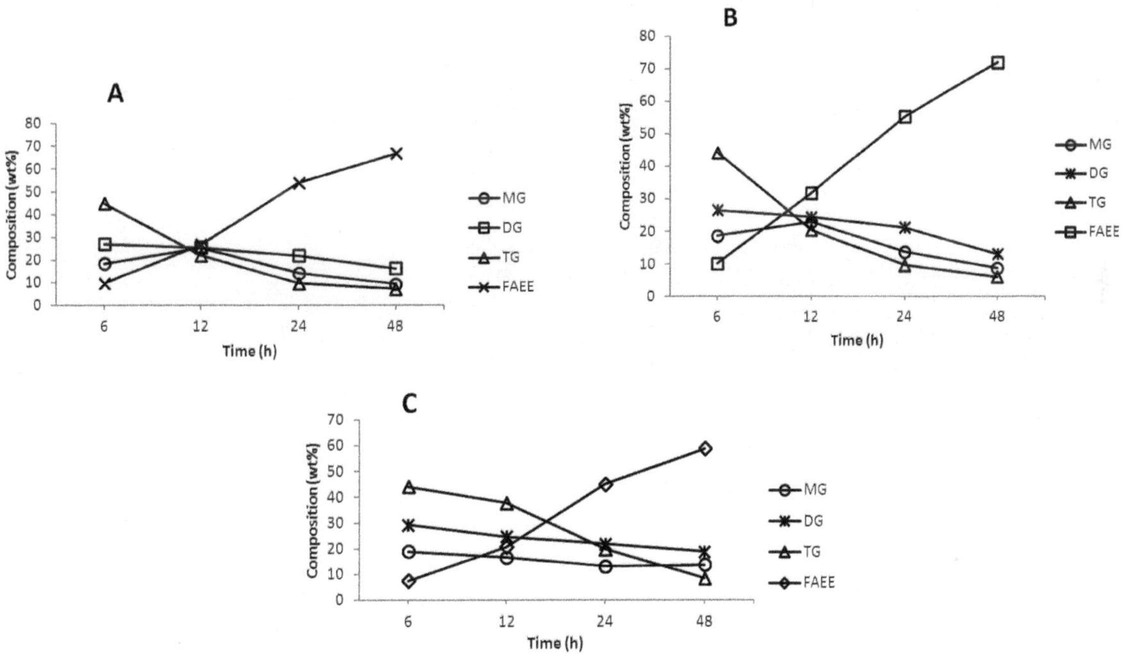

Figure 2. Composition of product obtained by three step flow ethanolysis of Crude Jatropha by Lipase from Pseudomonas cepacia at different reaction periods for oil to ethanol ratio of 1:4. **A.** at 30 °C; **B.** at 40 °C and **C.** at 50 °C

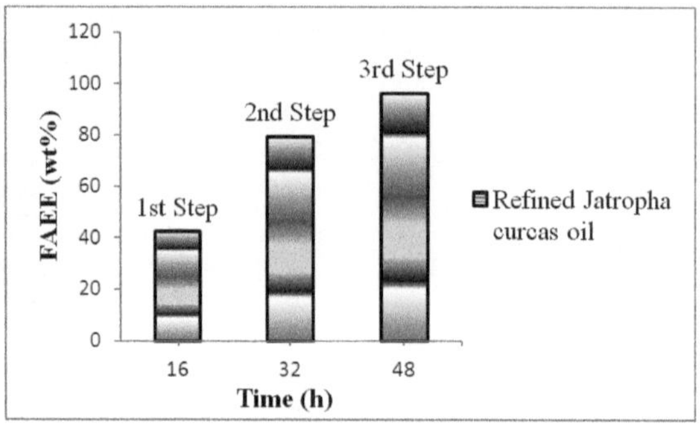

Figure 3. Three Stepwise Ethanolysis of Pre-treated *Jatropha curcas* Oil at 40 °C for 48 h using immobilized Lipase from *Pseudomonas cepacia* for oil to alcohol ratio of 1:4

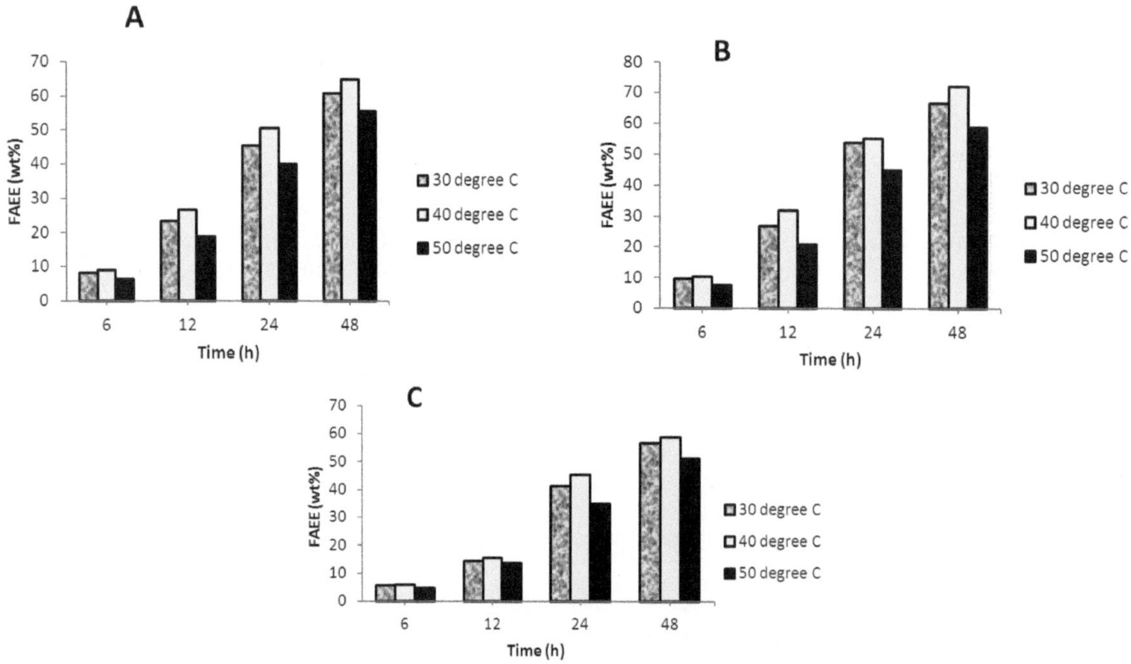

Figure 4. Effect of temperature on FAEE concentration against reaction time for crude *Jatropha curcas* oil using immobilized Lipase from *Pseudomonas cepacia.* **A.** oil-to-ethanol molar ratio of 1:3; **B.** oil-to-ethanol molar ratio of 1:4; **C.** oil-to-ethanol molar ratio of 1:6

3.3 The Immobilized Enzyme's Performance

The relative activity of the enzyme after 13 cycles was seen relative to the yield of the FAEE obtained. FAEE yield of 96.1 wt% was obtained at the 13th cycle after transesterification of the pre-treated Jatropha oil. This was a good evidence of the operational stability of the enzyme. This high yield and lipase stability after several uses may have resulted as a result of the hexane in the reaction, which notably lowered the adsorption of the glycerol on the enzyme.

3.4 Characteristics of the Biodiesel Obtained

The characteristics of the biodiesel obtained from the Nigerian *Jatropha curcas* oil, as well as its standard are presented in Table 1. It is noteworthy that the characteristics of the biodiesel produced were quite comparable with the standard.

4. Conclusion

This study presents the ethanolysis of both crude and pre-treated *Jatropha curcas* oil using immobilized lipase enzyme from *Pseudomonas cepacia.* The solvent used was n-hexane and ethanol was added stepwisely so as to hinder enzyme inhibition. The effects of oil-to-ethanol molar ratio and reaction temperature on the ester yield were studied. Maximum FAEE yields of 72.1% and 96.1% were obtained from ethanolysis of the crude and pre-treated *Jatropha curcas* oil at optimized conditions (Reaction temperature of 40 °C, oil to ethanol ratio of 1:4 and reaction period of 48 h). The use of ethanol is seen as a better alcohol for enzyme catalyzed reaction, as it can be synthesized from renewable sources and also cheaper when compared to methanol. The excellent performance of the enzyme is an indication that if the process is embarked upon, the potential to be efficient is high. This study therefore reveals that ethanolysis of non edible oil; *Jatropha curcas* oil using a stable immobilized lipase enzyme could lead to a green product; biodiesel. The results reported will serve as a motivation for other researchers involved in biodiesel production in not neglecting the use of enzyme as catalysts as it gives high yield and a green product. This study also has the potential to be economically feasible if the continuous process is embarked upon.

Acknowledgements

I will like to acknowledge the financial assistance rendered by the Education Trust Fund in Nigeria. Big thanks also to National Research Institute of Chemical Technology, Zaria for giving me the working space to analyse my work.

References

Abigor, R. D., Uadia, P. O., Foglia, T. A., Haas, M. J., Jones, K. C., Okpefa, E., ... Bafor, M. E. (2000). Lipase-catalysed production of biodiesel fuel from some Nigerian lauric oils. *Biochemical Society Transactions, 28*(6), 979-982. http://dx.doi.org/10.1042/0300-5127:0280979

Akbar, E., Yaakob, Z., Kamarudin, S. K., Ismail, M., & Salimon, J. (2009). Characteristic and composition of Jatropha curcas oil seed from Malaysia and its potential as biodiesel feedstock feedstock. *European Journal of Scientific Research, 29*(3), 396-403.

Akintayo, E. T. (2004). Characteristics and composition of Parkia biglobbossa and Jatropha curcas oils and cakes. *Bioresource Technology, 92*(3), 307-310. http://dx.doi.org/10.1016/S0960-8524(03)00197-4

Akoh, C. C., Chang, S. W., Lee, G. C., & Shaw, J. F. (2007). Enzymatic approach to biodiesel production. *Journal of Agricultural and Food Chemistry, 55*(22), 8995-9005. http://dx.doi.org/10.1021/jf071724y

Al-Zuhair, S., Ling, F. W., & Jun, L. S. (2007). Proposed kinetic mechanism of the production of biodiesel from palm oil using lipase. *Process Biochemistry, 42*(6), 951-960. http://dx.doi.org/10.1016/j.procbio.2007.03.002

Aransiola, E., Betiku, E., Layokun, S., & Solomon, B. (2010). Production of biodiesel by transesterification of refined soybean oil. *International Journal of Biological and Chemical Sciences, 4*(2), 391-399.

Aransiola, E. F., Daramola, M. O., Ojumu, T. V., Aremu, M. O., Layokun, S. K., & Solomon, B. O. (2012). Nigerian Jatropha Curcas Oil Seeds: Prospect for Biodiesel Production in Nigeria. *International Journal of Renewable Energy Research (IJRER), 2*(2), 317-325.

Aransiola, E. F., Betiku, E., Ikhuomoregbe, D. I. O., & Ojumu, T. V. (2012). Production of biodiesel from crude neem oil feedstock and its emissions from internal combustion engines. *Afr. J. Biotechnol, 11*(22), 6178-6186. http://dx.doi.org/10.5897/AJB11.2301

Aransiola, E. F., Ojumu, T. V., Oyekola, O. O., & Ikhuomoregbe, D. I. O. (2012). A Study of Biodiesel Production from Non-Edible Oil Seeds: A Comparative Study. *The Open Conference Proceedings Journal, 3*, 18-22.

Berchmans, H. J., & Hirata, S. (2008). Biodiesel production from crude Jatropha curcas L. seed oil with a high content of free fatty acids. *Bioresource Technology, 99*(6), 1716-1721. http://dx.doi.org/10.1016/j.biortech.2007.03.051

Darnoko, D., & Cheryan, M. (2000). Kinetics of palm oil transesterification in a batch reactor. *Journal of the American Oil Chemists' Society, 77*(12), 1263-1267. http://dx.doi.org/10.1007/s11746-000-0198-y

Dizge, N., & Keskinler, B. (2008). Enzymatic production of biodiesel from canola oil using immobilized lipase. *Biomass and Bioenergy, 32*(12), 1274-1278. http://dx.doi.org/10.1016/j.biombioe.2008.03.005

Dubé, M. A., Zheng, S., McLean, D. D., & Kates, M. (2004). A comparison of attenuated total reflectance-FTIR spectroscopy and GPC for monitoring biodiesel production. *Journal of the American Oil Chemists' Society, 81*(6), 599-603. http://dx.doi.org/10.1007/s11746-006-0948-x

Jain, S., & Sharma, M. P. (2010). Prospects of biodiesel from Jatropha in India: A review. *Renewable and Sustainable Energy Reviews, 14*(2), 763-771. http://dx.doi.org/10.1016/j.rser.2009.10.005

Kumari, A., Mahapatra, P., Garlapati, V.K ., & Banerjee, R. (2009). Enzymatic transesterification of Jatropha oil. *Biotechnol Biofuels, 2*(1), 1-6. http://dx.doi.org/10.1186/1754-6834-2-1

Martín, C., Moure, A., Martín, G., Carrillo, E., Domínguez, H., & Parajó, J. C. (2010). Fractional characterisation of jatropha, neem, moringa, trisperma, castor and candlenut seeds as potential feedstocks for biodiesel production in Cuba. *Biomass and Bioenergy, 34*(4), 533-538. http://dx.doi.org/10.1016/j.biombioe.2009.12.019

Modi, M. K., Reddy, J. R. C., Rao, B. V. S. K., & Prasad, R. B. N. (2007). Lipase-mediated conversion of vegetable oils into biodiesel using ethyl acetate as acyl acceptor. *Bioresource Technology, 98*(6), 1260-1264. http://dx.doi.org/10.1016/j.biortech.2006.05.006

Moreira, A. B. R., Perez, V. H., Zanin, G. M., & de Castro, H. F. (2007). Biodiesel synthesis by enzymatic transesterification of palm oil with ethanol using lipases from several sources immobilized on silica–PVA composite. *Energy & Fuels, 21*(6), 3689-3694. http://dx.doi.org/10.1021/ef700399b

Noureddini, H., Gao, X., & Philkana, R. S. (2005). Immobilized Pseudomonas cepacia lipase for biodiesel fuel production from soybean oil. *Bioresource Technology, 96*(7), 769-777. http://dx.doi.org/10.1016/j.biortech.2004.05.029

Ognjanovic, N., Bezbradica, D., & Knezevic-Jugovic, Z. (2009). Enzymatic conversion of sunflower oil to biodiesel in a solvent-free system: process optimization and the immobilized system stability. *Bioresource Technology, 100*(21), 5146-5154. http://dx.doi.org/10.1016/j.biortech.2009.05.068

Salis, A., Pinna, M., Monduzzi, M., & Solinas, V. (2005). Biodiesel production from triolein and short chain alcohols through biocatalysis. *Journal of Biotechnology, 119*(3), 291-299. http://dx.doi.org/10.1016/j.jbiotec.2005.04.009

Shah, S., & Gupta, M. N. (2007). Lipase catalyzed preparation of biodiesel from Jatropha oil in a solvent free system. *Process Biochemistry, 42*(3), 409-414. http://dx.doi.org/10.1016/j.procbio.2006.09.024

Shah, S., Sharma, S., & Gupta, M. N. (2004). Biodiesel preparation by lipase-catalyzed transesterification of Jatropha oil. *Energy & Fuels, 18*(1), 154-159.

Shimada, Y., Watanabe, Y., Samukawa, T., Sugihara, A., Noda, H., Fukuda, H., & Tominaga, Y. (1999). Conversion of vegetable oil to biodiesel using immobilized Candida antarctica lipase. *Journal of the American Oil Chemists' Society, 76*(7), 789-793. http://dx.doi.org/10.1007/s11746-999-0067-6

Shimada, Y., Watanabe, Y., Sugihara, A., & Tominaga, Y. (2002). Enzymatic alcoholysis for biodiesel fuel production and application of the reaction to oil processing. *Journal of Molecular Catalysis B: Enzymatic, 17*(3), 133-142. http://dx.doi.org/10.1016/S1381-1177(02)00020-6

Tsai, W. T., Lin, C. C., & Yeh, C. W. (2007). An analysis of biodiesel fuel from waste edible oil in Taiwan. *Renewable and Sustainable Energy Reviews, 11*(5), 838-857. http://dx.doi.org/10.1016/j.rser.2005.05.005

Wang, R., Hanna, M. A., Zhou, W. W., Bhadury, P. S., Chen, Q., Song, B. A., & Yang, S. (2011). Production and selected fuel properties of biodiesel from promising non-edible oils: Euphorbia lathyris L., Sapium sebiferum L. and Jatropha curcas L. *Bioresource Technology, 102*(2), 1194-1199. 10.1016/j.biortech.2010.09.066

Wang, X., Liu, X., Zhao, C., Ding, Y., & Xu, P. (2011). Biodiesel production in packed-bed reactors using lipase-nanoparticle biocomposite. *Bioresource Technology, 102*(10), 6352-6355. http://dx.doi.org/10.1016/j.biortech.2011.03.003

Watanabe, Y., Shimada, Y., Sugihara, A., Noda, H., Fukuda, H., & Tominaga, Y. (2000). Continuous production of biodiesel fuel from vegetable oil using immobilized Candida antarctica lipase. *Journal of the American Oil Chemists' Society, 77*(4), 355-360. http://dx.doi.org/10.1007/s11746-000-0058-9

Yucel, Y. (2011). Biodiesel production from pomace oil by using lipase immobilized onto olive pomace. *Bioresource Technology, 102*(4), 3977-3980. http://dx.doi.org/10.1016/j.biortech.2010.12.001

Characterization of Inorganic Elements in Woody Biomass Bottom Ash from a Fixed-bed Combustion System, a Downdraft Gasifier and a Wood Pellet Burner by Fractionation

Adrian. K. James[1], Steve S. Helle[1], Ronald W. Thring[1], Gurkaran S. Sarohia[1] & P. Michael Rutherford[1]

[1] Environmental Science & Engineering, University of Northern British Columbia, 3333 University Way, Prince George, BC, V2N 4Z9, Canada

Correspondence: Adrian. K. James, Environmental Science & Engineering, University of Northern British Columbia, 3333 University Way, Prince George, BC, V2N 4Z9, Canada. E-mail: ajames@unbc.ca

Abstract

The direct combustion of biomass residues produces large quantities of bottom ash. Environmental sustainable management requires that ash recycling should be carried out whenever possible. Suitable applications of bottom ash are based predominantly on its chemical properties. The presence of major ash forming and trace elements along with other intrinsic properties unique to bottom ash, suggest its potential as a soil additive. But, ash quality must be of a high standard to prevent environmental pollution. This comparative study characterizes bottom ash obtained from three types of bioenergy systems - a fixed-bed boiler, a downdraft gasifier and a wood pellet burner. The chemical properties were analyzed and discussed for each bottom ash, together with their respective particle fractions that were obtained by sieving. The pH of the starting ash samples for the gasifier, boiler and pellet burner were 10.36, 12.49 and 13.46, respectively. Ni with a concentration of 229 mg/kg in the pellet burner ash, exceeded the maximum limit for soil amendments (in British Columbia, Canada) within the particle size fraction $\geq$ 850 μm but < 2000. All samples were significantly enriched in both Ca (50-61%) and K (10-26%). The elements Mg, Al, Mn, Fe, P and Na each contributed 10% or less to the inorganic portion of the ash. Concentrations of inorganic contents varied with particle size. Water soluble phosphates were very low in the samples. The results suggest that size fraction separation can be a useful method to isolate fractions containing higher (or lower) amounts of some metals. This method may be a useful technique for managing ash that contains elements exceeding environmental limits.

Keywords: bottom ash, soil amendment, major elements, trace elements, particle size fraction, ash utilization

1. Introduction

The employment of wood combustion technologies have resulted in a rapid increase in the use of woody-biomass residues for energy production. Boilers, gasifiers and pellet burners are but some of the systems available for wood energy production. A disadvantage of using biomass for energy production is that large amounts of residual ash are generated (Dahl, Nurmesniemi, Poykio, & Watkins, 2010). Ash from these combustion processes vary in quality depending on the fuel type, operating conditions of the system and the type of combustion system (Obernberger, Biedermann, Widmann, & Riedl, 1997). Ideally, biomass ash should be recycled whenever possible, but large amounts of wood ash are typically landfilled (Picco, February 2010). Countries such as Sweden and the United States of America (USA) use landfills to dispose majority of their ash (Clarholm, 1994; Vance, 1996; Demeyer, Voundi-Nkana, & Velow, 2001). The quality of the ash in part determines recycling options; some options such as land spreading require low trace element concentrations to prevent environmental pollution (Pels, de Nie, & Kiel, 2005). The application of wood ash to soils may counteract soil acidification and correct for nutrient deficiencies; this is widely used in regions that carry out extensive forest harvesting, such as Northern Europe and parts of North America (Demeyer et al., 2001; Stupak, Asikainen, Roser, & Pasanen, 2008; Aronsson & Ekelund, 2004). In Canada, biomass ash is mainly landfilled but land applications are more common in some provinces than others. In British Columbia the use of ash as for soil applications is relatively limited due to current provincial regulations. Until adequate research is carried out the majority of ash generated in B.C. may be limited to landfilling or other uses outside of soil applications.

Limiting factors for the use of wood ash in soils include trace element concentrations and potential organic pollutants (Sarenbo, 2009; Knapp & Insam, 2011). Elevated concentrations of trace elements in ashes may limit utilization of ash in soils and therefore also the recycling of nutrient elements (e.g. Ca, Mg, K, P), too. The handling and application of ash is improved by hardening or aggregation; but, ash with a high organic C content does not harden properly (Narodoslawsky & Obernberger, 1996).

The physical and chemical characteristics of wood ash may vary with particle size fraction (Dahl et al., 2010; Liao, Wu, & Yan, 2007; James, Thring, Helle, & Rutherford, 2013). Particle size fractions high in undesirable trace element concentrations could be separated from some ashes, thereby improving the overall quality of the residual ash material (and therefore improving recycling options). Previous research shows that arsenic concentrations in fly ash exceeded the Finnish environmental limits for the particle size fraction less than 0.125 mm (Dahl et al., 2010). Particle separation may be carried out by sieve fractionation and may influence variables such as trace element concentrations, pH, organic carbon concentration, bulk density and other chemical-physical properties.

Should the properties vary significantly in particle size fractions, greater applications of ash as a soil amendment or as a raw material for products requiring specific properties could be employed. Ash-producing industries may find this particularly useful when considering disposal or utilization methods for ash.

The objective of this research was to characterize ash from the three systems. A comparative study of the chemical properties of each bottom ash, as well as their respective fractions, obtained by sieving, was carried out. This paper will identify the inorganic elemental distribution of specific particle size fractions, pH and anion distribution of boiler, gasifier and wood pellet burner bottom ash. It is hoped that these results will assist in finding specific applications for bottom ash.

2. Experimental

Wood ash from from three types of combustion systems, a fixed-bed boiler (Canfor Pulp Mill), a downdraft gasifier (University of Northern British Columbia - UNBC) and a wood pellet burner (UNBC) were used in the study. The downdraft fixed bed gasifier is capable of producing power of 5 MW, the wood pellet burner 0.4 MW and an industrial scale fixed bed boiler, 27 MW. The fuel for the gasifier and boiler is hog fuel comprised predominantly of softwood sawmill waste and is primarily derived from pine wood. The pellet burner utilizes wood pellets made locally from soft-wood saw dust. Two ash samples were collected from each system. The boiler ash sample was collected on April 27, 2012 – BA. The gasifier ash sample was collected July 18, 2012 – GA. Pellet burner samples was collected March 5, 2010 – PA. Henceforth, the boiler sample is denoted as BA, gasifier sample GA and pellet burner sample, PA. The ash samples were separated into different particle fractions as follows.

2.1 Particle Size Distribution

Approximately 100 g of air-dried starting ash original sample obtained from boiler, gasifier, or pellet burner was separated into different size fractions using a stack of 5 sieves (2000 μm, x850 μm, 425 μm, 250 μm, 150 μm), arranged in decreasing diameter openings (Hermann & Baker, 2002). Ash was poured on the top sieve (largest opening), which was then covered; the sieve stack was then placed on an automatic shaker for 15 minutes after which the stack was removed. Each sieve with the retained material was gently tapped on the sides before being removed from the stack and weighed. Retained ash was removed and stored, while the sieve trays were thoroughly cleaned and reweighed to obtain the mass of ash retained on tray. The analyses were repeated twice and an average of the results taken. The following retained fractions were used for subsequent analyses: starting wood-ash, ≥ 2000 μm, ≥ 850 μm but < 2000 μm, ≥ 425 μm but < 850 μm, ≥ 250 μm but < 425 μm, ≥ 150 μm but less than 250 μm and <150 μm.

For reporting data purposes, the fractions will be denoted as 2000 μm, 850 μm, 425 μm, 250 μm, 150 μm and < 150 μm. Each of the retained fractions were weighed and then stored for further analysis.

2.2 pH Analysis

The pH of ash fractions were determined following methods described in (Kalra & Maynard, 1991). The pH was measured potentiometrically using the pH of saturated paste method. A 400 ml beaker was half-filled with ash, then sufficient deionized water was added to saturate sample. The sample was mixed and subsequently left to sit for 1 hour after which pH readings were taken. Analysis of all fractions were repeated twice and an average of the results taken.

2.3 Concentration and Distribution of Trace Elements and Major Ash-Forming Elements

Selected elements within each fraction were determined by ICP-MS at UNBC. 300 mg of sample was mixed with 10 mL conc. HNO_3 and then diluted to 50 mL with deionized water. This sample mixture was then subjected to microwave digestion, using a Milestone MLS 1200 Mega digestion system. Metal characterization was done by an inductively coupled plasma (ICP-MS) on an Agilent 7500 ICPMS machine. This was used to determine the alkali, alkali earth and trace elements.

2.4 Anion Analysis

Anion concentrations were determined using the Dionnex IC-5000 system. Ten (10) mL of deionized water was added to ~ 400 mg of samples and shaken for 12 hrs. The samples were then removed from shaker and centrifuged for 15 min, after which ~ 2 mL of sample was obtained for analysis using Ion chromatography. All results shown are considered to be only water soluble amounts. Duplicate analyses were conducted and an average concentration reported.

3. Results

3.1 Particle Size Distribution of Ash

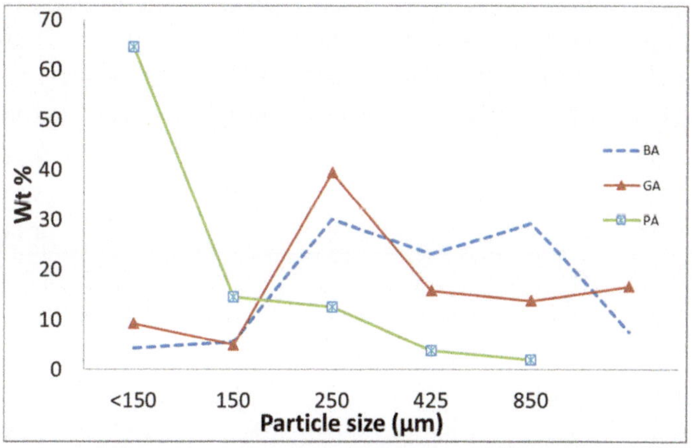

Figure 1. Retained Weight as a function of particle size distribution for boiler ash (BA), gasifier ash (GA) and pellet burner ash (PBA) samples (Data referenced) (James et al., 2013)

Table 1. Total carbon contents and pH of bottom ash samples of boiler, gasifier and pellet burner

	*Carbon			pH		
Particle size / μm	BA wt.%	GA wt.%	PA wt.%	BA	GA	PA
<150	17.17	3.67	5.75	12.22	10.43	13.53
150	13.85	6.11	36.08	12.56	10.41	13.36
250	21.56	3.24	19.16	12.12	10.36	13.20
425	52.76	5.76	6.62	10.74	10.28	12.13
850	66.41	3.71	5.94	10.49	10.05	11.16
2000	70.73	3.07		10.41	9.69	
Starting ash	51.79	5.13	5.64	12.49	10.36	13.46

*Data referenced from (Hermann & Baker, 2002) and on a dry basis.

3.2 pH Analysis

Table 1 shows the distribution of the pH for the PA, GA and BA samples. Ash from all three systems was high in pH but recorded different pH readings. Ash from the gasifier was the least alkali, followed by the boiler then the pellet burner. The starting ash samples for the gasifier, boiler and pellet burner were 10.36, 12.49 and 13.46 respectively. It was observed that the pH increased for all ash samples as the particle size fraction decreased, with the exception of the particle size fractions < 150 μm for the for BA. The increase in pH may have been due to the increase in concentrations of the alkali earth metals (refer to Table 4) as the particle size fractions decreased potentially leading to higher concentrations base-forming metal salts.

Table 2. Concentration and distribution of trace elements for samples of boiler, gasifier and pellet burner bottom ash (dry basis)

Particle Size (μ)m	Cr mg/kg	Cu mg/kg	Ni mg/kg	Pb mg/kg	V mg/kg	Co mg/kg	Mo mg/kg	As mg/kg	Cd mg/kg
PA									
850	49.58	61.23	229.59	7.22	14.70	6.74	3.13	0.49	0.50
425	47.32	108.91	175.40	3.55	15.03	6.90	3.55	0.80	1.91
250	79.91	135.53	109.48	4.17	14.30	7.67	4.01	0.85	3.69
150	52.09	153.60	69.70	2.07	12.85	7.54	4.52	0.84	4.45
<150	52.38	170.72	40.65	2.74	10.18	7.93	5.24	0.98	9.46
Starting ash	54.13	132.47	54.50	0.44	10.55	7.20	4.82	0.92	7.71
Gasifier ash									
2000	23.40	16.66	20.00	0.63	29.37	5.29	0.48	1.39	0.03
850	12.81	37.75	19.09	7.36	21.95	4.17	1.15	1.76	0.04
425	18.54	31.30	27.23	1.72	20.71	4.63	0.74	1.40	0.07
250	35.33	60.97	68.13	5.12	29.65	7.35	4.30	6.31	0.30
150	44.54	66.37	67.74	4.10	28.57	7.04	3.33	4.85	0.31
<150	47.57	207.57	78.81	4.20	33.41	7.86	5.33	7.49	0.44
Starting ash	33.51	43.38	60.74	1.64	28.53	6.78	4.13	5.42	0.27
Boiler ash									
2000	5.84	34.01	6.89	3.34	3.50	1.56	1.51	0.52	2.21
850	4.64	81.52	6.31	9.59	4.14	1.86	1.40	0.69	2.85
425	10.36	50.42	11.06	7.09	12.16	3.23	1.61	1.14	4.11
250	19.87	77.59	19.05	9.99	18.54	5.17	2.29	1.82	6.18
150	30.00	107.49	25.07	9.00	24.49	6.70	2.71	2.07	6.18
<150	28.97	419.58	25.36	13.86	24.66	6.99	2.79	2.37	7.52
Starting ash	11.82	32.85	13.39	4.89	12.44	3.79	1.88	1.45	4.76
*Environmental Limits	1060	2200	180	500		150	20	75	20

*Environmental limit's for ash intended for land application in B.C., Canada (BC Ministry of Environment, 2008).

3.3 The Concentration and Distribution of Elements in Ashes and Anion Distribution

3.3.1 Trace Elements

The concentrations of elements for the starting ash and within specific particle size fractions of Cr, Cu, Ni, Pb, V, Co, Mo, As and Cd are shown in Table 3.2. Of the starting ash samples analysed, Cu from the Pellet burner system was highest in trace element concentration; 132 mg/kg for Cu (Table 3.2). Cu was enriched in pellet burner ash by ~ 3 and ~ 4 times that of the gasifier and the boiler samples, respectively. The elements Ni and Cr followed next in concentrations, ranging from 33 - 55 mg/kg for the pellet burner and gasifier starting ash. The boiler samples were ~ 5 times lower than the pellet burner ash in both Ni and Cr contents. In most results, the boiler samples were slightly lower in trace metal concentrations. All elements in the starting ash fractions of the 3 ash samples were well within the British Columbia soil amendment limits (BC Ministry of Environment, 2008).

Table 2 also shows the distribution of trace elements within particle size fractions for the respective ash. For most elements, a slight increase in concentration was observed as the particle size fraction decreased in all ash types. All fractions for each element were within the environmental limits for soil amendments in British Columbia, except the 850 μm fraction of Ni, obtained from the pellet burner ash; concentration of 229 mg/kg. While not exceeding the limit, the 425 μm fraction also had a high concentration of 176 mg/kg Ni. It is evident that some elements are more concentrated in specific fractions of ash. The results suggest that fraction separation can be a useful method to isolate fractions containing higher amounts of some metals. This method may be a useful technique for isolating elements exceeding environmental exposure limits and at the same time rendering the residual ash useful.

Table 3. Percent total metal distribution in boiler, gasifier and pellet burner ash

Elements	BA wt.%	GA wt.%	PA wt.%
Ca	60	61	55
K	15	10	21
Mg	6	6	11
Al	4	7	2
Mn	4	3	6
Fe	3	6	1
P	3	3	2
Na	2	2	1
Remaining elements	2	2	1

Table 4. The concentration of major ash forming elements within particle size fractions for boiler, gasifier and pellet burner on a dry basis

Particle Size (μ)m	Ca mg/kg	Al mg/kg	Mg mg/kg	Na mg/kg	K mg/kg	P mg/kg	Mn mg/kg	Fe mg/kg
PA								
850	136392	9266	25062	1607	61751	3880	11185	9973
425	195843	10538	35162	2756	99339	6606	16921	9055
250	239720	9758	46113	3459	112781	8109	22630	8883
150	253148	9193	51437	3414	105948	8264	25198	7993
<150	262728	7555	56533	2854	95660	8551	28115	6168
Starting ash	246186	7689	50910	2851	95232	8075	25845	6659
GA								
2000	20910	11888	4966	4966	5006	932	854	12652
850	19848	9897	4213	4213	9120	1083	1018	10707
425	27422	11904	5363	5363	12933	1382	1509	11911
250	140745	16083	13992	13992	20778	6955	7267	12732
150	112986	16250	12220	12220	20677	5569	6072	12903
<150	167030	15936	15733	15733	18430	8178	8463	12814
Starting ash	124946	14852	12308	12308	20077	6026	6522	12339
BA								
2000	46393	2427	4518	2456	25997	2086	2693	1558
850	51562	2728	4946	2548	25342	2419	2971	1797
425	62728	6208	6450	2466	22548	3057	3615	4732
250	122267	10015	12189	2616	22393	6481	7229	7613
150	153376	12080	14939	2786	20800	8259	9084	10131
<150	166787	12079	16256	2788	21529	9144	10084	9756
Starting ash	92564	6763	8974	2507	23354	4846	5394	5186

3.3.2 Major Elements General Trends (Starting Ash)

Table 3 displays the percent of total ash forming elements in each ash sample. These percentages were obtained by taking the sum of all the total metal content found in 1 kg (d.b) of ash. All samples were significantly enriched in both Ca (50-61 wt.%) and K (10-26 wt.%). The high weight percent metal content found in these ash are expected for woody biomass ash. The elements Mg, Al, Mn, Fe, P and Na each contributed 10% or less to the total elemental portion of the ash. The other inorganic elements present in ash, though not mentioned in the table cumulatively contributed 2% or less to the total metal content of the ash under study and are not discussed in this paper. All analysis were based on the top eight most concentrated elements Ca, K, Mg, Al, Mn, Fe, P and Na.

The elements Al, P, Mg, Mn, Na, Fe, K, and Ca are shown in Table 4. The table shows that the concentration of P, Mn, Mg, K and Ca were found to be higher in pellet burner ash when compared to the other two ash types. The elements Al, Fe, and Na were present in higher quantities in the gasifier ash when compared to the other two ash types. Additionally, the pellet burner ash contained ~ 4 times higher concentrations of K, Mg and Mn, than the other ash types. However, Ca was only ~2 times higher in pellet ash. Some degree of variation in concentrations is expected due to the likely variation of fuel types, temperatures and other factors. Research has shown that calcium concentrations increase at temperatures below 900 °C which is primarily due to the decomposition of calcium carbonates (Misra, Ragland, & Baker, 1993). Above 900 °C, the Ca concentration also increases due to the dissociation and volatilization of potassium oxide formed after dissociation. The variation in oxide and/or carbonate concentrations bound to the ash will affect the concentration of the elements in samples when concentration is calculated on a weight basis.

3.3.3 Major Elements General Trends (within Fraction)

Within fractions, it was observed that P, Mg, Mn, Na, and Ca either increased with decreasing particle size or remained relatively constant. Ca was highest in concentration within each fraction and for all ash types (Table 4). The highest values reported for Ca were seen in Pellet burner ash, from ~136,000 to ~262,000 mg/kg moving from fractions 850 μm to < 150 μm. Additionally, K was found to be highest in the pellet burner ash, containing concentrations ranging from ~62,000 to ~112000 mg/kg when moving from 850 μm to 250 μm fractions. While most of the elements were present in higher quantities in pellet burner ash as compared to other ash types, Al, Na and Fe were present in higher concentrations in gasifier ash.

The concentration of each element not only varies across different systems but within fractions as well. This variation in concentrations poses a difficulty in the general application of bottom ash to soils due to the lack of standardization in ash quality. According to Obernberger and Supancic (2009), ash recycling to agricultural lands can help reduce the use of artificial fertilizers and close the natural mineral cycle. This is largely due to the presence of N, P and K in biomass ash. However, Pels et al. (2005) suggests that the use of biomass ash as a soil fertilizer is limited due to a number of deficiencies in ash. The author noted that biomass ash can only be a source of K because ash from thermal sources is low in N and the P present has a very low solubility at soil conditions. In previous research on the same ash, the ultimate analysis showed N content all less than 1 wt.% on a dry basis for all ash samples (James et al., 2013). Wood naturally has low N contents. Furthermore, due to the conversion of most of the wood N to NH_3, NO_x and/or N_2 during the combustion of the wood, it is expected that the nitrogen content be low (Misra et al., 1993; Steenari, Karlsson, & Lindqvist, 1999; Knapp & Insam, 2011).

Table 5. The concentration (dry ash basis) of water soluble phosphates relative to particle size distribution for boiler gasifier and pellet burner ash

sieve size/ μm	BA mg/kg	GA mg/kg	PA mg/kg
<150	-	0.23	0.66
150	-	0.1	1.87
250	-	0.22	1.67
425	0.35	0.32	3.64
850	2.57	0.15	0.29
2000	2.29	0.74	
AR	-	0.08	0.83

While the percent mass of phosphorus in the metal contents of the samples ranged from 1-3%, the amount that is available for plant uptake or that is water soluble must be considered, this was measured in the form of phosphates (PO^{-3}_4). Phosphorous, may exist in both organic and inorganic forms (BC Ministry of Environment, 2008). The total and plant available phosphorous may vary. Water soluble phosphorus is the amount of phosphorus that is readily available to plants (Fuhrman, Zhang, Schroder, & Davis, 2005; Nurmesniemi, Manskinen, Poykio, & Dahl, 2012). Research conducted on wood based biomass in an incinerator reported that the water soluble phosphorus content in bottom ash sample is negligible (Nurmesniemi et al., 2012). The research concluded that the poor water solubility of phosphorus was a draw back in the application of wood ash to soils since only a small portion P is extractable and available for plants when used in forest fertilizers. Our research showed some amounts of water soluble phosphorus present in the form of phosphates, though low in concentrations (Table 5). The starting ash for the pellet burner and gasifier were 0.83 mg/kg and 0.08 mg/kg, respectively, while the boiler ash was below the detection limit. Concentrations varied within particle size fractions. The pellet ash ranged from 0.66 – 3.6 mg/kg; lowest observed in particle size fraction < 150 μm and the highest in 425 μm fraction. The gasifier ash ranged from 0.1 – 0.74 mg/kg for the different fractions. The water soluble phosphate were only detected within the three largest particle size fractions for the boiler ash and ranged between 0.35 – 2.57 mg/kg. While the three largest fractions of the boiler samples had higher concentrations of phosphorus, water soluble phosphorus were not detected for these fractions. It does suggest that the phosphorus present in these fractions may be insoluble in water.

The presence of relatively high Ca and Mg, typically in the form of oxides, hydroxides and carbonates, suggests use of these ashes as potential liming agents. This could help in controlling soil pH. As the particle size fraction decreased, the concentrations of Ca and Mg increased and so did the pH. Therefore, specific fractions could be used to achieve appropriate liming requirements.

3.3.4 Inorganic Relationship to Total Carbon Content

The results were also analysed to identify the possibilities of any relationship that may exist between the elemental distribution of major inorganic components and the total carbon content (James et al., 2013). Figures 3a and 3b show a graphical display of the elements Ca, and P versus total carbon. These two elements showed a linear correlation for boiler sample; most of the carbon (over 60%) in this sample was organic carbon. A strong correlation was obtained with R^2 value of 0.9272 and 0.9218 for Ca and P, respectively. The equations are displayed in Equations 1 and 2. As a general trend, as the total carbon content increases the elemental concentrations for Ca and P decrease. These equations may help in determining the approximate amount of certain nutrients that may be present in ash. It would be worth removing as much carbon as possible to increase the concentration of nutrients.

$$y = -1969C + 182159 \qquad R^2 = 0.9272 \tag{1}$$

$$y - 113.92C + 9973.5 \qquad R^2 = 0.9218 \tag{2}$$

No major trend or correlation was observed when samples from the gasifier and pellet burner were analysed. This may have been due to the very low carbon distribution within the particle size fractions for each sample.

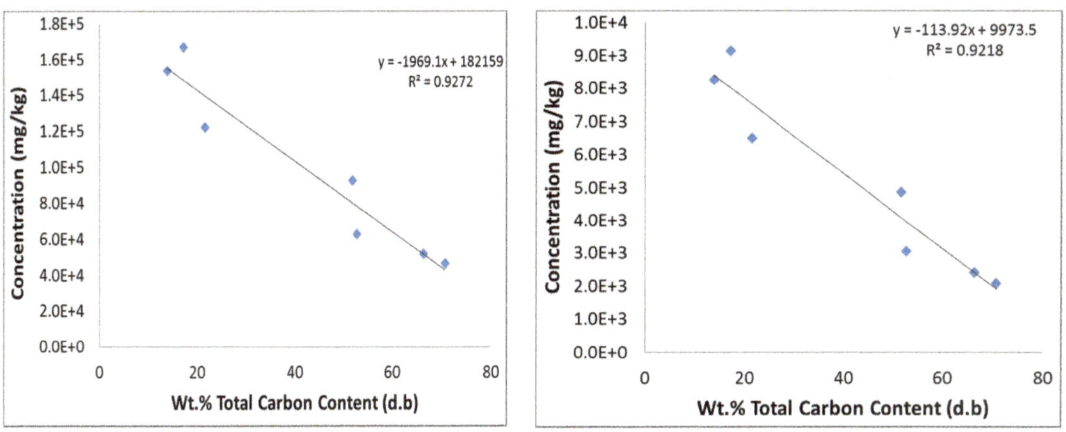

(a) (b)

Figure 3. The correlation of calcium (Ca) and phosphorus (P) concentrations, respectively shown by (a) and (b), as a function of total carbon found in boiler ash

The presence of organic carbon in ash, if recycled to forest soils presents a challenge. Unburned carbon creates a dilution effect which reduces the concentration of inorganic nutrients available in the ash (Sarenbo, 2009). This is evident from the two correlations above where the metal content increases with the decreasing carbon contents. If a high concentration of carbon is present in ash, greater amounts of ash will be required to fulfill the nutrients demand by the soil. It should be noted that high carbon ash has also been associated with elevated concentrations of environmentally harmful products of incomplete combustion, such polycyclic aromatic compounds (PAH) (Sarenbo, 2009; Bundt, Krauss, Blaser, & Wilcke, 2001); other work with these ashes has shown very low concentrations of PAHs, dioxins and furans (data not shown). The presence of carbon also reduces the ability of the ash to harden, which is integral to improving the handling of ashes and reducing its solubility (Sarenbo, 2009). This property limits its application to soil. Perhaps, if suitable methods for recirculating and reburning high carbon ash are determined, higher nutrient concentrations could be obtained while at the same time tackling some of existing ash related issues.

On the other hand, should high carbon ash possess similar properties as biochar then it could be used to obtain positive results in soil applications. Research has shown that biochar addition to soil may enhance properties, increasing plant growth nutrient availability, cation exchange capacity, water retention and with a reduction in the release of some greenhouse gases (Robertson, Rutherford, Lopez-Gutierrez, & Massicotte, 2012; Laird, 2007; Lehmann, 2007). Further studies are needed to determine if high carbon ash could also have these positive effects.

4. Conclusion

The inorganic element contents, pH and phosphate distribution of wood ash samples obtained from a fixed-bed boiler, fixed-bed gasifier and pellet burner were characterised. The samples showed variations in pH values both within particle size fractions and across the three systems. The starting ash samples for the gasifier, boiler and pellet burner ash had a pH of 10.36, 12.49 and 13.46, respectively. The pH increased with decreasing particle size fractions. This increase in pH may have been due to the increase in concentrations of the alkali earth metals as the particle size fractions decrease, forming higher concentrations of base-forming metal salts.

The trace element contents for the starting ash samples of the three ash types were all within the environmental limits for soil amendments in British Columbia, Canada. However, when analyzed within particle fractions, Ni with a concentration of 229 mg/kg from the pellet burner ash, exceeded the limit within the particle size fraction ≥ 850 μm but < 2000. The starting ash samples were significantly enriched in both Ca (50-61 wt.%) and K (10-26 wt.%) on a total metal composition. The highest values reported for Ca were seen in Pellet burner ash, from ~136,000 to ~262,000 mg/kg. Very low concentrations of water soluble phosphates were obtained for all ash types in some case below the detection limit. A strong correlation was developed between the total carbon present and the metal contents for Ca and P.

These results suggest that fraction separation can be a useful method to isolate fractions containing higher amounts of some metals. This method may be a useful technique for elements exceeding environmental exposure limits.

However, the variation in concentrations poses a difficulty in the general application of bottom ash to soils due to the lack of standardization in ash quality. As shown from the research, the inorganic content of the ash samples varied across systems and within fractions. In addition to known factors such as fuel type and temperature, the research showed that other factors such the carbon content and particle size could play a role in determining the concentrations of the available inorganics. It is therefore difficult to determine standard compositions of ash.

Acknowledgments

We would like to thank the following organizations for their support in conducting this research Canfor Pulp Mill and the University of Northern British Columbia Central Equipment Laboratory. Also, we acknowledge funding support from National Science and Engineering Research Council (NSERC) and Canfor Pulp Limited Partnership (CPLP).

References

Aronsson, K., & Ekelund, N. (2004). Biological effects of wood ash application to forest and aquatic ecosytems. *Journal of Environmental Quality, 33*, 1595-1605. http://dx.doi.org/10.2134/jeq2004.1595

BC Ministry of Environment. (2008). *Land application guidelines for the organic matter recycling regulation and the soil amendment code of practice.* New Westminister, BC: SYLVIS Environmental.

Bundt, M., Krauss, M., Blaser, B., & Wilcke, W. (2001). Forest fertilzation with wood ash: effect on the distribution and storage of polycyclic aromatic hydrocarbons (PAHs) and poly choloronated biphenyls (PCBs). *Journal of Environmental Quality, 30*, 1296-1304. http://dx.doi.org/10.2134/jeq2001.3041296x

Clarholm, M. (1994). Granulated wood ash and a N-free fertilizer to a forest soil-effects on P availability. *Forest Ecology and Management, 66*, 127-136. http://dx.doi.org/10.1016/0378-1127(94)90152-X

Dahl, O., Nurmesniemi, H., Poykio, R., & Watkins, G. (2010). Heavy metal concentrations in bottom ash and fly ash fractions from a large sized (246MW) fluidized bed boiler with repect to their Finnish forest fertilizer limit values. *Fuel Processing Technology, 91*, 1634-1639. http://dx.doi.org/10.1016/j.fuproc.2010.06.012

Demeyer, A., Voundi-Nkana, J., & Velow, M. (2001). Characteristics of wood ash and influence on soil properties and nutrient uptake: An overview. *Bioresource Technology, 77*, 287-295. http://dx.doi.org/10.1016/S0960-8524(00)00043-2

Fuhrman, J., Zhang, H., Schroder, L., & Davis, R. (2005). Water-soluble phosphorus as affected by soil to extract ratios, extraction times and electrolyte. *Communication in soil science and plant analysis, 36*, 925-935. http://dx.doi.org/10.1081/CSS-200049482

Hermann, T., & Baker, S. (2002). *Evaluating particle size.* Kansas: Kansas State University Agricultural Experiment Station and Cooperative Extension Service.

James, A., Thring, R., Helle, S., & Rutherford, M. (2013). Characterization of biomass bottom ash from an industrial-scale fixed bed boiler. *Energy and Environment Research, 3*(2), 21-32. http://dx.doi.org/10.5539/eer.v3n2p21

Kalra, Y., & Maynard, D. (1991). *Methods manual for forest soil and plant analysis.* Alberta: Forestry Canada.

Knapp, B., & Insam, H. (2011). Recycling of biomass ashes: current technologies and future research needs. In H. Insam & B. A. Knapp (Eds.), *Recycling of Biomss Ashes.* New York: Springer. http://dx.doi.org/10.1007/978-3-642-19354-5_1

Laird, D. (2007). The charcoal vision: a win-win-win scenario for simultaneously producing bioenergy, permanent sequestering carbon, while improving soil and water quality. *Agron. Journal, 100*, 178-180. http://dx.doi.org/10.2134/agrojnl2007.0161

Lehmann, J. (2007). A handful of carbon. *Nature, 447*, 143-144. http://dx.doi.org/10.1038/447143a

Liao, C., Wu, C., & Yan, Y. (2007). The characteristics of inorgani elements in ashes from a 1MW CFB biomass gasification power plant. *Fuel Processing Technology, 88*, 149-156. http://dx.doi.org/10.1016/j.fuproc.2005.06.008

Misra, M., Ragland, K., & Baker, A. (1993). Wood ash composition as a funtion of furnace temoperature. *Biomass and Bioenergy, 4*(2), 103-116. http://dx.doi.org/10.1016/0961-9534(93)90032-Y

Narodoslawsky, M., & Obernberger, I. (1996). From waste to raw material - the way of cadmium and other heavy metals from biomass to wood ash. *Journal of Hazardous Mateials, 50*(2-3), 157-168. http://dx.doi.org/10.1016/0304-3894(96)01785-2

Nurmesniemi, H., Manskinen, K., Poykio, R., & Dahl, O. (2012). Forest fertilizer properties of the bottom ash and fly ash from llarge-sized (115MW) industrial powerplant incinerating wood-based biomass residues. *Journal of the Univeristy of Chemical Technology and Metallurgy, 1*, 43-52.

Obernberger, I., Biedermann, F., Widmann, W., & Riedl, R. (1997). Concentrations of inorganic elements in biomass fuels and recovery in different ash fractions. *Biomass and Bioenergy, 12*, 211-224. http://dx.doi.org/10.1016/S0961-9534(96)00051-7

Obernberger, I., & Supancic, K. (June/July 2009). Possibillities of ash utilisation from biomass combustion plants. Proceedings of the 17th European Biomass conference & Exhibition. Hamburg: ETA Renewable Energies.

Pels, J., de Nie, D., & Kiel, J. (2005). *Utilization of ashes from biomass combustion and gasification.* Paris, France: Published at 14th European biomass conference & exhibition.

Picco, D. (February 2010). *Technical assistance for the development and improvement of technologies methodologies and tools for enhanced use of agricultural biomass residues.* Goriza, IT: C.E.T.A- Centro di Ecologia Teorica ed Applicata.

Robertson, S., Rutherford, M., Lopez-Gutierrez, J., & Massicotte, H. (2012). Biochar enhances seedling growth and alterss root symbioses and properties of sub-boreal forest soils. *Canadian Journal of Soil Science, 92*, 329-340. http://dx.doi.org/10.4141/cjss2011-066

Sarenbo, S. (2009). Wood ash dilemma - reduced quality due to poor combustion performance. *Biomass and Bioenergy, 33*, 1212-1220. http://dx.doi.org/10.1016/j.biombioe.2009.05.007

Steenari, B., Karlsson, L., & Lindqvist, O. (1999). Evaluation of the leaching characteristics of wood ash and the influence of ash agglomeration. *Biomass and Bioenergy, 16*, 119-136. http://dx.doi.org/10.1016/S0961-9534(98)00070-1

Stupak, I., Asikainen, A., Roser, D., & Pasanen, K. (2008). Review of recommnedation for forest energy harvesting and wood ash recycling. In D. Roser, A. Asaikainen, K. Raulund-Rasmussen & I. Stupak (Eds.), *Sustainable Use of Forest Biomass for Energy*. Heidelberg: Springer. http://dx.doi.org/10.1007/978-1-4020-5054-1_7

Vance, E. (1996). Land application of wood-fired and combustion boiler ashes: an overview. *Journal of Environmental Quality, 25*, 937-944. http://dx.doi.org/10.2134/jeq1996.00472425002500050002x

10

Remediation of Nickel from Crude Oil Obtained from Bomu Oil Field Using Cassava Waste Water Starch Stabilized Magnetic Nanoparticles

Joshua L. Konne[1] & Kenneth Okpara[1]

[1] Department of Chemistry, Rivers State University of Science & Technology, Port Harcourt, Nigeria

Correspondence: Joshua L. Konne, Department of Chemistry, Rivers State University of Science & Technology, Port Harcourt, Nigeria. E-mail: konne.joshua@ust.edu.ng

Abstract

Starch Stabilized Magnetic Nanoparticles (SSMNPs) were prepared by reacting 4.4 g of $FeCl_3.6H_2O$ with 1.98 g of $FeCl_2.4H_2O$ in 61 mL of 0, 0.025, 0.5, 0.1 and 0.5% w/v of cassava waste water starch solution. Deionised water (0% w/v) was used as the control. Powder X-ray diffraction analysis of the first four samples confirmed that all the samples contained magnetite nanoparticles with estimated sizes of 16.99, 15.73, 11.48 and 17.73 nm respectively. The SSMNPs were then applied in the removal of nickel ions from crude oil obtained from Bomu oil field. The samples were divided into batches A and B. Both samples were shaken for 3.3 h but only batch A was centrifuged before analysis. 0.2 mL of each sample was siphoned and dissolved in *n*-hexane for analysis. The results obtained showed that the sample with the least starch concentration (15.73 nm) recorded the best performance in both cases with 56 and 93% of Ni (II) complexes removal respectively. All measurements were done using a UV-VIS spectrophotometer set at λ_{max} of 400 nm predetermined from a scan of Ni (II) ions solution.

Keywords: starch, cassava-waste-water, magnetic nanoparticles, nickel, crude oil

1. Introduction

Nickel, a ferromagnetic metal is one of the trace metals found in crude oil and particularly in Bomu oil field crude as reported by Onojake, Oforka and Osuji (2011). Trace metals from parent rocks got incorporated into crude oil in the form of porphyrin complexes which originated from the chlorophyll components of decayed biomass. This is responsible for higher concentrations of nickel found in oils formed from sedimentary marine organic matter than those from other terrestrial biomass. Bomu crude and terrain fell in this category (because of salt water coastline). High nickel concentration of 8.93 ppm has been reported for Bomu crude by the same author (Onojake et al., 2011).

Similarly, other authors have reported different concentrations of nickel in some Nigeria crude oils. These include: - Bonny light crude 0.13–0.17 ppm (Odebunmi & Adeniyi, 2004), Onshore/off-shore Nigeria crude oil 210–1620 ppb (Muhammad et al., 2013) and 351–11000 ppb also cited in the same paper. The different reports showed marked variation in nickel concentrations of Nigeria crude oil due to factors such as source of the crude oil, sample handling and the method of analysis (Odebunmi & Adeniyi, 2004). The analytical techniques used in the two reports were HPLC and AAS respectively.

The presence of nickel in crude oil is generally undesirable because of the following reasons; it accelerates the corrosion of distillation towers and turbines (Odebunmi & Adeniyi, 2004), catalytic poisoning during refining processes with resultant undesirable side reactions that compromised efficiency (Welz & Vale, 2012). In addition the inhalation of nickel dust or aerosols generated during the combustion of crude oil is also harmful to humans as a potential carcinogen. Other ailments and disease conditions such as bronchitis, skin reactions, stomach and blood disorder, asthma, lung and heart disorders could also be linked to prolonged exposure.

Coking and hydrotreatment are some pre-treatment methods of processing heavy crude in order to remove metal ions, sulphur and nitrogen compounds (Reynolds, 2004). However, due to the avoidance of high carbon, coke is usually applied in limited quantity which doesn't guarantee efficient removal of trace metals from crude.

In a drive for more effective and cheaper processes, Reynolds (2004) treated surrogate metal ions bound to petroporphyrin fractions isolated from heavy crude with some chemical agents followed by washing in aqueous solutions to remove the metals. The author monitored the process using UV-VIS spectrophotometer (λ_{max} = 410 nm) and found that maleic acid in dimethyl formamide completely removed vanadyl porphyrin while both nickel and vanadyl porphyrin removals were achieved by the use of superacids (fluorosulfonic acid and trifluoromethane sulfonic acid).

In a similar approach, Wen, Yang, Xu and Gao (2013) removed nickel and vanadium from crude oil using microwave-chemical method with 35% efficiency in the presence of 500 mg/L demetallizers (synthesized from polyethylene and polyamides) at 70 °C and 5 min reaction time. The efficiency was increased to 43.41 and 41.63% for nickel and vanadium respectively at 30 min reaction time. The nickel and vanadium concentrations were measured using infrared spectroscopy.

Starch Stabilized Magnetic Nanoparticles (SSMNPs) have been applied recently in the remediation of arsenic contaminated soil with higher efficiency recorded for the starch-stabilized adsorbents than the non SSMNP (Zhang, Pan, Zhao, & He, 2011; An, Liang, & Zhao, 2011). Arsenate adsorption on the SSMNPs was found to be generally starch concentration dependent except for the 0.5% concentration that behaved somewhat different. The authors attributed the success to greater adsorption surface and stronger adsorption affinity of SSMNPs to arsenate ions. It would be expected that nickel (II) porphyrin complex in crude oil would require adsorbent with high surface area due to its bulky nature hence, the use of SSMNPs with high surface area and strong affinity could be promising. In addition magnetic nanoparticles would exert stronger affinity for ferromagnetic elements like nickel from crude oil. Secondly, SSMNPs are more environmentally benign and cheaper adsorbents than most of the methods in use.

The use of UV-VIS spectrophotometer in monitoring the depletion of nickel from Bomu crude oil would also provide some quantitative data from this technique for the first time. To date, there seemed to be no report on the use of starch-stabilized magnetic nanoparticles (SSMNPs) obtained from cassava waste water for the remediation of nickel from crude oil in the available literature. Cassava waste water is the major by-product of cassava processing and an abundant waste in Niger Delta region of Nigeria. This report presents the first application of SSMNPs in the remediation of nickel from Bomu crude.

2. Experimental

2.1 Preparation of Magnetic Iron Oxide Particles

The preparation of magnetic iron particles was done by the method reported in literature with slight modifications (Zhang et al., 2011). 4.4 g of $FeCl_3.6H_2O$ and 1.98 g of $FeCl_2.4H_2O$ were dissolved in 61 mL of de-aerated water to precipitate the iron oxide. The solution was purged for 30 mins with nitrogen gas in order to create the inert environment needed to prevent the oxidation of Fe^{2+} species. 143 mL of 0.7 M NaOH was added drop wise to the mixture with the aid of a syringe with further influx of N_2 for additional 10 mins and continuous shaking for 1 h. The mixture was allowed to stand for 24 h. The colour of the mixture changed from brown to dark brown and finally to black during the first 7 h. The NaOH increased the pH to 11 which supplied the basified environment for magnetite formation according to the following equation:

$$2Fe^{3+}{}_{(aq)} + Fe^{2+}{}_{(aq)} + 8OH^{-}{}_{(aq)} \rightarrow Fe_3O_{4(s)} + 4H_2O_{(l)} \quad (1)$$

The clearer supernatant of the mixture was decanted away followed by confirmatory magnetization test of the bottom black portion with a bar magnet. The black magnetic iron oxide particles were attracted to the sides of the sample bottle when a bar magnet was brought closer to it. The resultant precipitates (black magnetic iron oxide particles) were allowed to stay in contact with the mother liquor in capped bottles in order to minimised further oxidation of the product.

2.2 Preparation/Characterization of Starch Stabilized Magnetic Nanoparticles (SSMNP)

The SSMNPs were prepared by the dissolution of 0, 0.025, 0.05, 0.10 and 0.5% (w/v) dried starch powder in 100 mL of deaerated deionized water with vigorous stirring until a uniform mixture was obtained. 61 mL of the mixture was used as solvent for the preparation of ten (10) SSMNPs (two replicates for each concentration) following the same method described for magnetic iron oxide particles. Four of the samples (0 to 0.10) were dried at 120 °C in the laboratory oven and sent for powder X-ray analysis at Sheda Science and Technological Complex (SHESTCO) Abuja, Nigeria. The instrument used was A PAnalytical Pro MPD XRD facility on a Reflection-Transmission Spinner configuration. Minimum step size of 2-Theta: 0.001 and CuKα radiation source was used for the analysis. Particle size estimations were done by feeding the generated full width at half maximum (FWHM) from the reflection patterns to the Scherrer formula (Patterson, 1939).

$$L = \frac{\kappa\lambda}{\beta \cos\theta} \quad (2)$$

Where L = Linear particle dimension, κ = shape, factor = 0.93 Å, β = FWHM in degrees and θ in degrees respectively.

2.3 Application of SSMNP for Nickel Removal

The ten SSMNPs samples were divided into two batches. 0.5 g/v of the different SSMNPs and the control (0%) were mixed with 10 mL of crude oil and shaken for 3.3 h at 800 rpm using a 65800 Thermolyne Maxi shaker. Batch A samples were centrifuged at 2000 rpm for 30 mins while batch B samples were allowed to stand before analysis. 0.2 ml of each sample was siphoned with a micro syringe and dissolved in 10 mL of *n*-hexane for spectrophotometric analysis of nickel.

2.4 Wavelength Selection and Calibration of UV-VIS Spectrophotometer

The wavelength of maximum adsorption (λ_{max}) of nickel ions was determined by a scan of 0.02 M nickel sulphate hexahydrate solution in a 1 mL cuvette from 350 to 500 nm range using a UV-VIS spectrophotometer, model 2500 LaboMed, Inc. This was followed by a calibration curve of varying concentrations of nickel (II) ions at the same wave length. The slope (molar absorptivity) obtained was used in the determination of the concentrations of the absorbing nickel ions in the samples.

3. Results and Discussion

The preparation of starch-stabilized magnetic nanoparticles resulted in the formation of the desired products according to Equation 1. Preliminary confirmation tests were the observed black colour (Figure 1a) of the products and subsequent attraction to a bar magnet. The magnetic particles (control) crude oil-mixture can be seen in the middle of the four test samples (SSMNPs-crude oil-mixtures) of Figure 1b. The powder X-ray reflection patterns of the four samples (Figure 2) were all indexed to the Fe_3O_4 phase in the following order: - control (A) and Fe_3O_4 + 0.025 (B), Fe_3O_4 + 0.05 (C) and Fe_3O_4 + 0.10 (D) starch concentrations. The peak positions and intensity were similar to literature data (Kim, 2004). The plots were done without background correction and could be responsible for the noisy background.

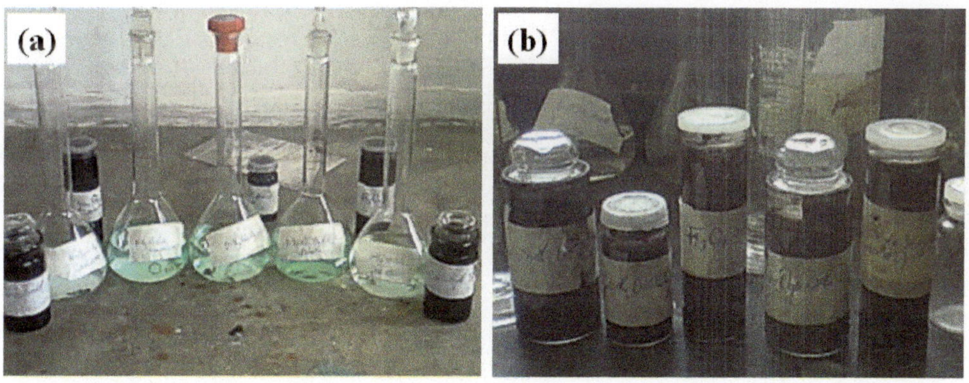

Figure 1. Varying concentrations of (a) SSMNPs (black) in small sample bottles beside Ni^{2+} ions in volumetric flasks and (b) Magnetic particles (Fe_3O_4)-middle and SSMNPs + crude oil samples

However, few impurity phases marked (●) were also observed for the samples. Peak broadening due to decrease in particle size and increase in amorphous content (starch particles) were observed with the starch loaded magnetic nanoparticles. The sizes were estimated to be 16.99, 15.73, 11.48 and 17.73 nm for samples A-D respectively from Scherrer formula. Particle size was found to decrease with increase in the starch concentration except for D which showed an abnormal increase. Similar observation has been reported for a 0.05% starch concentration and this was attributed to the NaOH/starch ratio that would result in higher starch hydrolysis at higher starch concentrations.

The λ_{max} obtained from a scan of 0.02 M nickel ions solution between 350 to 500 nm intervals was determined as 0.19 at 400 nm (Figure 3). The result compared closely to 401 and 410 nm for nickel (II) ions and complex respectively, reported previously (Kumar, Reddy, & Reddy, 2007; Okoye, Chukwuneke, Ekere, & Ihedioha,

2013). The molar absorptivity obtained from the calibration curve (plot of absorbance vs concentration of 0.02, 0.04, 0.06, 0.08 and 1.0 M nickel (II) ions) at 400 nm was 5.23 M^{-1} cm^{-1}. The aqueous light green solutions of the Ni^{2+} ions are shown in Figure 1a. A good linear regression (R^2) of 0.9958 implied that the result was in good agreement with Beer-Lambert's law. However, slight deviations from Beer-Lambert's law at concentrations greater than 0.01 M are normal (Harris, 2003). The calibration plot showed these deviations at 0.06 and 0.1 M nickel (II) ions (Figure 4).

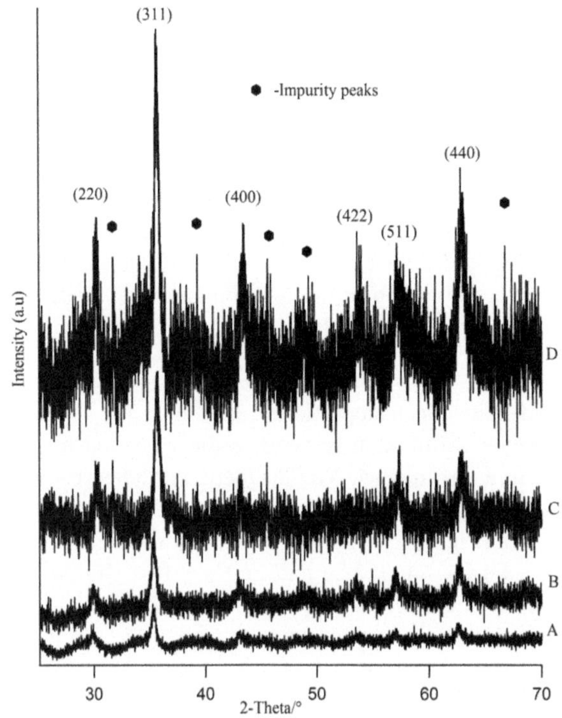

Figure 2. XRD of A- Fe_3O_4(control), B- Fe_3O_4 + 0.025 starch, C- Fe_3O_4 + 0.05 starch and D- Fe_3O_4 + 0.10 starch concentrations showing indexed magnetite patterns and some impurity peaks (●)

Table 1 showed the calculated concentrations and percentages of nickel (II) complexes in crude oil (c), crude oil with magnetic particles (c + m) and crude oil with varying concentrations of SSMNPs (c + m + 0.025… c + m + 0.1). The results showed that the magnetic particles effectively removed 55 and 24% concentration of nickel from the Bomu crude oil for batches A and B respectively. This result was greater than 53% previously reported for the removal of Ni from petroporphyrin-rich fractions and asphaltenes by biocatalytic method (Mogolloñ, Rodríguez, Larrota, Ortiz, & Torres, 1998).

Table 1. Samples, calculated concentrations and percentages of [Ni^{2+}] left after removal

Samples	Conc-A ($Moldm^{-3}$)	A-%[Ni^{2+}]	Conc-B ($Moldm^{-3}$)	B-%[Ni^{2+}]
c	0.1091	100	0.1091	100
c+m	0.0491	45	0.0829	76
c+m+0.025	0.0484	44	0.0076	7
c+m+0.050	0.1072	98	0.0359	33
c+m+0.10	0.1206	111	0.2771	254
c+m+0.50	0.1084	99	0.2209	202

Table 1 shows the concentrations of nickel (II) ions left in samples and their relative percentages to the original concentration in crude before the removal process.

The higher efficiency recorded in batch A (c + m) showed that centrifugation increased the sticking probability of the adsorbates to the magnetic particles and quicker coalescing of nickel porphyrin complexes to the bottom of the centrifuge tubes due to centrifugal force than those of batch B where interactions were limited to surface adsorption on the magnetic layer. This was responsible for the observed differences between batch A and B (c + m) results.

However, the efficiency in batches A and B increased to 56 and 93% respectively when SSMNPs of 0.025% w/v concentrations were applied. The use of starch in the preparation of magnetic particles effectively reduced particle aggregations that led to slight particle size reduction (16.99 to 11.48 nm), provided functionalised particle surfaces and stabilizing effect. The particle sizes of the samples produced agreed with previous report by Zhang et al. (2011), howbeit, the control sample (A) was smaller. Therefore increase in surface area of the SSMNPs (adsorbents) was responsible for the observed increase in efficiency when SSMNPs were involved.

The overall results showed that the least starch concentration produced the best result in both cases studied. Batch B also produced a second higher efficiency of 67% at 0.05% w/v starch concentration. This means that higher surface coverings reduced the attractions of the magnetic core of SSMNPs to the outermost layers (over certain critical length scales) and created some loosely bound layers which might have caused the interference that resulted in loss of efficiency. Hence there was virtually no reduction observed at higher starch concentrations rather anomalous concentration values above 100% were obtained particularly for batch B. But when the system was centrifuged, the loosely bound layers coalesced and settled to the bottom of the centrifuge tubes reducing interference as observed in batch A results. The higher efficiencies (67 and 93%) observed in batch B can also be linked to the fact that the presence of Ni^{2+} in crude as porphyrin complexes (central metal ion complex) favoured layer by layer attraction and adsorption. Therefore more layers settled to the surfaces of the adsorbent at lower starch concentrations under gravity/magnetic attraction in B than in a more agitated batch A system.

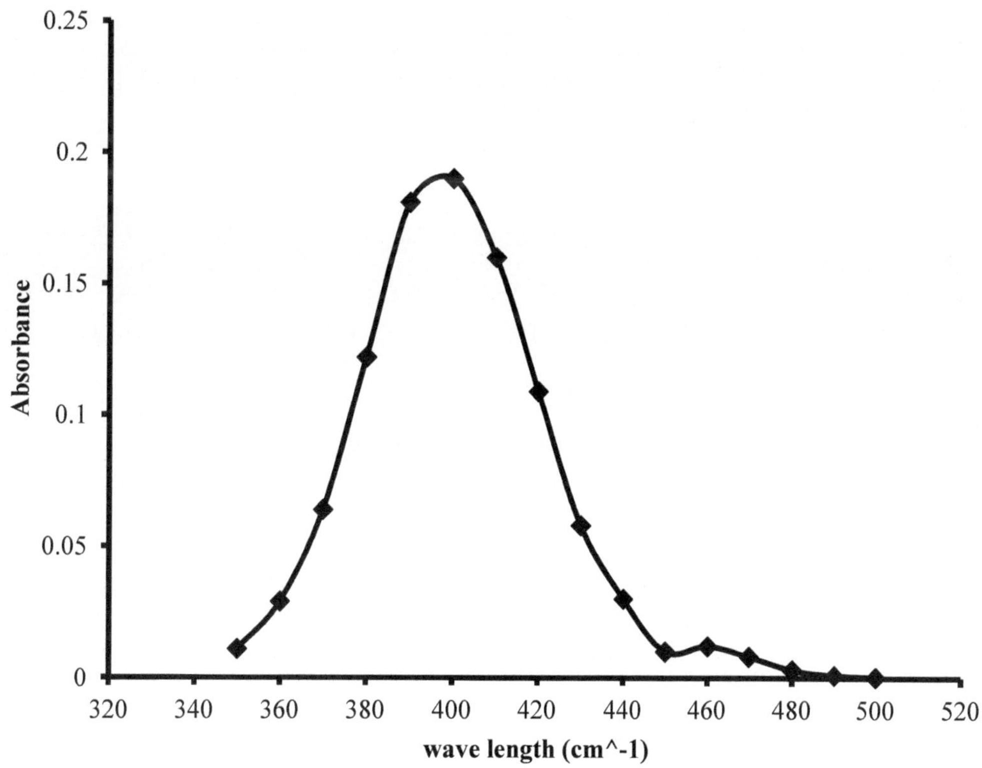

Figure 3. Absorbance vs wavelength of nickel (II) ions

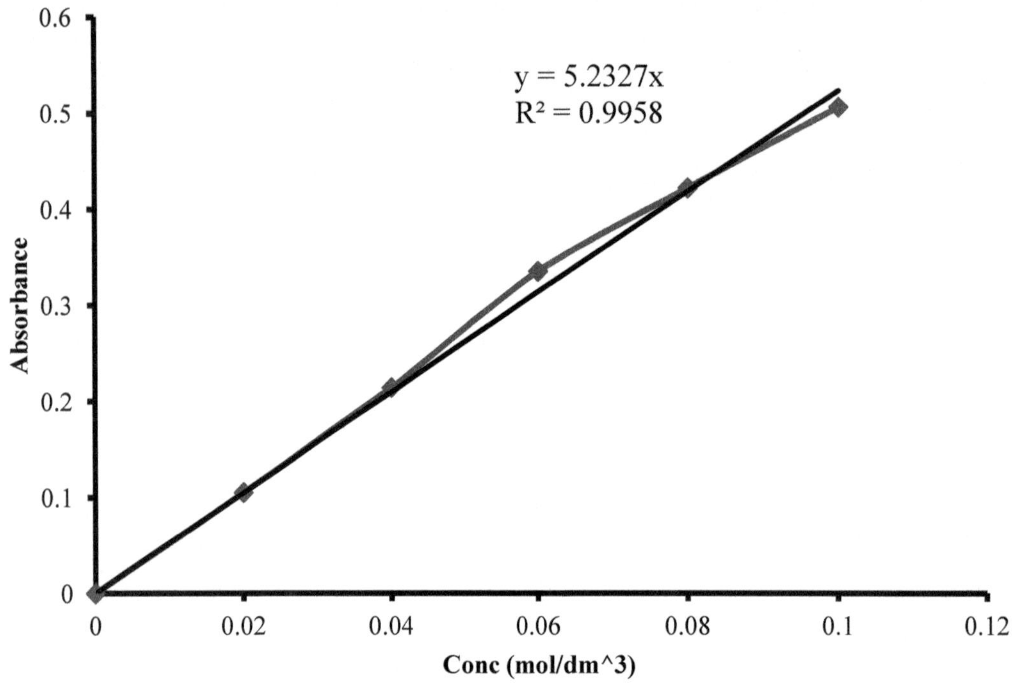

Figure 4. Absorbance vs concentration plot of Ni (II) ions at 400 nm

4. Conclusion

SSMNPs have been tested and shown to be more effective in the removal of nickel porphyrin complexes from Bomu crude oil than any other methods reported previously. The results showed that the sample with the least starch concentration exhibited the strongest attraction and surface affinity for nickel complexes in crude oil with up to 93% of nickel complexes removal. The use of SSMNPs in adsorption studies are currently being investigated in our laboratories. More work is needed on studying the effectiveness of SSMNPs in the removal of other trace metals in crude oil.

Acknowledgements

The authors wish to thank Mr. Bob Manuel of the Department of Chemistry, Rivers State University of Science and Technology Port Harcourt, Nigeria for his technical support on the UV-VIS spectrophotometer. We also wish to acknowledge Sheda Science and Technological Complex (SHESTCO) Abuja, Nigeria for the Powder X-ray analysis.

References

An, B., Liang, Q., & Zhao, D. (2011). Removal of arsenic (V) from spent ion exchange brine using a new class of starch-bridged magnetite nanoparticles. *Water research, 45*(5), 1961-1972. http://dx.doi.org/10.1016/j.watres.2011.01.004

Harris, D. C. (2003). *Quantitative Chemical Analysis* (6th ed.). W. H. Freeman and Company.

Kim, C. S. (2004). Crystallographic and Magnetic Properties of Iron Oxide Nanoparticles for Applications in Biomedicine. *Journal of Magnetics, 9*(3), 83-85.

Kumar, A. P., Reddy, P. R., & Reddy, V. K. (2007). Spectrophotometric determination of Nickel (II) with 2-hydroxy-3-methoxybenzaldehyde thiosemicarbazone. *Indian J. Chem., 46A*, 1625-1629. http://dx.doi.org/10.4283/JMAG.2004.9.3.083

Mogolloń, L., Rodriguez, R., Larrota, W., Ortiz, C., & Torres, R. (1998). Biocatalytic removal of nickel and vanadium from petroporphyrins and asphaltenes. *Applied biochemistry and biotechnology, 70*(1), 765-777. http://dx.doi.org/10.1007/BF02920187

Muhammad, I., Ikeh, P. O., Usman, B. B., Shehu, B. D., Salawu, K., & Mikailu, D. A. (2013). Determination of Vanadium, Nickel, Copper and Iron as Complexes of Bis-Acetylpivalyl Methane (Ethylene Diamine) in Nigerian onshore and offshore Crude Oils Using HPLC. *Journal of Natural Sciences Research, 3*(8), 105-111.

Odebunmi, E. O., & Adeniyi, S. A. (2004). Characterization of crude oil and Petroleum products for trace elements. *J. Chem. Soc. Nigeria, 29*, 149-154.

Okoye, C. O. B., Chukwuneke, A. M., Ekere, N. R., & Ihedioha, J. N. (2013). Simultaneous ultraviolet-visible (UV–VIS) spectrophotometric quantitative determination of Pb, Hg, Cd, As and Ni ions in aqueous solutions using cyanidin as a chromogenic reagent. *International Journal of Physical Sciences, 8*(3), 98-102.

Onojake Mudiaga, C., Oforka Nicolas, C., & Osuji Leo, C. (2011). Trace Metals Geochemistry of Crude Oils from Umutu/Bomu Fields in South West Niger Delta Nigeria. *Energy and Environment Research, 1*, 139-146. http://dx.doi.org/10.5539/eer.v1n1p139

Patterson, A. L. (1939). The Scherrer formula for X-ray particle size determination. *Physical review, 56*(10), 978. http://dx.doi.org/10.1103/PhysRev.56.978

Reynolds, J. G. (2004). Removal of nickel and vandium from heavy crude oils by exchange reactions. *Prepr. Pap.-Am. Chem. Soc., Div. Fuel Chem, 49*, 79.

Welz, B., & Vale, M. G. R. (2012). Determination of Total Nickel and Vanadium and Their Volatile and Thermally Stable Species in Crude Oil Using High- Resolution Continuum Source Graphite Furnace Atomic Absorption Spectrometry. *Analytical Instrumentation, 14*.

Wen, Z., Yang, J., Xu, X., & Gao, J. (2013). Removal of Nickel and Vanadium from Crude Oil by Microwave-Chemical Method. I. Chemical Method. *Petroleum Science and Technology, 31*(9), 991-999. http://dx.doi.org/10.1080/10916460903330023

Zhang, M., Pan, G., Zhao, D., & He, G. (2011). XAFS study of starch-stabilized magnetite nanoparticles and surface speciation of arsenate. *Environmental Pollution, 159*(12), 3509-3514. http://dx.doi.org/10.1016/j.envpol.2011.08.017

11

Comparative Study on Charcoal Yield Produced by Traditional and Improved Kilns: A Case Study of Nyaruguru and Nyamagabe Districts in Southern Province of Rwanda

Alphonse Nahayo[1], Isaac Ekise[1] & Alphonsine Mukarugwiza[1]

[1] Higher Institute of Agriculture and Animal Husbandry (ISAE-Busogo), Department of Forestry and Nature Conservation (FNC), Musanze, Rwanda

Correspondence: Alphonse Nahayo, Higher Institute of Agriculture and Animal Husbandry (ISAE)-Busogo, Department of Forestry and Nature Conservation, P. O. Box 210, Musanze, Rwanda. E-mail: nahayo1@yahoo.fr

Abstract

Deforestation and shortage of wood are serious environmental issues in Southern Province of Rwanda. This is likely to happen due to inadequate strategies and capacity to produce and utilize wood for energy on a sustainable basis. Furthermore, the production of charcoal in rural areas is done through the earth mound kilns causing more pressure on forests due to increased demand of charcoal. The main purpose of this study was to compare the charcoal yield produced from improved and traditional methods. This study was conducted in Nyabimata sector of Nyaruguru District and Tare sector of Nyamagabe District in June and July 2012. Improved charcoal and traditional charcoal were produced in order to determine the best method to be used. Data were analyzed using the Gen Stat Discovery 4th Edition. The results revealed that improved techniques can increase the charcoal production and reduce the air pollution where one can obtain at least 3 bags of charcoal in 1 m^3 of wood and 15 liters of tar collected from the chimney containing the major elements responsible for green house gases emission. The yields of charcoal obtained according to the weight of wood used were less in traditional earth mound kiln (T_1) techniques with the percentage of 7.5% than what improved earth mound kiln (T_2) and casamance kiln (T3) techniques produced with 19% and 20% respectively. Measures should be taken in order to increase the level of improved charcoal making adoption, such as encouraging people to invest in improved charcoal production, organizing the trainings to the charcoal makers and planting more trees.

Keywords: improved earth mound kiln, traditional earth mound kiln, Casamance kiln, charcoal yield

1. Introduction

Charcoal is a fuel commonly used by households for cooking food and heating in certain parts of the developing world. Biomass users prefer charcoal over other biomass fuels such as wood, residues and dung because charcoal has a higher energy density than other biomass fuels. It also has excellent cooking properties because it burns evenly, for a long time, and can be easily extinguished and reheated (Kammen & Lew, 2005).

The Food and Agriculture Organization (FAO) has estimated that the total charcoal production in 1992 was 24 million tons. About the half of the world's charcoal use is in Africa, where traditional production techniques lead to low conversion efficiency (Kammen & Lew, 2005).

In Kenya, concerning the urban household energy use, 66% is supplied by charcoal and 18% by fuelwood. For rural household energy, 90% is from fuelwood and only 5% from charcoal (O'keefe, Raskin, & Bernow, 1984).

In Tanzania, around 91% of all energy consumed is woodfuel. Miombo woodland is the source of about 70% of the annual consumption (Monela et al., 1993). It accounts for 97.6% of the total wood products consumed in the country (MNRT, 2001). The estimated national annual fuelwood consumption in Tanzania is 44.8 million m^3 (Kaale, 2005). The major consumer centres are households for cooking (95.4%), rural industries (2.8%) and agriculture (1.4%). The rural industries and agriculture use exclusively fuelwood. On the other hand, the rural households use almost fuelwood exclusively.

In year 2000, the charcoal consumption was estimated in three African cities to about 140,000 tons for Maputo (Mozambique), about 314,000 tons for Dar es Salaam (Tanzania) and about 245,000 tons for Lusaka (Zambia).

In these countries, the households used certain fuel energy sources for cooking certain foods or for other kind of activity requiring energy such as lighting or heating (CHAPOSA, 2002). The same author reported that in Lusaka (Zambia), 65% of the households used charcoal as the only energy source while the rest of the households used charcoal in combination with firewood (23%), kerosene (17%) and electricity (1%). As for Dar es Salaam, 69% of the households used charcoal for cooking as their first choice.

Charcoal is essentially produced by heating fuelwood (or any other raw biomass) in some types of kilns with limited access to air, a process called carbonization. Charcoal making kilns can vary greatly in structure and size, from simple earth mound kiln to semi permanent brick ovens to large and permanent metal structures. Carbonization creates a fuel of higher quality than the original fuelwood because of inherent inefficiency in the process (Kokou, Nuto, & Atsri, 2009).

Wood collection is the main cause of deforestation. Therefore, the improved charcoal making decreases the quantity of wood used for charcoal production through the improvement of technique and the utilization of chimney helps reduce the pollution of atmosphere through the smoke. With improved methods of charcoal making, a greater quantity of charcoal could be produced. There is no much waste, air pollution in the burning process as with more traditional means. Improved kiln for charcoal making also plays a role in reducing exposure to indoor smoke, and it is the cheapest source of energy we can use in our country in order to reduce deforestation, cooking time, smoke in the kitchen and to economize the cooking materials because a given quantity of wood in improved kiln produces about twice as much charcoal as the same quantity of wood in traditional kiln, also a given quantity of charcoal produces approximately twice as much heat as the same weight of wood. Burning charcoal produces temperature in excess of 1,000 degrees Celsius and it is relatively easy to make it (Gerald, 1986).

In Sub-Saharan Africa, the production and consumption of charcoal lead to high associated Green house gas emissions which are the result of three factors (Müller, Michaelowa, & Eschman, 2011):

1) An unsustainable supply of biomass in which forests are being depleted for the production of the fuel.
2) The use of inefficient technologies to convert wood into charcoal with yields as low as 10% observed in certain countries (10 kg of wood required to produce 1 kg of recovered charcoal).
3) The use of specific technologies in which the conversion of wood into charcoal leads to a high level of methane emissions

The potential for reducing green house gas emissions by promoting the application of improved kiln technology is tremendous, not only due to higher charcoaling efficiencies, but also due to the application of green house gas reducing technologies (e.g. destruction of the methane stream) and the possibility of electricity co-generation as stated by Sander, Hyseni and Haider (2011). These opportunities can refer either to the technology and practices for charcoal making or to a decrease in non-renewable share of biomass used.

International Center for soil Fertility and agriculture Development (IFDC) through its department of value chain has undertaken different activities aiming at improving sustainable energy in Rwanda. Private forests owners have been, since the beginning, among the targeted groups with IFDC and charcoal makers have worked with them. With the contribution of IFDC, the sensitization and training about the use of improved kiln for charcoal making in Nyaruguru and Nyamagabe districts started in 2010.

Charcoal production is one of the primary causes of deforestation in Rwanda. Changing behavior of farmers through the adoption of improved kiln will play an important role in environmental protection by reducing deforestation and air pollution. The main objective of this study was to compare the charcoal yield produced from improved and traditional charcoal making techniques in Nyaruguru and Nyamagabe Districts in Southern province of Rwanda. The specific objectives were (i) to determine between improved and traditional kiln the best technique to be used in order to produce much charcoal by using less wood; (ii) to determine between the traditional and improved kilns the technique to be used in order to produce the charcoal intended for domestic uses of high consistency in terms of weight. The following hypotheses were formulated: (1) There is no significant difference in the charcoal yield among the three charcoal making methods (traditional earth kiln, improved earth kiln and casamance kiln) at p-value of 5%; (2) There is no significant difference in weight of charcoal measured by each bag from three kilns of charcoal making at p-value of 5%.

2. Material and Methods

2.1 Study Area Description

The District of Nyaruguru (1010.267 km^2) is located in Southern Province of Rwanda. It has 14 administrative sectors which are Munini, Kivu, Cyahinda, Nyagisozi, Ngoma, Ngera, Rusenge, Muganza, Ruheru, Busanze, Mata, Ruramba, Kibeho and Nyabimata. The study was conducted in Nyabimata sector. The altitude varies between 2000 and 2500 m a.s.l. The Eastern part of the District is characterized by the central plateau with an altitude varying between 1500 m to 2000 m a. s. l. The annual average temperature is around 17 °C in the region of high altitude and 20 °C in the region of average altitude.

The District of Nyamagabe lies between Huye and Rusizi in the South-West of Rwanda. It also contains the half of Eastern Nyungwe National Park, a popular tourist destination, being one of the last remaining forest areas of Rwanda and home to chimpanzees and many other species of primate and birds. Nyamagabe District is divided into 17 sectors: Buruhukiro, Cyanika, Gatare, Kaduha, Kamegeli, Kibirizi, Kibumbwe, Kitabi, Mbazi, Mugano, Musange, Musebeya, Mushubi, Nkomane, Gasaka, Tare and Uwinkingi. The study was done in Tare sector. Nyamagabe District has an average altitude varying from 1800 to 2700 m a.s.l. It has uneven altitude with some summits going beyond 3000 m high. Nyamagabe District relief is characterized by irregular slopes ranging from 60° to 120° making the soil susceptible to soil erosion and degradation. Nyamagabe District enjoys a humid tropical climate moderated by the effect of high altitude. Annual rainfall varies from 1300 to 1450 mm with average temperature of 18 °C.

2.2 Material

Table 1. Materials used

Materials	Use
Wood of *Eucalyptus maidenii*	The Eucalyptus trees are the most suitable for charcoal making and suitable to increase the consistency of charcoal.
Panga and axe	For cutting wood
Hoe	For digging and land leveling
Spade	For digging and land leveling
Taper	For measuring
Balance	For weighing wood or charcoals
Rake	For collecting charcoal in the kiln
Pitch fork	For collecting charcoal
Hygrometer	For measuring humidity

2.3 Methods

Experimentation design.

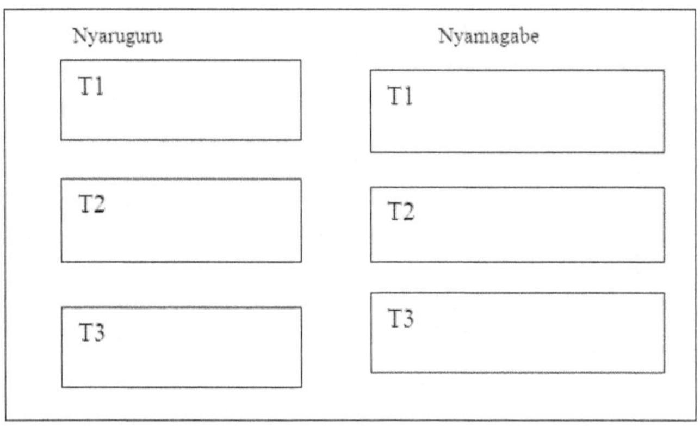

Figure 1. Experimental layout

Experimental kilns were laid out by using randomized complete block design (RCBD) with 3 treatments and 2 replications. The treatments were: T1= Traditional earth kiln; T2= Improved earth kiln; T3= Casamance kiln.

3. Preparation of Kilns and Carbonization Follow Up

3.1 Cutting down Trees, Site Selection and Loading the Kiln

The wood was cut for three weeks and dried in solar energy before the carbonization operation begins; this is necessary to reduce the amount of water content in order to minimize the amount of heat energy wasted during the drying stage. As the rate of carbonization is sensitive to many factors, the wood used had the same moisture content and the same size of length of 2 m each. The site was selected from the dry places, free from stumps; small vegetation was removed and the land was leveled roughly to produce square or circular platform. The transport of wood was reduced to minimum level by making the kiln in or near the forest.

Before charging the kiln, the pit was covered with a layer of branch litter. This is needed during carbonization process because the soil takes the heat and continues to burn the charcoal. Then, the charcoal burns completely and produces high content of ash and carbon. This layer was put between soil and wood as it had the importance of collecting charcoal from clear ground. The first layer of wood in the kiln is composed of small wood in length playing the role of circulating smoke in the kiln and preventing the soil from touching the trees. Trees were loaded into the kilns with the aim to not lift any place between trees. Small branches were used to fill the empty spaces between trees. The installation of chimneys was done before covering the kiln with the soil. The chimneys were placed in the middle of the kiln on the upper side.

Figure 2. The chimney installation in improved kiln

After loading the kiln, the wood was covered with the leaves of *Eucalyptus maidenii*. The earth layer was sealed without any hole in the kiln.

3.2 Care Taken during Carbonization and Discharging Charcoal

The labor was available during the day and night in order to control the kilns. Then the supply of air to the kiln was carefully controlled and all air entries were completely excluded during carbonization.

The kiln was kept tightly and sealed in order to reduce the danger of the whole load to burst into flame and to be lost. It was prohibited to walk on the top of the kiln for avoiding the fire. The end of this process was shown by the reduction of the smoke and the appearance of the bluish color from the chimney. It consisted in the removal of the heated soil layer on the top of the kiln and the addition of another wet soil without disturbing the leaf litter. Once it was exposed to the open air, charcoal was left for a period of about 24 hours. During this time, the charcoal was cooled to air temperature. The danger of spontaneous combustion disappeared and the charcoal was ready for packing.

Figure 3. Liquid products in casamance kiln

The smoke from the kiln was left in the chimney in the form of water vapor or gas. Reddish water and mixture of tars and other unknown products were collected. Note that the composition of those substances is not known, but research is under process. This liquid product was sticky with strong and pungent smell which lingers on anything with which it is in contact. All these liquids, if there were no chimneys, they would have gone in the atmosphere to pollute and damage the environment.

The Table 2 below summarizes the methodology used to produce charcoal in Traditional kiln (T1), Improved kiln (T2) and Casamance kiln (T3).

Table 2. Charcoal production in Traditional kiln (T1), Improved kiln (T2) and Casamance kiln (T3)

Characteristics	**Traditional kiln (T1)**	**Improved kiln (T2)**	**Casamance kiln (T3)**
Size of wood used	2m	2m	Wood with the same moisture content and the wood size was 2m; 1.8m;1.7m; 1.5m; 1.3m; 1.1m; 0.9m; 0.7m and 0.5m in order to respect the circular plat form of the kiln.
Pit kiln measurements	2.5 m for both length and width	2.5 m for both length and width	2.5 m of radius
Materials	The same materials used: See Table 1 above		
Site selection	Level land roughly to produce irregular form	Level land roughly to produce rectangular plat form	The site was selected from dry places, free from stumps and vegetation was removed. The site was also leveled roughly to produce circular plat form of 2.5m of radius

Characteristics	Traditional kiln (T1)	Improved kiln (T2)	Casamance kiln (T3)
Loading the kiln	Layers of loading the kiln: disorganized loaded wood was the first layer in the pit kiln	The layers of loading the kiln: On the pit kiln, put a layer of branch litter followed by wood and put the layer of branch litter again and finally a layer of the soil to cover all kilns. For this method, the first woods loaded in the kiln were composed of small wood in length and diameter with the role of circulating the smoke in the kiln and preventing the soil from touching the trees. Small woods were loaded with the aim of not leaving any space between trees.	The layers of loading the kiln were as follows: a layer of branches and a big layer of wood followed by a layer of branches: all were covered with a layer of soil
Internal capacity	$17m^3$	$17m^3$	$17m^3$
Chimney	Not used: The kiln was not covered completely so as to simplify the aspiration of smoke in the kiln	The chimney was installed in the middle part of the kiln, on the upper direction. After, the kiln was covered completely and the air entry was opened and closed when needed	Three chimneys were used: one was installed on upper side, other on the left side and the last was installed on the right side
Smoke	The smoke emerges everywhere on the top	The smoke was released through one chimney	The smoke was released through three chimneys
Care during carbonization	The labor was available during the day and night in order to control the kiln	The kiln was kept tightly and sealed in order to reduce the danger of the whole load to burst into flame. The labor was available during the day and night in order to control the kiln.	The same as for improved kiln
Methods used to cool charcoals	The end of this process was shown by the reduction of the smoke. This phase consisted of removing the heated soil layer on the top of the kiln and the addition of wet soil. Charcoal was exposed to open air for about 4 hours. During this time, the charcoal was cooled to air temperature. The danger of spontaneous combustion disappeared and the charcoal was put into different packages	The end of this process was shown by the reduction of the smoke and its bluish color from the chimney. This phase consisted of removing the heated soil layer from the top of the kiln and the addition of another wet soil without disturbing the leaf litter. Charcoal was then exposed to open air for a period of about 4 hours to be cooled to air temperature. The danger of spontaneous combustion disappeared and the charcoal was put into packages and the tar was collected.	The same as for the traditional and improved kilns

3.3 Data Collection and Analysis

The mass of the wood loaded into each kiln was determined either by direct weighing or from the wood volume where each kiln had $17m^3$ in Nyaruguru District and $17m^3$ in Nyamagabe District. The moisture content of wood was determined by hygrometer. During the charcoal making process, the emission was released from the kiln through one or more chimneys at a time. In order to test the degree of emissions released from different kilns, we have collected all liquids (tar) released in the chimney from each kiln and measured them in liter. After the end of firing, each kiln was sealed and allowed to cool. After cooling, charcoal was removed and weighed. The quantitative data obtained were analyzed by means of Gen Stat Discovery Edition 4 software.

4. Results and Discussion

Means of charcoal making in three treatment areas of T1 (Traditional method), T2 (Improved method), T3 (Casamance kiln) in two different Districts, Nyaruguru and Nyamagabe, are presented in the Table 3 below.

Table 3. Means of charcoal making in three treatment areas of Nyaruguru and Nyamagabe Districts

Parameters	T_1	T_2	T_3	P<0.05	L.S.D	C.V.%
Carbonization in days	12.500^a	5.000^b	5.000^b	<0.001	1.299	5.4
Bags/m^3 of wood	2.000^c	3.000^b	3.250^a	<0.001	0.1299	1.5
Humidity in %	76.5	76.5	76.5	ns	42.76	17.6
Bags for each kiln	29.50^b	52.50^a	55.00^a	<0.001	3.182	2.2
Charcoal in Kg	930^b	2362^a	2475^a	<0.001	203.5	3.3
Charcoal in %	7.5^b	19.0^a	20.0^a	<0.045	9.75	19.8
Tar in Liter	0.00c	15.00b	15.75a	<0.001	0.6496	2.0

Values followed by the same letters in a row are not significantly different at $P < 0.05$; n.s. = not significant, L.S.D= Least significance difference, C.V= Coefficient of variation in percentage.

The Table 3 shows that the charcoal yields obtained by using casamance kiln (T_3) and improved traditional earth mound kiln (T_2) are significantly different in two Districts at $p < 0.05$. The charcoal yields obtained using the improved kilns were higher than the ones obtained by using traditional earth mound kiln (T_1) with the mean value of 55.00^a, 52.50^a and 29.50^b respectively.

The duration of carbonization in days, bags of charcoal obtained from 1m^3 of wood, bags of charcoal obtained from 17 m^3 of wood, weight of charcoal, charcoal in % and tar collection were very significantly different among three types of kilns at $p<0.05$. The number of bags of charcoal obtained in 1m^3 of wood from casamance kiln (T3) technique was different from the one obtained by using traditional (T_1) and improved traditional earth mound kiln (T_2) techniques with the mean value of 3.350 got from casamance kiln (T3). Other techniques presented the mean value of 3.000 and 2.000 respectively. Even if the weight of bags used for packing were the same in each technique, the charcoal weight obtained in three techniques was totally lower in traditional earth mound kiln (T_1) (mean weight of charcoal: 930kg =29.50 bags) than casamance kiln (T3) (mean weight of charcoal: 2475 kg= 55.00 bags), and improved traditional earth mound kiln techniques (T_2), (mean weight of charcoal: 2362 kg= 52.50 bags).

The yields of charcoal obtained in percentage according to the weight of wood used was less in traditional earth mound kiln (T_1) techniques with the percentage of 7.5% than what improved traditional earth mound kiln (T_2) and casamance kiln (T3) techniques produced with 19% and 20% respectively. The weight of charcoals had a significant difference ($p<0.05$) in terms of bags of charcoal obtained from the 1m^3 of wood and in terms of weight of charcoals obtained from each kiln of 17m^3 of wood while the yield of charcoal in percentage was significantly different at p- value <0.05. The days spent for carbonization by using traditional earth mound kiln (T_1) was higher than the improved traditional earth mound kiln (T_2) and casamance kiln (T3) techniques with the means of 12.5 and 5.00 respectively.

There was no difference of moisture content in wood used among three types of kilns. Concerning the quantity of tars collected during wood burning for three techniques, results showed that there was no tar produced in traditional earth mound kiln (T_1). However, there was a higher amount of tars produced by improved traditional earth mound kiln (T_2) and casamance kiln (T3) techniques with the means of tars equal to 15 and 15.75 liters respectively. This means that the improved kilns reduce the air pollution by retaining the harmful chemicals into the tar filtered by the chimney. The tars collected in improved (T_2) and casamance kiln (T3) presented a significant difference at $p<0.05$.

The chimney plays an important role in reducing the air pollution by filtering the smoke. The tar can be considered as a secondary product of charcoal production because it is often used as an insecticide or in painting the houses. Traditional kiln techniques contribute much to the environmental degradation because it uses a huge quantity of wood and it produces gases which are not filtered in the chimney as it is done for improved earth mound kiln techniques. The tar produced through the chimney is a significant production of products of incomplete combustion (PIC). The PIC emitted during the charcoal making process include carbon monoxide (CO), methane (CH_4), total nonmethane organic compounds (TNMOP), and particulate matter. Nitrogen oxides (NO, NO_2, and N_2O) as well as CO_2, CH_4 are important greenhouse gases (GHG) emitted (David & Kilk, 2001).

The consistency and the quality of charcoal for casamance kiln (T3) and improved traditional earth mound kiln (T2) are also generally higher than traditional earth mound kiln. Casamance kiln (T3) was able to consistently deliver the yields in the range of 3.25 bags per $1m^3$ of wood used. The loss of charcoal in traditional method is caused by a higher air content entering the kiln and the long time of carbonization. The charcoals are also burnt completely and therefore, the quantity and quality of charcoal decreased. In addition, the moisture content in the wood has an impact on charcoal production. The quality of charcoal made from traditional earth mound kiln and improved earth mound kiln are significantly different in terms of time of carbonization and weight of charcoal. Improved charcoal making requires small quantity of wood to produce the same quantity of charcoal produced in the traditional method (Kokou, Nuto, & Atsri, 2009). The improved earth mound kiln reduces the emission of harmful volatile substances into the atmosphere up to about 75 % (Bill, 1981).

5. Conclusion

The traditional earth mound kiln is used to make charcoal and this technique persists in Southern Province of Rwanda mainly because it is flexible in use and cheap. However, they often produce very low yields (typically 1 or 2 bags of 30 Kg each of charcoal from $1m^3$ of wood), inconsistent quality (because it is difficult to maintain uniform carbonization) and environmental pollution from the released tar and poisonous gases. But the efforts can be made to improve the traditional charcoal making by equipping earth mound kilns with chimneys. With good practice, yield of 3 bags of 45 Kg or above of charcoal from 1 m^3 of air-dried wood was possible. The quantity of charcoal made by improved earth mound kiln is greater than the one produced by traditional earth mound kiln of the same size.

Acknowledgements

The authors acknowledge the contribution of Fabien Kayitare, Ir. Théogene Ndayambaje and Jeannette Mukamabano to the completion of this study.

References

Bill, S. (1981). *Improved wood, Waste and Charcoal burning stoves: A practitioner's manual, Netherlands*. p. 229.

CHAPOSA. (2002). *Charcoal potential in Southern Africa. Inco_Dev: International cooperation with developing countries (1998-2002)*. Final report. p. 88.

David, M. P., & Kilk, R. S. (2001). Emissions of greenhouse gases and other airborne pollutants from charcoal making in Kenya and Bresil. *Journal of Geophysical Research, 106*, 24-155.

Gerald, F. (1986). *Charcoal Making in Developing Countries* (17-76). London, UK.

Kaale, B. K. (2005). *Baseline study on biomass energy conservation in Tanzania, SADC Programme for Biomass Energy Conservation (ProBEC)* (p. 55).

Kammen, D. M., & Lew, D. J. (2005). *Renewable and Appropriate Energy Laboratory Report.*

Kokou, K., Nuto, Y., & Atsri, H. (2009). Impact of charcoal production on woody plant species in West Africa: A case study in Togo. *Scientific Research and Essays, 4*(9), 881-893.

MNRT. (2001). National Forestry Program in Tanzania. Ministry of Natural Resources and Tourism. United Republic of Tanzania.

Monela, G. C., Oktingáti, A., & Kiwele, P. M. (1993). Socioeconomic aspects of charcoal consumption and environmental consequences along Dar es Salaam - Morogoro highway, Tanzania. *Journal of Forest Ecology and Management, 58*, 249-258. http://dx.doi.org/10.1016/0378-1127(93)90148-G

Müller, N., Michaelowa, A., & Eschman, M. (2011). Proposal for a new standardized baseline for charcoal projects in the clean development mechanism. Zurich. p. 86.

O'Keefe, Phil, Raskin, Paul et Bernow, Steve (Ed.) Energy and Development in Kenya: Opportunities and Constraints. Uppsala – Stockholm, The Scandinavian Institute of African Studies – The Beijer Institute, Coll. Energy, Environment and Development in Africa, no. 1, 1984, p. 202.

Review of Technologies for the Production and Use of Charcoal. Energy and Resources Group & Goldman School of Public Policy. University of California, Berkeley, CA 94720-3050; National Renewable Energy Laboratory, 1617 Cole Boulevard, Golden, CO 80401-3393. p. 19.

Sander, K., Hyseni, B., & Haider, W. (2011). Wood-Based Biomass Energy Development for Sub-Saharan Africa, World Bank, Washington. Issues and approaches. p. 64.

12

Design and Realization of a 300 W Human Power Energy Generator System on a Bicycle

Ming Chun Hsieh[1] & David King Jair[2]

[1] Department of Electronic Engineering, Kun Shan University, Tainan, Taiwan, R.O.C.

[2] Department of Mechanical Engineering, Kun Shan University, Tainan, Taiwan, R.O.C.

Correspondence: Ming Chun Hsieh, Department of Electronic Engineering, Kun Shan University, Tainan, Taiwan, R.O.C. E-mail: michelhsieh@gmail.com

Abstract

In this study, we discuss a new type of human power energy generator (HPEG) system that has been designed to work with bicycles. To analyze the magnetic circuit, the finite element analysis was used to simulate the magnetic flux density in the axial flux permanent magnet（AFPM）generator. The performance of the designed HPEG was tested by mounting the E-bike in the training stand and pedaling at different speeds. Based on the experiments, the maximum output of current and voltage were 1.4 A and 215 V, respectively. The output power of the developed HPEG system had the maximum output power 300 W at the rotation speed of 790 rpm. On the road, the HPEG system is able to generate excess power to recharge the battery while pedaling, braking and coasting downhill, which turns the extra momentum into electricity to charge the battery. When the speed of the bike is above 25 km/hr, it is easy to generate approximately 100 watts of power by pedaling. It can provide a 10% increase in range. Furthermore, the developed HPEG system can also provide a method of generating electricity by means of a modified exercise bike for use in energy storage. Energy stored in battery form can act as a supplemental energy source for battery banks that may already be used for household appliances.

Keywords: planetary, axial-flux, permanent-magnet, power generator, HPEG

1. Introduction

Access to inexpensive energy in useful forms has been one of the most important issues for mankind throughout history. The fear of forthcoming energy shortages has even been the cause of wars. Recently, there has been the introduction of renewable energy, such as small hydro, wind, solar, geothermal, and biofuel energy. The most interesting of these energies are wind power and solar energy. However, the use of solar energy technologies has a shortcoming of very high cost. The limited life of solar panels is about 10-20 years and the production requires the use of a large amount of silicon, germanium, and boron, which may cause pollution in other areas (Nelson, 2003). On the other hand, the wind power energy has high installation cost, big noise, locations and wind farm requirements (Lei et al., 2005; Bansal et al., 2002). Due to various factors mentioned above, there is low interest in renewable energy investment in the private sector. Therefore, the use of renewable energy tends to be limited to government and education and is not prevalent in the public. When selecting a suitable renewable generator, consumers focus on low cost, easy installation, minimal limitation due to topography, height and size, realizable and low maintenance rather than the volume of electricity generated. Hence, bicycles become one of the best choices. Because of their mobility and convenience, bicycles turn out to be a useful tool for people traveling in the city. Moreover, bicycles do not require fossil fuel or generate any air pollution or noise. By combining bicycles and generators, the public would be able to recycle the long ignored human energy, significantly decreasing electricity bills and better protecting the environment. Therefore, it is generally recognized that there is a need for new methods of affordable, non-polluting personal generator. In this paper we provide the most primitive clean energy that has been used for thousands years – human power energy. Human power energy is also more abundant and less dependent on environmental conditions, such as climate, than solar energy and alternative sources of energy.

It is obvious that the bicycle is a well-understood and often used machine, but the electrical versions have had little commercial success over the past century. In the past, some types of bicycle generators have been developed, such as drum generators, sidewall generators, and hub generators. Sidewall generator is the most common type of the bicycle generator. It consists of a steel frame holding a permanent magnet direct current generator. The bicycle tire rubs against the drum of the generator that turns the generator shaft to create electrical power. Drum generator basically like sidewall generator, except that their roller is designed to rub against the top of the tire instead of the side (Heine & Oehler, 2005). Since all the friction drums of these generators have to make contact with the bicycle tire. This will lead the friction that occurs at the drum/tire interface and cause the energy loss. Hub generator is built into the hub of a bicycle wheel, thus make it more effective. However, it is less powerful than the other types due to the heavy weight and is difficult to install. Therefore, these generators can only produce 3-6 W of power. The efficiency of the bicycle generator has not yet reached the level that the public demands.

In this study, a new type of high power planetary axial flux permanent magnet (AFPM) generator is proposed for the construction of the human power energy generator (HPEG) system installed on a bicycle to generate electric power while the wheel in operation. The advantage of AFPM generator is that it performs efficient cooling and can be manufactured in slim styles. As the rotors of the AFPM generator are outer-rotors, they can be directly attached for direct drive without any gears (Gieras et al., 2004). High-energy permanent magnets using Nd-Fe-B structures allowed formation of relatively large airgaps to make the production of coreless types possible. The noted advantages of coreless type are the relatively high density in the volume of coil and less loss of energy from the cogging torque (Parlikar et al., 2012). Since the generator is non-contact with the wheel, the rotor does not rely on friction for the transfer of rotation, it can be easily employed for drive. The design, fabrication and the measurement of the HPEG are described in detail.

2. Method

The principle of power generation is very simple, that is Faraday's law. The law states that the induced electromotive force (EMF) in any closed circuit is proportional to the rate of change in the magnetic flux through the circuit (Sadiku, 2007). Design of permanent magnet machines, such as the permanent magnet axial flux generator, is a procedure involving iterative computations based on performance requirements and tradeoffs. Typically, empirical relations or simplified models are used in the early stages of the design, followed by finite element analysis (FEA) lately. The unique pancake profile of AFPM machines has made them ideal candidates for in-wheel generator applications as they inherently match vehicle rim aspect ratios of large diameter and thin width. Flux in an AFPM machine flows through the airgap axially. AFPM machines have discs for the rotor and the stator geometry. The proposed AFPM in-wheel machine configuration in this paper is shown in Figure 1.

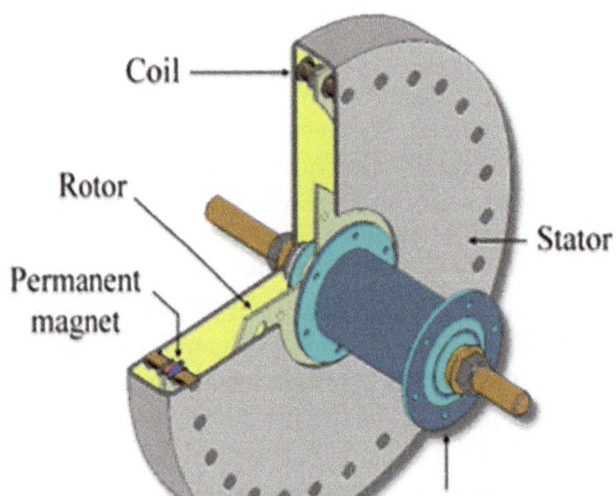

Figure 1. The scheme of the developed HPEG system

As illustrated in Figure 1, permanent magnets are glued to the surfaces of solid mild steel rotor discs. Particular care should be paid to the systems for jointing permanent magnets to the rotor. We used standard cylindrical axially flux magnetized Neodymium-Iron-Boron (NdFeB) permanent magnets for the field excitation in the rotor (Neji et al., 2006). The permanent magnets were encased in the rotor disc. The rotor has a diameter of 200 mm and 18 NdFeB permanent magnets were arranged in a N-S-N-S pattern around the circumference of the rotor.

The rotor disc was bolted to the rear wheel hub, and the stator was attached to the frame. Since the generator is non-contact with the wheel, the rotor do not rely on friction for the transfer of rotation, therefore, it can be easily employed for drive.

By using symmetry in electrical machines, most FE solutions can be obtained by 2D modeling, which is less computer processing intensive than 3D modeling. Even though 2D modeling is a somewhat approximation of the full model, it has been proved accurate. Figure 2 shows the magnetic flux lines of reaction field.

Figure 2. Magnetic flux lines in the machine

The attractive force between the stator and rotor in single-sided AFPM is a serious drawback for the topology. This force causes bending of the rotor and stator support structures and must be evaluated. The attractive forces can be determined analytically by Equation 1 (Patterson & Spée, 1995):

$$A_F = \frac{B_g^2 S_m \alpha_m}{2\mu_0} \tag{1}$$

Where A_F is the attraction force, B_g is the flux density in the airgap, S_m is the airgap surface area, α_m is the percentage of magnet to airgap area, and μ_0 is the permeability of free space. From the analysis, the attraction force of $A_F = 6645$ N was calculated for the rotor thickness to be such that the maximum deflection of the rotor disk is a fraction of the air gap length. Care must be taken in the resultant bending in the rotor back yoke (Δr) and the stator back support (Δb). Finite element analysis was once again used to determine the maximum deflection in the structure so as not to violate the mechanical clearance. The maximum combined reduction in clearance was calculated to be 0.08 mm which is within acceptable limits.

The prototype of the small power generation system is shown in Figure 3. Figure 3 shows the arrangement of magnet poles glued onto a rotor surface. The diameter and thickness of the NdFeB permanent magnet were 20 and 15 mm, respectively. The stator was ring shaped and consists of 18 induction coils with respect to the permanent magnets of the rotor. To reduce the cost, we inserted the screw into the coil as iron core instead of silicon steel. One important mechanical aspect with regard to AFPM machines pertains to the attraction forces between the stator and the rotor; in particular single-sided AFPM machines as they present the worst case. In order to counteract the attraction forces, non-ferromagnetic holders were designed and constructed. A set of paired bars was selected to handle the force.

Figure 3. The HPEG system has been installed on a commercial bicycle

In this research, the HPEG system was installed on a commercial bicycle and tested under different rotational speeds and wire configuration. Since the space between the hub and frame of the commercial bicycle is limited, only one stator was attached to the frame and the rotor plate was bolted to the rear wheel hub. Besides, without the contact between the generator and rear tire, the rotor, which does not rely on friction for the transfer of rotation, can be easily employed for drive without energy loss. Finally, the performance of the generator was obtained with experimental tests. It is expected that by driving the generator at high speeds, higher output power is obtained.

3. Results and Discussion

The system is not optimized in weight and dimension: it is only a first low cost prototype, useful to demonstrate the technology of HPEG system and to obtain direct information on operating tests. The experiments were conducted to observe the performance of the HPEG system. The HPEG system performance was tested by mounting the bike in the training stand and pedaling at different speeds. For the health-improving or energy-preserving activities applications, we are interested in the speed range from 20 km/hr to 100 km/hr. Figure 4 shows the output voltage and current as a function of speed. As shown in Figure 4, the maximum output current and voltage in parallel configuration were 0.7 A and 430 V.

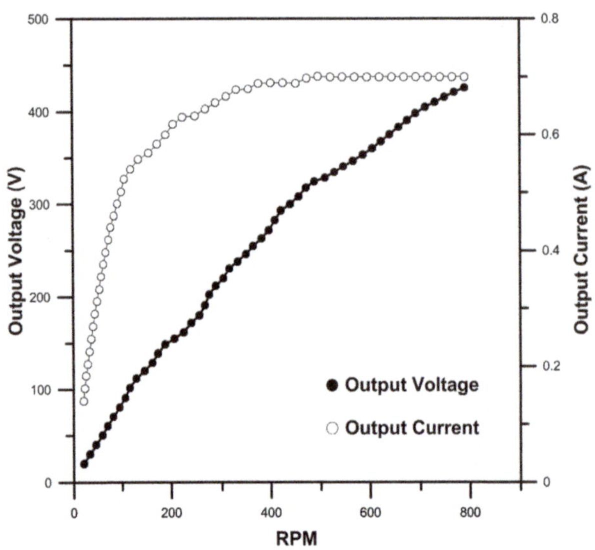

Figure 4. The output voltage and current as a function of speed in parallel

The output power of the developed HPEG in parallel configuration is illustrated in Figure 5. As shown in Figure 5, the developed HPEG had the maximum output power 300 W at the rotation speed of 790 rpm. We also found that at the normal speed 500 rpm (corresponds to downhill speed 60 km/hr) it is easy to achieve 215 W, even if in road travel at 200 rpm (about 25~30 km/hr), it can also achieve 100 W.

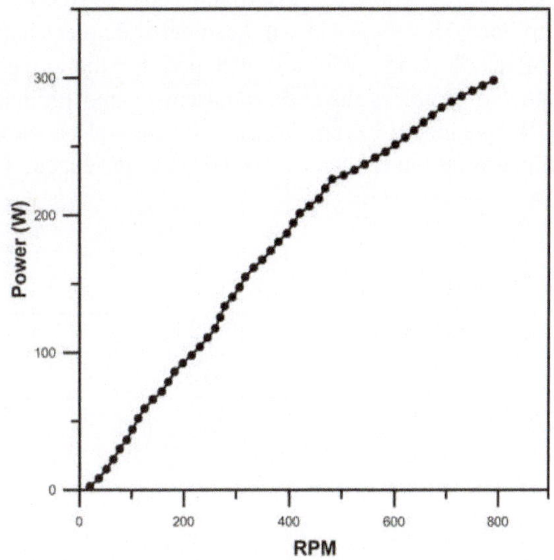

Figure 5. The output power of the developed HPEG

Furthermore, the developed HPEG system can also provide a method of generating electricity by means of a modified exercise bike for use in energy storage and running household appliances. The output alternating current (AC) is converted into a direct current (DC) by battery charger. The energy can be stored in various types of lead-acid batteries. Energy stored in battery form can act as a supplemental energy source for battery banks that may already be used for household appliances. If AC appliances are in place then an inverter must be used to transfer the 12 volts of DC current into the standard 110 volts of AC current for usage by these appliances. Appliances that could be powered include radios, televisions, lights, power tools and other appliances that pull relatively low amounts of energy for their usage.

In order to increase the driving range, we can install the HPEG system on the hub of the rear wheel of the electric bicycle, which can create excessive power to recharge the battery while pedaling, braking and coasting on a downhill. In this work, the Lithium battery was adopted for its light weight and long duration power. The power system consisted of a battery (36 V, 9 Ah) and an electric brushless DC motor (350 W). Figure 6 illustrates schematically the prototype of the electric bicycle with HPEG system. The system can transfer the kinetic energy to electric power and recharge the battery by use of a planetary AFPM generator, which can extend the range by 10-20 miles.

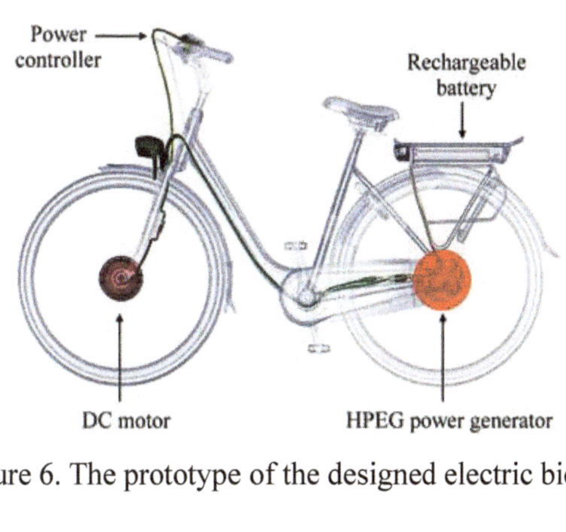

Figure 6. The prototype of the designed electric bicycle

The daily and hourly energy demand has been calculated. It is possible to calculate the rate of recharge that is needed to replenish the battery. We are examining the system with a 12 volt, 100 Ah battery. It would take 6 hours of pedaling to full charge the battery by mounting the bike in the training stand and pedaling at speed of 400 rpm. Though the time needed to generate electricity seems too long for one person to put out, it would probably be most appropriate for households with multiple people. Table 1 lists the appliances that the full charge battery could be powered. Currently the average installation cost is 10,000 US dollars for a 1,000 watt solar panel. Moreover, the amount of hours that the solar panel can generate electricity depends on the amount of sunlight received. The majority of small wind mills are installed by the government and education, not residential, due to unstable wind power, 6 meters distance requirement and limited space. Compared to solar panels or wind mills, HPEG is more convenient because it can be installed on bicycles. The cost of HPEG is approximately 300 US dollars and could be lower once it can be mass-produced. Therefore, HPEG is cheaper than other renewable energy devices.

Table 1. The appliances that the full charge battery could be powered

	Radio	Fluorescent lighting	37-inch LCD TV	Toaster	Air conditioner
Power(W)	20	20	150	600	2000
In use time(hr)	60	60	8	2	0.6

4. Summary

The purpose of the paper is to introduce an electric generator, which easily recycles kinetic energy, allowing the public to conserve energy and reduce carbon consumption. In this study, the prototype of the HPEG was designed, fabricated and tested under different rotational speeds and wire configuration. Human power is a readily available source of renewable energy. It is user-friendly, portable, and available whenever you need it. The power output of the designed HPEG was directly proportional to the effort of the person pedaling. Experiments showed that the maximum output power of the developed HPEG could achieve 300 W at the speed of 790 rpm and 100 W at the normal speed of 25~30 km/h during road test. It can be also employed as an emergency power generation and battery charging for household appliances and exercise systems that do useful work generating power while exercising. The prototyped generator is relatively small and cheap. The information gained with current experience can be employed as a useful guidance for further developments on this technology.

Acknowledgements

The work was financially supported by the National Science Council under contract NSC 98-2221-E-168-040.

References

Bansal, R. C., Bhatti, T. S., & Kothari, D. P. (2002). On some of the design aspects of wind energy conversion systems. *Energy Conversion and Management, 43*, 2175-2187. http://dx.doi.org/10.1016/S0196-8904(01)00166-2

Gieras J. F., Wang, R., Kamper, M. J. (2004). *Axial Flux Permanent Magnet Machines.* The Netherlands: Kluwer Academic Publishers.

Heine, J., & Oehler, A. (2005). Testing the efficiency of generator hubs. *Vintage Bicycle Quarterly, 3*(4), 27-30.

Lei, C., Zuoxia, X., & Nan, L. (2005). Environmental value of wind power generation. *Renewable Sources of Energy, 15*, 47-49.

Neji, R., Tounsi, S., & Sellami, F. (2006). Ontribution to the Definition of a Permanent Magnet Motor with Reduced Production Cost for the Electrical Vehicle Propulsion. *European Trans. On Electrical Power, 16*, 437-460. http://dx.doi.org/10.1002/etep.95

Nelson, J. (2003). *The Physics of Solar Cells.* Imperial London, UK: College Press. http://dx.doi.org/10.1142/p276

Parlikar, V. V., Kurulkar, P. M., Rathod, K. P., & Kumari, P. (2012). An Axial-Flux Permanent Magnet (AFPM) Generator for Defence Applications-Paradigm Shift in Electrical Machine. *International Journal on Electrical & Power Engineering, 3*(1).

Patterson, D., & Spee, R. (1995). The design and development of an axial flux permanent magnet brushless DC motor for wheel drive in a solar powered vehicle. *Industry Applications, IEEE Transactions on, 31*(5), 1054-1061. http://dx.doi.org/10.1109/28.464519

Sadiku, M. N. (2007). *Elements of electromagnetics*. Oxford university press.

13

Green Technology for Biodiesel Production From *Mesua Ferrea* L. Seed Oil

Mohammad Aslam[1], Prashant Saxena[2] & Anil K. Sarma[1]

[1] Sardar Swaran Singh National Institute of Renewable Energy, India

[2] DR BR Ambedkar NIT Jalandhar, India

Correspondence: Anil K. Sarma, Sardar Swaran Singh National Institute of Renewable Energy, India. E-mail: anil_tu@yahoo.co.in

Abstract

In this paper, the non-edible *Mesua ferrea* L. (MFL) seed oil with an acid value 18.8 mg KOH/g of oil, was tranesterified in the presence 1 MPa hydrogen pressure and also without the application of external pressure at 250–275 °C for 1 hour. The heterogeneous 1–5 wt.% *Musa balbisiana* Colla underground stem (MBCUS) ash catalyst was used during transesterification. The catalyst has a nano-metric dimension and has a versatile composition that consist of several alkali metal oxides, alkaline earth metal oxides, non-transition metal oxides and nonmetal oxides. The resultant biodiesel was separated using rotary vacuum evaporator and purified using molecular sieve and carbon columns. The biodiesel so obtained from MFL seed oil have been characterized as per available methods and found that the fuel properties are in conformity to ASTM and EN standards. 5 wt.% catalyst was found more effective as compared to other composition during ethanolysis. The effect of hydrogen pressure upon tranesterification was found negligible. Ethanol is a renewable liquid because it is made from renewable precursor such as biomass. It is expected that the use of whole renewable precursor during the process will fulfil the demand of 100% Green Technology for biodiesel production.

Keywords: biodiesel, tranesterification, ethanolysis, *Mesua ferrea L.*, green technology

1. Introduction

Due to the limited resources of fossil fuel, increasing prices of crude oil and environmental concerns over the past couple of decades, biodiesel has recently gained significant attention (Lukic, Krstic, Jovanovic, & Skala, 2009). Biodiesel refers to a vegetable oil or animal fat-based diesel fuel consisting of alkyl (methyl, propyl, or ethyl) esters obtained by chemical reaction of the lipids (vegetable oil, animal fat) with an alcohol. Chemically, biodiesel consists of alkyl esters instead of the alkanes and aromatic hydrocarbons of petroleum derived diesel (Adebayo, Ameen, & Abass, 2001). More than 350 oil-bearing crops have been identified, among which mainly jatropha, sunflower, safflower, soybean, cottonseed, rapeseed, and peanut oils are considered to be a potential alternative fuels for diesel engines. Nonetheless, other unknown oleaginous crops, which are being grown in less-favoured countries, could perform well as an adequate fuel with chemical and physical properties similar to those of diesel fuels (Dorado, Ballesteros, Lopez, & Mittelbach, 2004). As India is deficient in edible oils, non-edible oils become the primary choice for biodiesel. Generally, a blend of 5% to 20% is proposed in India (Dwivedi, Jain, & Sharma, 2011).

Mesua ferrea L. (MFL) is a timber plant that grows naturally in the northeastern parts of the Himalayan regions of India and in Srilanka. The oil seed contains 55–57 wt.% nonedible, reddish-brown-colored oil (the shelled kernel contains > 75 wt.% oil), which had been traditionally used as fuels (Sarma & Konwer, 2005). It is used in folk medicine for the treatment of fever, dyspepsis and renal diseases (Dennis, Kumar, & Srimannarayana, 1988). Biodiesel (i.e., the fatty acid alkyl ester) can be processed through the conventional tranesterification process. However the most commonly used methanolysis cannot be regarded as a green technology because methanol is mostly derived from the fossil fuel sources. Under such conditions, vegetable oils can be ethanolysed under high pressure, temperature conditions using some renewable catalyst (Zhou, Konar, & Boocock, 2003).

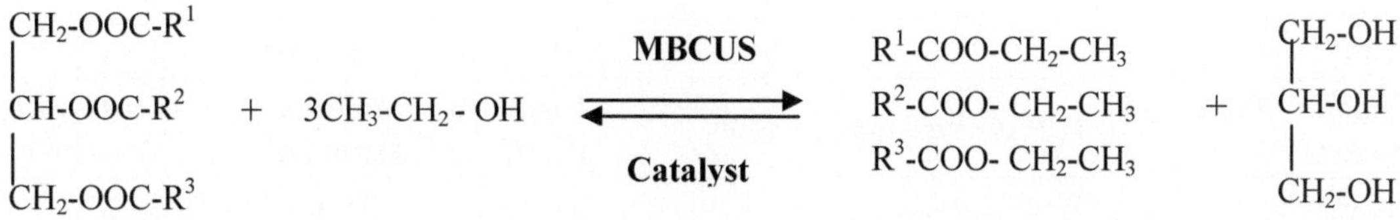

Figure 1. General tranesterification reaction of MFL to FAEE with MBCUS nano-catalyst

The commonly used homogeneous base catalysts are NaOH, KOH and their alkoxides. Homogeneous alkali-catalyzed transesterification is much faster than acid-catalyzed transesterification (Vicente, Martınez, & Aracil, 2005). However, a large amount of water is needed to transfer the catalysts from the organic phase to a water phase after the reaction. So, it is considerably more costly to separate the catalyst from the produced solution (Ebiura, Echizen, Ishikawa, Murai, & Baba, 2005; Kim et al., 2004; Dorado, Ballesteros, Lopez, & Mittelbach, 2004).

Heterogeneous base catalysts have many advantages: they are noncorrosive, environmentally benign and present fewer disposal problems. Meanwhile, they are much more easily separated from the liquid products and can be designed to give higher activity, selectivity and longer catalyst lifetimes (Gryglewicz, 1999; Tanabe & Holderich, 1999).

Musa balbisiana Colla underground stem (MBCUS) derived ash has been employed as a heterogeneous catalyst in this work for the first time for ethanolysis of MFL seed oil, which was also covered recently for methanolysis of Jatropha oil having high percentage of free fatty acids (Sarma, Kumar, Aslam, & Chouhan, 2014).

Thus a single step transesterification process with ethanol in a high pressure- high temperature (HPHT) batch process under various temperature ranges has been refined and discussed in this study. Moreover, the product obtained has been analysed for different fuel properties as envisaged for biodiesel as per ASTM standards. The details of the catalyst characterization and use during transesterification of Mesua Ferrea L oil having high percentage of free fatty acids have been expounded in the articles.

2. Experimental

2.1 Materials

Mesua ferrea Linn. (A national tree of Sri Lanka) seeds was procured from the Assam (India) and the seeds were transported to the Institute for physico-chemical analysis and further studies.

MBCUS was employed as catalysts. Petroleum ether -60–80 °C AR (Loba Chemie),Chloroform (Merck, 99.8%), Ethanol (Merck, 99.9%), Methanol extra pure (Loba Chemie, 99.5%), Iodine mono chloride/Wijs Solution (Merck), Phenolphthalein (Himedia, 98%), Potassium hydroxide pellets GR (Merck, 85% minimum), Sodium thio sulphate penta hydrate GR (Merck, 100%), Potassium iodide AR (Himedia, 99.8%), Sodium sulphate anhydrous AR (Himedia, 99%), Starch soluble AR (Himedia, 99%), Hexane GC grade (Sigma Aldrich), Heptane GC grade (Sigma Aldrich, 99.5%), Methanol GC grade (Sigma Aldrich), Methyl heptadecanoate GC grade (Merck), Molecular sieve etc. were used as AR grade chemicals in this study.

2.2 Feedstock Characterization

Mesua ferrea L. (MFL) seed kernels were ground, using a mechanical grinder, and extracted in a soxhlet apparatus, using petroleum ether (boiling point 60–80 °C). The extracted lipid was obtained by filtering the solvent-lipid mixture to get rid of the solid from solvent before the petroleum ether was removed using rotary evaporator 45–50 °C and 2.25 × 10^3 Pa. The extracted seed oil was stored in freezer at -2 °C for subsequent physicochemical analysis. The fatty acid profile of the oil was determined from GC analysis at PBTI, Mohali, India; an NABL accredited laboratory as shown in Table 1. The morphology of MFL seed oil was represented in the Figure 2.

Table 1. Fatty acid profile of *Mesua ferrea* L. seed oil

S. No.	Common Name of Fatty Acids	IUPAC/Systematic Name	Lipid No. (C:D)	Molecular Formula	Fatty Acid (%)
Saturated Fatty Acids					
1	Caproic Acid	Hexanoic acid	6:0	$C_6H_{12}O_2$	0.022
2	Myristic Acid	Tetradecanoic acid	14:0	$C_{14}H_{28}O_2$	0.047
3	Palmitic Acid	Hexadecanoic acid	16:0	$C_{16}H_{32}O_2$	13.635
4	Heptadecanoic Acid	Heptadecanoic acid	17:0	$C_{17}H_{34}O_2$	0.046
5	Stearic Acid	Octadecanoic acid	18:0	$C_{18}H_{36}O_2$	13.646
6	Arachidic Acid	Icosanoic acid	20:0	$C_{20}H_{40}O_2$	1.3
7	Behenic Acid	Docosanoic acid	22:0	$C_{22}H_{44}O_2$	0.351
8	Lignoceric Acid	Tetracosanoic acid	24:0	$C_{24}H_{48}O_2$	0.186
Mono Unsaturated Fatty Acids					
9	Palmetoleic Acid	Hexadec-9-enoic acid	16:1	$C_{16}H_{30}O_2$	0.087
10	Oleic Acid	(9Z)-Octadec-9-enoic acid	18:1	$C_{18}H_{34}O_2$	50.713
11	Cis-11-Eicosenoic Acid	(Z)-Eicos-11-enoic acid	20:1	$C_{20}H_{38}O_2$	0.18
Poly Unsaturated Fatty Acids					
12	Linoleic Acid	(9Z,12Z)-9,12-Octadecadienoic acid	18:2	$C_{18}H_{32}O_2$	19.47
13	Linolenic Acid	all-cis-6,9,12-octadecatrienoic acid	18:3	$C_{18}H_{30}O_2$	0.316
Summarized Fatty Acids Analysis					
Total Saturated Fatty Acids (%)		29.233			
Total Unsaturated Fatty Acids (%)		70.766			
Grand Total		**99.999**			

MFL Tree (National Tree of Sri Lanka)

MFL Leaves

MFL Seeds

MFL Kernels

Figure 2. Morphology of *Mesua ferrea* L. (MFL)

The acid value of the oil and biodiesel was measured as per AOCS official method Ca 5a–40 as stated in Equation (1). The chemicals required for titration were freshly prepared caustic KOH (0.25 N) and phenolphthalein indicator (1% in 95% ethanol). The percentage of free fatty acids was calculated as oleic acid using Equation (1).

$$Free\ Fatty\ Acids\ of\ Oleic\ \% = \frac{ml\ of\ Alkali \times N \times 56.1}{wt\ of\ Sample} \quad (1)$$

The acidity (AD) of the crude MFL was 9.4% with an acid value (AV) 18.8 mg KOH/g. For transesterification reaction analytical grade methanol (purity 99.5%, Loba Chemicals) and ethanol (99.9%, Merck) were used without further treatment.

The chemicals required for iodine value estimation were purchased from Merck and prepared as follows: reagent grade iodine monochloride solution, potassium iodide solution (15%), sodium thiosulphate (0.1 N), glacial acetic acid, chloroform and starch (1%). The Iodine values of the oil and biodiesel were estimated as per AOCS official method Cd 1–25 using freshly prepared Wijs solution and starch indicator from the Equation (2).

$$The\ Iodine\ Value = \frac{[(B-S) \times N \times 12.69]}{Weight\ of\ Sample} \quad (2)$$

Where B = titration of blank; S = titration of sample; N = normality of sodium thiosulphate solution.

2.3 Catalyst Preparation

MBCUS was cut into thin pieces of the dimension 1.5 cm × 15 cm and sun dried in March, 2011 for about 15 days, which were later burnt traditionally in an open environment upon a tin sheet for ash production. These ashes obtained were further processed in a furnace in the laboratory at 550 °C for one hour. For material balance, a piece of the fresh MBCUS was collected at sampling plastic bag of known weight and studied for proximate analysis. It was found that the fresh MBCUS contain 91.34% moisture at 105 °C, 7.45% volatile matter at 350 °C and 0.14% carbon at 550 °C. While the ash obtained at 550 °C was 1.07% of the fresh MBCUS (Figure 3).

MBC trunk MBCUS Root MBCUS Ash

Figure 3. Morphology of MBCUS catalyst

2.4 Catalyst Characterization

The physico-chemical characteristics of the MBCUS ash (Baked at 550 °C) were analysed to estimate the catalytic activity and basic composition as per AOAC methods. The catalyst was analysed in a TGA instrument manufactured by Perkin Elmer, model STA 6000 for thermal and oxidative stability. The weight losses of the catalyst (prepared @ 550 °C), temperature range 30–1000 °C, heating rate 10 °C/min and constant flow of air and nitrogen (20 ± 0.5 ml/min) were recorded. The contents of the catalyst were ascertained from appropriate AOAC, TGA, XRF analysis and SEM EDX pattern. The basic strength of the catalysts (H-) was estimated using Hammett indicator method. The catalyst sample (100 mg) was shaken along with suitable amount of Hammett indicators diluted with methanol and left to equilibrate for about two hours upto no colour change observed. The titrate color was then recorded. The following Hammett indicators were used: neutral red (H^- = 6.8), bromothymol blue (H^- = 7.2), phenolphthalein (H^- = 9.3), 2,4-dinitroaniline (H^- = 15.0), 4-nitroaniline (H^- =

18.4). The basic strength of the catalysts was defined as being stronger than the Hammett indicator if it showed a colour change but weaker if no colour changes observed.

The structural study of the catalyst was done with the XRD pattern using X-ray diffractometer (XRD). Nitrogen adsorption isotherms were obtained at -196 °C using a Quantachrome NovaWin - Data Acquisition and Reduction (Quanta chrome Instrument) system. These isotherms were used to see the BET surface area, pore volume and pore size distribution.

The SEM images and subsequent EDX analysis of the samples were carried out using FE-SEM QUANTA 200 FEG of FEI Netherlands make. The EDX pattern showed the composition and different crystalline materials present in the MBCUS ash catalyst. The transmission electron microscope images of the MBCUS samples were taken using a Tecnai G2 20 of FEI Neitherlands make transmission electron microscope (TEM).

The nature, composition and detail physicochemical properties of the MBCUS nano-catalyst have been accepted for publication in a reputed journal recently (Sarma, Kumar, Aslam, & Chouhan, 2014).

2.5 Transesterification in Batch Reactor

The tranesterification reactions were carried out in a HPHT Bench Top reactor of 1.8 L capacity (Amar equipments Company, Mumbai, India). The reactor contains to heating mantle, cooling system and pressure-temperature control and recording devices.

The average molecular weight was calculated from the triglyceride composition of MFL seed oil and was found to be 866. The oil to alcohol molar ratio in the reaction mixture was kept at 1:9 and the catalyst concentration was kept at 1%, 3% and 5% of the initial oil mass. The reaction was carried out under 250 °C and 275 °C for easy conversion of the free fatty acids to ethyl ester and also for resisting, oxidation of the reactants. This was accomplished by stripping the reaction mixture inside the reactor. All the experiments were performed with a total oil, alcohol (ethanol) and catalyst volume not exceeding 1.5 L. The initial hydrogen pressure inside the reactor was maintained at 1.0×10^5 Pa (atmospheric) which rises up to 4.4×10^6 Pa with increasing internal vapour pressure (IP) at 275 °C. The reactions were also conducted in the absence of catalyst under the same set of experimental condition. In this instance, the internal vapour pressure raised up to 6.6×10^6 Pa at 275 °C. The mixture was vigorously stirred at 650 revolutions per minutes. Samples were removed by opening the flash bottom valve after completion of the reaction. The samples collected from the reactor were allowed to settle in a ventilator before analysis. Since triglycerides (TGs), diglycerides (DGs), monoglycerides (MGs) and free fatty acids (FFAs) analysis relates only with the oily liquid phase of the sample and the results are presented as percentage of the total oily mass, the oily mass was purged of impurities that might affect the final measurements. So, the oily phase was subsequently transferred to the rotary vacuum evaporator (Heydolph, Germany Make) and the unreacted alcohol was separated in 45–50 °C and 2.25×10^3 Pa. Then, the samples were centrifuged to separate the catalyst, glycerine (GL) and oily phase. The upper layer of the centrifuged sample contained the oily phase (biodiesel), the lower one of the glycerine, while in the middle, between the oil and the glycerine phase appears the catalyst. The catalyst was later washed using petroleum ether and methanol to remove the bound ester and glycerol.

2.6 Fuel Property Analysis of the FAEE

The crude MFL seed oil was initially analyzed using GC FID detector as per AOCS Official method 1998, Ce 1–62 and Ce 2–26 at Punjab Biotechnology Incubator (Mohali) an NABL accredited laboratory of India. The fuel properties of the FAEE obtained at optimum conditions were analysed using appropriate ASTM D6751 and EN 14214 standard procedures specified for B100 biodiesel stocks.

3. Results and Discussion

3.1 Structural Analysis and Porosity Analysis of MBCUS Catalyst

The XRD analysis ascertains the presence of crystalline phases the in MBCUS catalyst, which have been reported earlier (Sarma, Kumar, Aslam, & Chouhan, 2014). It has been found that the surface area of the catalyst is 38.710 m^2/g, total pore volume 4.233×10^{-8} m^3 /g for pores smaller than 796.2×10^{-11} m (Radius) at P/P_o equal to 0.98780 and pore size expressed as average pore radius is equal to 21.8681×10^{-11} m. These values are significantly better than earlier reported literature (Sharma, Khan, Puri, & Tuli, 2012; Kumar et al., 2010). All these attributes show that the catalyst chosen could be an excellent one for transesterification reaction.

3.2 SEM-EDX and TEM Analysis of MBCUS Catalyst

The SEM-EDX analysis of the catalyst were conducted (Figure 4) and confirmed the presence of C ($CaCO_3$), O (SiO_2), Mg (MgO), Al (Al_2O_3), Si (SiO_2), P (GaP), Cl (KCl), K (MAD-10 Feldspar), CA (Wollastonite), as depicted in the Figure 4. The typical elemental composition as evaluated from Figure 4, that depicts the mass fraction composition as C (9.86%), O (44.01%), Mg (3.92%), Al (1.21%), Si (12.03%), P (1.59%), Cl (2.69%), K (16.83%), Ca (7.19%). Earlier studies of heterogeneously catalysed tranesterification reactions of vegetable oil have focused on single metal oxides, including CaO, MgO, Al_2O_3, zeolites, silicates etc.

Functionalized compounds such as calcium bentonite functionalised with potassium (Hoffmann & Conradi, 1988), porous silica and pumice granules, mesoporous silica and metallosilicate materials of SBA-15 types (Sanz et al., 2012) were also shown very good activity during tranesterification as reported. Therefore, the versatile composition, including several alkali and alkaline earth metal oxides, carbonate and silica makes MBCUS ash a perfect heterogeneous catalyst for transesterification reaction (Sarma, Kumar, Aslam, & Chouhan, 2014).

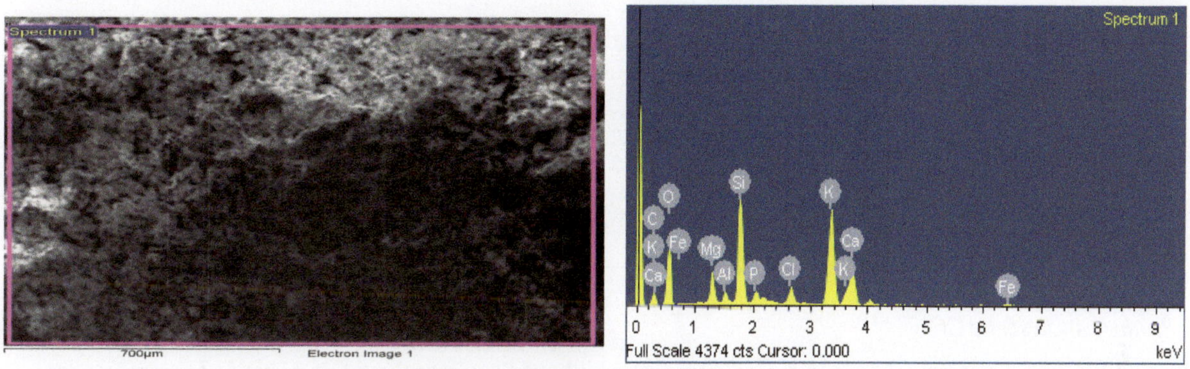

Figure 4. SEM-EDX spectrum and composition

The structural information of the catalyst was studied from the TEM images as shown in the Figure 5. It is clearly visible that the structural dimensions are similar (100 nm length) and brick like (Approx. 70 nm × 20 nm × 15 nm). The nano size structural dimension of MBCUS catalyst provided effective surfaces to their catalytic efficacy during tranesterification of triglyceride, irrespective of the free fatty acids present in the oil. Similar results were also covered by Boz, Degirmanbasi and Kalyon (2009) for the tranesterification of canola oil using nano-γ-Al_2O_3 loaded with KF that showed 97.7% ester yield with 3% catalyst loading.

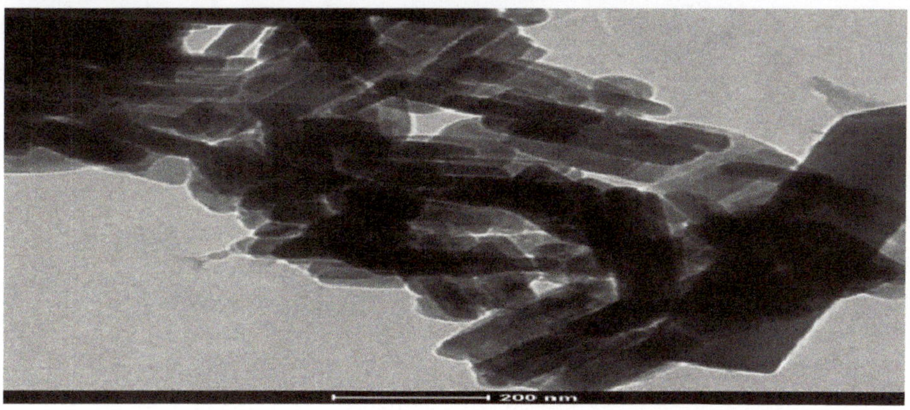

Figure 5. TEM images and structural information of MBCUS-catalyst

3.3 Basicity Analysis of MBCUS Catalyst

The basicity of the catalyst was determined within the range of $9.3 < H^- < 15.0$. The base strength of MBCUS is thus stronger than the weakest indicator phenolphthalein ($H^- = 9.8$) that exhibited a colour change from pink to colourless. However, it is weaker than the strongest indicator i.e. 2, 4-dinitroaniline ($H^- = 15.0$) as no colour change was observed.

3.4 Metallic & Non-Metallic Contents in MBCUS Catalyst

Nonmetallic concentration in the MBCUS ash catalyst was found out using appropriate standard methods as applicable discussed in Table 2.

Table 2. Metallic & non-metallic contents in MBCUS catalyst

S. No.	Parameters	Concentration	Specification
1	Chloride	0.60 wt.%	AOAC 915.01
2	Carbonate	ND(MDL: 2 mg/kg)	
3	Sodium	0.34 wt.%	
4	Potassium	27.70 wt.%	AOAC 975.03
5	Lead	5.19 mg/kg	
6	Cobalt	1.97 mg/kg	
7	Carbon	4. 5 wt. %	
8	Silica (various forms)	37.00 wt.%	TGA method

3.5 Product Analysis by GC

Important Physico-Chemical Characteristics of MFL seed oil have been represented in Table 3. The process for the production was optimized with regard to the conversion of the oil to ester @ 250 °C and @ 275 °C. The biodiesel was produced from MFL seed oil under the specified experimental condition and found in conformity to ASTM standards with regard to fuel property while the reaction was carried out at 275 °C. It has been observed that the acid value of MFL oil was 18.8 mg KOH/g oil (which is equivalent to about 9.4 wt.% FFA content), the acid value of the end product was 1.6 mg KOH/g oil (Figure 6). This has been observed even without externally applied hydrogen pressure. The experiment was repeated further for MFL oil at 275 °C without H_2 pressure. The tranesterification reaction was carried out under 250 °C and 275 °C for the conversion of FFAs and triglycerides to fatty acid ethyl-ester. This was accomplished by stripping the reaction mixture inside the reactor. All the experiments were performed with a total oil, alcohol (ethanol) and catalyst volume not exceeding 1.5 L. The initial hydrogen pressure inside the reactor was maintained at 1.0×10^5 Pa (atmospheric) which raises up to 4.4×10^6 Pa during the reaction along with increasing internal vapour pressure (IP) at 275 °C due to the presence of ethanol. The reactions were also conducted in the absence of catalyst under the same set of experimental condition. In this instance, the internal vapour pressure raised up to 6.6×10^6 Pa at 275 °C. This clearly demonstrated that the presence of hydrogen does not affect tranesterification.

Table 3. Physico-chemical characteristics of *Mesua ferrea* L. seed oil

Parameters	MFL Oil	MFL–FAEE (B100)
Density at 40 °C (g/cm^3)	0.90059	0.873
Specific gravity at 20 °C (g/cm^3)	0.90762	0.879
API density at 15 °C (g/cm^3)	0.91735	0.890
Distillation @ 360 °C (wt%)	--	90
Flash point (°C)	115.0	113
Calorific value (MJ kg^{-1})	36.0	35
Viscosity @ 40 °C (cSt)	34.10	5.525
Acid Value (mg KOH/g)	18.8	1.8
Free fatty acids (%) as oleic	9.40	0.8
Iodine Value (g I_2/100g)	131.31	113.24
Rams bottom carbon residue (wt%)	0.99	0.250
Copper strip corrosion	1a	1a

The iodine value decreases with increasing the catalysts percentage with respect to biodiesel production. It was observed that the unsaturation in the biodiesel decreased as iodine values decreased from 130 to 112 as shown in Figure 7. Iodine number basically represents the degree of unsaturation, the larger is the number higher is the unsaturation. While tranesterification took place some of the unsaturation present in the fatty acid chain got saturated with the proton shift that takes place in situ. This also indirectly indicated the conversion of oil to FAEE.

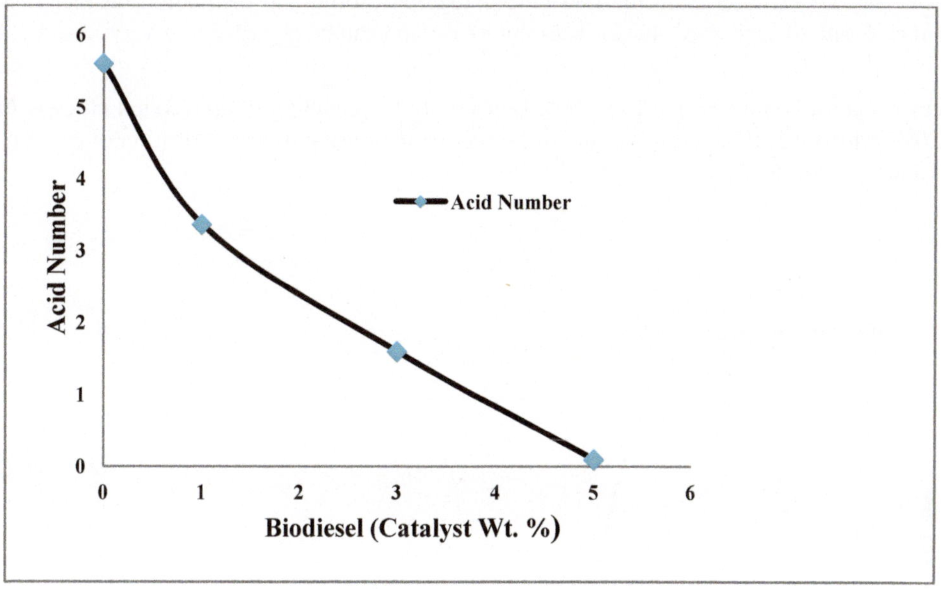

Figure 6. FFA trend using MBCUS

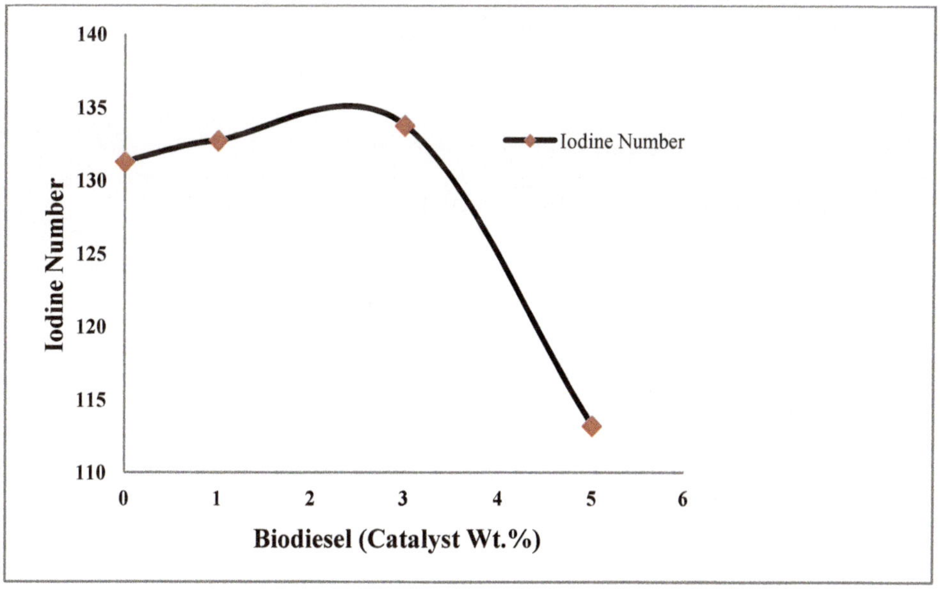

Figure 7. Iodine index variation MF oil to FAEE

It has been observed that the FFA level drastically decreased to near about 1.8 mg KOH/g from 18.8 mg KOH/g oil in 1 h. This clearly demonstrated the evidence of efficacy of the catalysts and furthers no effect of hydrogen pressure upon FFA conversion at higher temperature. MBCUS is a very good catalyst due to high reactivity and efficacy during the transesterification and the negligible cost of production. The catalyst and product separation were much easier as compared to other available methods (two step transesterification). MBCUS catalysts can be commercialized for the bio-diesel production for the development of society and economic growth of the Country as well.

The TGA profile of the FAEE was conducted under a constant flow of N_2 20 ml/min in 1 h constant heating rate and observed that over 90% of the FAEE was boiled off at 350 °C confirming the conversion of MFL to FAEE (Figure 8). The GC profile of the FAEE of MFL was recorded and found that 95% of the oil was converted to FAEE at 275 °C and one hour reaction time in a batch process in 1:9 molar ratio of oil: alcohol as specified in Figure 9 in the absence of H_2 pressure. The same was also confirmed from the vacuum distillation of the FAEE obtained and 90% recovery was found at 360 °C as per ASTM D 6751-07 b. The distillation characteristics are regarded as the most important parameter of any liquid fuel, which represents volatility during combustion and formation of proper air-fuel mixture. Both the Gas chromatograph profile and distillation characteristics, are therefore in accordance with the specified standards of the FAEE and can be regarded as a very safe fuel for our purpose.

The MBCUS nano - catalyst could be reused for three times with the reduction of conversion efficacy of oil to FAEE by about 20%, which was reported in our previous publication during tranesterification of Jatropha oil to FAME (Sarma, Kumar, Aslam, & Chouhan, 2014).

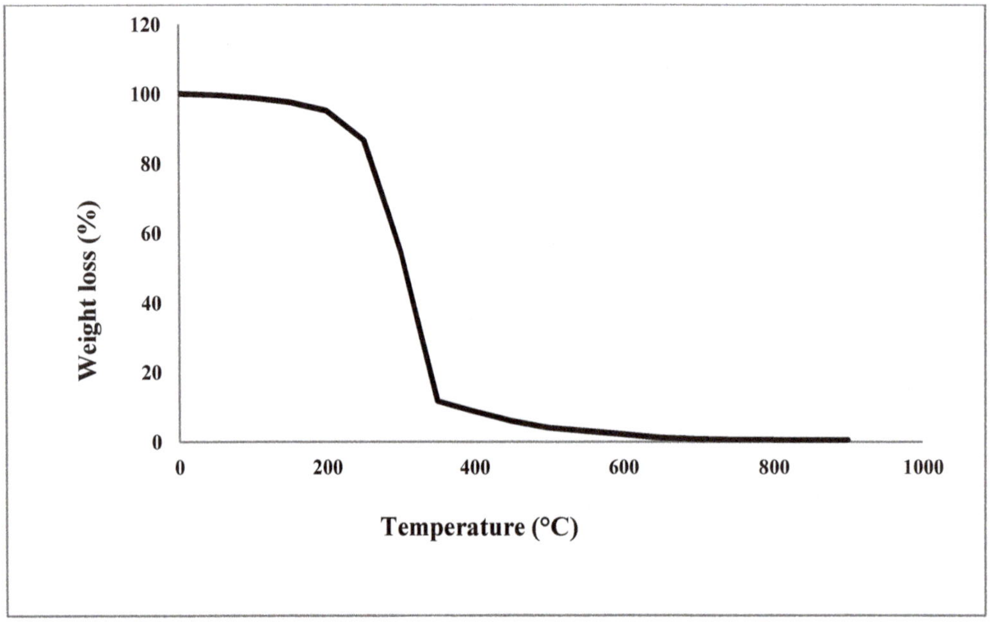

Figure 8. TGA profile of the MF FAEE obtained

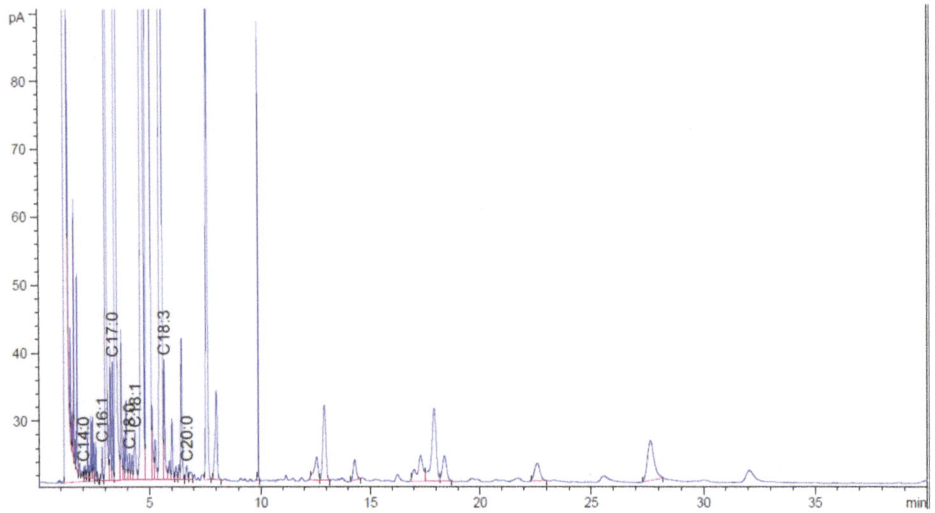

Figure 9. GC Profile of the MF FAEE obtained

4. Conclusion

High free fatty acid containing MFL seed oil could be transesterified with ethanol (1:9 M ratio) in a single batch process using 5 wt.% MBCUS and 1 with or without external pressure in closed reactors.The vapour pressure inside the reactor increased upto 3.8 MPa even without external pressure. The experiment was not conducted beyond 300 °C because at that stage poly unsaturated fatty acids would have been cracked when the reaction time exceeds 15 min. The MBCUS nano-catalyst performed well during the transesterification reaction with good reactivity and reusability. This technology is applicable to most types of oil having high percentage of free fatty acids. The industrialist may take an interest in using such materials while setting up biodiesel production, refinery in the near future for all types of high free fatty acid containing oil.

Acknowledgements

Finally, the authors acknowledge the Ministry of New and Renewable Energy, Govt of India, New Delhi for financial assistance in the form of project grants 7/152/2010-BF date 13/09/2011 for carrying out the research and APCSEET2013 for funding the travel grants during the presentation and covering the publication fees.

References

Adebayo, G. B., Ameen, O. M., & Abass, L. T. (2011). Physico-chemical properties of biodiesel produced from Jatropha Curcas oil and fossil diesel. *Journal of Microbiology and Biotechnology Research, 1*(1), 12-16.

Boz, N., Degirmenbasi, N., & Kalyon, D. M. (2009). Conversion of biomass to fuel: Transesterification of vegetable oil to biodiesel using KF loaded nano-γ-Al_2O_3 as catalyst. *Applied Catalysis B: Environmental, 89*(3), 590-596. http://dx.doi.org/10.1016/j.apcatb.2009.01.026

Dennis, T. J., Kumar, K. A., & Srimannarayana, G. (1988). A new cyclo hexadione from Mesua ferrea. *Phytochemistry, 27*(7), 2325-2327. http://dx.doi.org/10.1016/0031-9422(88)80153-5

Dorado, M. P., Ballesteros, E., López, F. J., & Mittelbach, M. (2004). Optimization of alkali-catalyzed transesterification of Brassica C arinata oil for biodiesel production. *Energy & Fuels, 18*(1), 77-83. http://dx.doi.org/10.1021/ef0340110

Ebiura, T., Echizen, T., Ishikawa, A., Murai, K., & Baba, T. (2005). Selective transesterification of triolein with methanol to methyl oleate and glycerol using alumina loaded with alkali metal salt as a solid-base catalyst. *Applied Catalysis A: General, 283*(1), 111-116. http://dx.doi.org/10.1016/j.apcata.2004.12.041

Gaurav, D., Siddharth, J., & Mahendra Pal, S. (2011). Pongamia as a Source of Biodiesel in India. *Smart Grid and Renewable Energy, 2011.*

Gryglewicz, S. (1999). Rapeseed oil methyl esters preparation using héterogeneous catalysts. *Bioresource Technology, 70*(3), 249-253. http://dx.doi.org/10.1016/S0960-8524(99)00042-5

Hoffmann, M. M., & Conradi, M. S. (1998). Hydrogen exchange reactions in supercritical media monitored by in situ NMR. *The Journal of Supercritical Fluids, 14*(1), 31-40. http://dx.doi.org/10.1016/S0896-8446(98)00094-1

Kim, H. J., Kang, B. S., Kim, M. J., Park, Y. M., Kim, D. K., Lee, J. S., & Lee, K. Y. (2004). Transesterification of vegetable oil to biodiesel using heterogeneous base catalyst. *Catalysis Today, 93*, 315-320. http://dx.doi.org/10.1016/j.cattod.2004.06.007

Kumar, S., Pulikottil, A. C., Sharma, A., Sarkar, M., Kagdiyal, V., Yadav, A., ... Kumar, A. (2010). A catalyst composition for transesterification of organically/naturally derived oils and fats to produce alkyl esters and process for preparing the same. *WO, 20998*, A2.

López-Sanz, J., Pérez-Mayoral, E., Soriano, E., Sturm, M., Martín-Aranda, R. M., López-Peinado, A. J., & Čejka, J. (2012). New inorganic–organic hybrid materials based on SBA-15 molecular sieves involved in the quinolines synthesis. *Catalysis Today, 187*(1), 97-103. http://dx.doi.org/10.1016/j.cattod.2011.12.015

Lukić, I., Krstić, J., Jovanović, D., & Skala, D. (2009). Alumina/silica supported K_2CO_3 as a catalyst for biodiesel synthesis from sunflower oil. *Bioresource technology, 100*(20), 4690-4696. http://dx.doi.org/10.1016/j.biortech.2009.04.057

Sarma, A. K., Konwer, D., & Bordoloi, P. K. (2005). A comprehensive analysis of fuel properties of biodiesel from Koroch seed oil. *Energy & Fuels, 19*(2), 656-657. http://dx.doi.org/10.1021/ef049754f

Sarma, A. K., Kumar, P., Aslam, M., & Chouhan, A. P. S. Preparation and Characterization of Musa balbisiana Colla Underground Stem Nano-material for Biodiesel Production Under Elevated Conditions. *Catalysis Letters*, 1-10.

Sharma, M., Khan, A. A., Puri, S. K., & Tuli, D. K. (2012). Wood ash as a potential heterogeneous catalyst for biodiesel synthesis. *Biomass and Bioenergy, 41*, 94-106. http://dx.doi.org/10.1016/j.biombioe.2012.02.017

Tanabe, K., & Hölderich, W. F. (1999). Industrial application of solid acid–base catalysts. *Applied Catalysis A: General, 181*(2), 399-434. http://dx.doi.org/10.1016/S0926-860X(98)00397-4

Vicente, G., Martınez, M., & Aracil, J. (2004). Integrated biodiesel production: a comparison of different homogeneous catalysts systems. *Bioresource Technology, 92*(3), 297-305. http://dx.doi.org/10.1016/j.biortech.2003.08.014

Zhou, W., Konar, S. K., & Boocock, D. G. (2003). Ethyl esters from the single-phase base-catalyzed ethanolysis of vegetable oils. *Journal of the American Oil Chemists' Society, 80*(4), 367-371. http://dx.doi.org/10.1007/s11746-003-0705-1

14

Large Scale Power Generation: Up-Skilling Welsh Industry

Elizabeth Locke[1] & Sally Hewlett[1]

[1.] School of Engineering, Cardiff University, Wales, UK

Correspondence: Elizabeth Locke, School of Engineering, Gas Turbine Research Centre – GTRC, Heol Cefn Gwrgan, Margam, SA13 2EZ, Wales, UK. E-mail: lockee@cardiff.ac.uk

Abstract

As the move to a low carbon economy presents new challenges for existing industry, there is increasing consensus on the need to address the green skills agenda in order to ensure that the transition is timely and effective. There is however little empirical evidence of the development and delivery of low carbon training courses, specifically in the area of Large Scale Power Generation (LSPG). This study examined the level and form of existing low carbon combustion training in and around Wales and the demand for such training amongst Welsh industry. This is with a view to developing training courses through the Welsh Energy Sector Training project (WEST). Demand for potential WEST courses was found to be positive; specific interest for course content included: The Nature of Fuels, Utilization of Waste, Energy Conversion Processes, Energy Conversion Technologies, Combustion Science, Improving Combustion Efficiency & Emissions and Combustion Risks & Hazards. Critically, geographical factors must be taken into account when assessing the most appropriate form of delivery; e-learning could be a useful tool for maximizing participant numbers. An important implication from the research arose; existing training is generally provided at master's degree level, whilst findings indicate a preference for courses at introductory undergraduate degree level. As such a close collaboration with participants is required if the training developed is truly to be of value to industry in Wales during the transition to a low carbon economy.

Keywords: low carbon, training, combustion, skills, Wales

1. Introduction

'*The transition to a low carbon economy will require sufficient appropriately qualified and skilled people to manufacture, install, operate and maintain the low carbon technologies and approaches.*' (Jagger et al., 2013:43).

The move to a low carbon economy will place increasing demands on industry as organisations and their employees face new challenges in terms of skills and knowledge. As Jagger et al. (2013) argue, the skills shortage in terms of manufacturing, installing and operating low carbon technologies could critically delay the low carbon transition and increase its cost and duration. It is further reasoned that more attention should be focused on skills and training issues given how fundamental they are to the transition to a low carbon economy. Indeed consideration must be given to those 'deep green' jobs which are critical to a greater uptake of renewable technology. These jobs are central to the "*manufacture, installation, and operation of low carbon technologies such as electricity generation, energy efficiency equipment, vehicles and manufactured products*" (Jagger et al., 2013:45).

Higher Education (HE) establishments could play a fundamental role in the low carbon transition through research and development programmes and through the provision of training. This will require a close relationship between academia and industry, the success of which will amongst other things, depend on cooperation and mutual benefits. Fairweather (1988:16) suggests that since the 1980's in which it may be considered the 'decade of industry-university liaison,' the higher education sector and industry have a number of drivers for engagement. These include the commercial motivation for a competitive edge and Government policies aiming to boost economic growth. This reflects a changing perception of Universities and Colleges as 'ivory towers' to one in which academic organisations are 'active participants in society.' Indeed this rhetoric underpins the objectives of the Welsh Energy Sector (WEST) project, which aims to develop the low carbon knowledge base in Wales through the development of industry training. WEST is a project within Cardiff

University's Low Carbon Research Institute (LCRI) Convergence Energy Programme. The project is funded by the European Regional Development Fund (ERDF) through the Welsh Government (with additional support from Welsh Universities and industry). It is intended that training will be delivered at level 4 and above of the Credit and Qualifications Framework Wales (CQFW) which is equivalent to the first year of an Undergraduate Degree (Hewlett, 2014).

There are 5 key research themes within the LCRI which WEST will develop and pilot training modules through; Large Scale Power Generation (LSPG), Hydrogen, Low Carbon Built Environment (LCBE), Marine and Solar PV. The focus of this research is the LSPG theme for which Cardiff University's Gas Turbine Research Centre (GTRC) in Port Talbot, South Wales is responsible. Two members of the team currently lecture on the Sustainable Energy and the Environment MSc, delivered by the School of Engineering and will play central roles in the delivery of training. The training will be offered as fully funded to those individuals who live or work in the Convergence Zone in Wales, illustrated in Figure 1 as the shaded areas. The Convergence Energy Programme aims to enable Welsh industry to lead the way through research and development, supporting job creation, reducing carbon emissions and improving the quality of life in Wales. Central to this is meeting industry needs and developing economic and social benefits through energy sector training.

Figure 1. Welsh Convergence Zone

Energy Policy in Wales emphasises the important changes required in the power generation sector:

"*...any new fossil fuel plants should be carbon capture ready with fully developed plans for carbon capture and storage; and that these plants maximise efficiency through use of waste heat and co-firing where appropriate*" (Wales Gov, 2010:16).

At present gas is the primary fuel source for electricity generation in Wales (approximately 50% of all electricity generation), coal and nuclear power have declined, with around a 1/5 of electricity generation from low carbon sources (Wales Gov, 2012). Growth in renewable generation, from 2.9% in 2004 to 5.1% in 2010, has to some extent negated the fall in electricity generated from nuclear over the same period (Wales Gov, 2012). The total electricity generation in Wales has fallen 10% over the period, primarily due to a decline in coal and nuclear generating plants (Wales Gov, 2012). Considering the UK as a whole Wales was responsible for 9% of electricity generation between 2004 and 2010. Policy indicates that, whilst Carbon Capture Storage (CCS) technology is being developed, the priority for Wales lies with improvements in combustion efficiency. This will be achieved through the utilization of waste heat and the firing of renewable and alternative fuels (in combination with fossil fuels). In the longer term, as a result of the long history of coal mining in Wales, deep coal seams are seen as one possible option for storage in CCS, although the viability of this approach is yet to be confirmed.

Wales' anaerobic digestion capacity, for the production of biogas for combustion, is set to expand to increase energy from waste (EfW) activities; the £5m per year sustainable land management scheme 'Glastir' offers a capital grant scheme with the aim of improving business and resource efficiency, and lowering the carbon equivalent emissions of agricultural and horticultural holdings (Wales Gov, 2010). Proposals have also been submitted for an energy from waste (EfW) combined heat and power (CHP) facility in Cardiff with the possibility of a District Heating (DH) network to serve homes and businesses with low carbon heat. According to Viridor Ltd (2013) there is the potential to generate around 30MW of electricity and offer DH for utilisation of low grade heat (Viridor Ltd, 2013). Cardiff City Council will now be examining the feasibility of developing a DH network in the city. Indeed DH has been viewed as a vital low carbon infrastructure by neighbouring Local Authorities, Bridgend, the City and County of Swansea and Neath and Port Talbot Council who have all undertaken feasibility studies for DH in the last three years. Developments in EfW technology specifically concerning emissions control, mean that modern plants are a far different technology from the 'incinerators' of the past, the legacy however can be difficult to dispel. If any further plant developments are to be approved scheme developers will need to work hard to win public and political support through education and consultation (Hewlett, 2014).

Welsh Energy Policy emphasises that low carbon solutions should be locally sustainable, further stressing the important role of industry and organisations:

"*....business and industry in particular will play a key role in driving the transition to a low carbon economy*" (Wales Gov, 2012:6).

In order to increase the proportion of renewable energy in the production mix, a greater level of awareness and understanding on the use of alternative fuels for power generation is required. The GTRC is undertaking research into the functionality of new combustion systems, components and fuels, under different temperature and operating conditions and pressures. The drive to utilise greater levels of alternative fuels as well as the cleaner combustion of existing fuels (for land based power generation [and aviation sectors]) underpin the research at the GTRC. This reflects the wider need described by Welsh Energy Policy to develop the skills required to build a "*competitive Welsh energy supply-chain that can help unlock our own energy resources...*" (Wales Gov, 2012:7). To realise this full potential, significant training and up-skilling will be required in order to provide the existing and future workforce with the skills necessary to sustain a competitive low carbon economy. Wales Gov (2012) emphasise that Wales is well placed for the transition to a low carbon economy given the considerable human resources capacity in the low carbon sector in Wales, as well as the natural domestic resources. REA (2012) also highlights green industry in Wales as providing a key example of the job potential in the new green economy; it does however suggest that skills shortages could prevent the UK reaching its renewable potential. In order for Wales to meet the challenges and opportunities that a low carbon economy presents, effort must be focused on developing a strong skills and knowledge base. Effective engagement with industry will be pivotal in tackling the green skills agenda.

The objective of this study is therefore to assess the level of industry interest in LSPG low carbon training in Wales, developed and delivered by the GTRC through the WEST project. In order to achieve this aim it was considered appropriate to employ a survey approach. Firstly investigating the level and form of existing training provision in and around Wales and hence any gaps in training provision, prior to targeting potential course participants to gauge their level of interest in combustion training. As Helm (2007:317) discussing the 'carbon lock-in' of the existing fossil-fuel based energy system, states "*the limits of technological change lie not with science and technology, which tend to evolve much faster than governing institutions but rather with the organisational, social and institutional changes that allow the diffusion of new technological solutions.*" As such significant change is required across these interrelated aspects of sociotechnical complexity. This paper will focus on one such aspect, the organisational change required, focusing on training and skills. The need to address potential skills shortages is widely acknowledged, there is however scant empirical evidence on the potential form and content of low carbon training specifically in the area of large scale power generation (LSPG). The paper is organised as follows: the methodology section provides detail on the main methods employed in the study. The results are then presented and discussed in relation to the low carbon agenda. The paper concludes with recommendations on the importance of a adopting a collaborative approach to ensure that industry training developed is both desired and effective.

2. Method

2.1 Investigating Existing Course Provision

Given the LSPG theme focus on alternative fuels and efficiencies in combustion with a special emphasis on gas turbine technology (GT), a review of relevant HE courses available inside and outside Wales was conducted. Of the courses outside Wales, only those institutions considered within practical travelling distance were considered for full-time (FT) and part-time (PT) courses, while institutions further afield were considered where delivery was conducted online (i.e. distance learning). Relevant projects were deemed to be those whereby *HE* institutions in *Wales* were offering training *direct to industry* in relation to the *large scale power generation* theme. It was also deemed valuable to establish the HE institutions delivering LSPG relevant training to traditional full time HE students from within or near to Wales. In addition training delivered by HE institutions and others outside Wales on the topic of Gas Turbine technology will also be examined.

It was considered important to assign a measure of relevance to the courses found to indicate a degree of overlap. The most useful and concise way to do this was deemed to be the Likert scale as employed by Hall and Allan (2014); as such a 1-5 score was assigned, with 1 illustrating a minor relevance to the topic and 5 showing significant overlap with course material. The benefits of using the Likert scale were considered to be its simplicity of construction, the ease with which participants can read and complete the scoring system and the likelihood of producing a highly reliable scale (Bertram, 2007). However the weaknesses of the approach have also been considered namely, the 'central tendency bias' with participants avoiding extremes of opinion in addition to 'acquiescence bias' in which participants may seek to 'agree' with the researcher, providing answers that are assumed to be desirable by the researcher (Bertram, 2007:8). In addition issues of reproducibility and validity have been considered.

2.2 Assessing Industry Appetite

In order to assess the level of demand for low carbon training within Welsh industry an existing database created by Cardiff University's Engineering School was utilised. The database consists of around 1000 organisations, from which a review process enabled the companies to be refined to 60 relevant target participants (i.e. those assumed to be using large scale power generation technology), located within the Convergence Zone. In addition efforts were made to locate the contact details of all power stations in Wales. The power stations were considered a primary target as the principal users of industrial gas turbines in Wales. As such they were viewed as a vital source of potential participants and whose opinion was critical in the design and development of a relevant GT educational program.

This enabled the database to grow to 75 companies who could potentially be interested in attending LSPG WEST training. The companies were targeted with a questionnaire sent to the Training Officer; this was deemed sensible given the nature of the project. In addition an event was delivered, the 'LCRI LSPG Seminar and Exposition,' in which low carbon power generation energy companies were invited to hear more about the training courses to be developed by the LSPG theme. As part of the event a workshop was held describing to attendees the nature and purpose of the WEST project. The presentation also proposed possible course topics and structure and attendees were invited to complete a questionnaire to register their interest in the different proposals. Two companies in attendance completed the questionnaire, bringing the total questionnaire responses to 9 companies. The questionnaire was also made available on the GTRC and WEST specific website, however no participants responded through this channel.

Descriptions of potential core topics within the theme of LSPG were devised and respondents were asked to rate their preference for the subjects as well as the preferred level of training. A Likert Scale was also utilized for the assessment of industry appetite employing a 1-4 rating for the level of interest in proposed topics (i.e. 1: of no interest, 2: of minor interest, 3: fairly interested, 4: definitely interested). For ease of comparison and in order to generate a holistic assessment of demand in light of the research objectives, data analysis consisted of generating scores for each topic based on total participant responses in a summative fashion. This enabled cross-topic comparison, allowing as it does for the weighting of respective topics. Data was displayed in the form of a bar chart, which was considered a clear and visually appealing way to illustrate survey findings.

3. Results

The following section provides an overview of the main empirical findings; firstly the investigation of LSPG related courses delivered in and around Wales (through the survey of Sector Skills Councils and HE establishments) prior to the results of the industry survey.

3.1 Existing Training Provision

3.1.1 Sector Skills Councils

Of the large number of sector skills councils considered, only a selection were deemed to be most relevant to LSPG related training. These included The Sector Skills Council for Science, Engineering and Manufacturing (SEMTA) and the Construction Industry Training Board (CITB). Sector Skills Councils were considered the most significant agents for sourcing training for industry. However upon investigating the training offered, predominantly through the websites of each respective Council, the relevance to LSPG related training was found to be minimal. Sector Skills Councils are engaged with their industries to train predominantly at level 3 of the CQFW and below, focusing on practical skills and mandatory qualifications. As such the potential for duplication with WEST training materials was deemed to be low.

Table 1. Wales & surrounding areas LSPG related HE course provision

Institution					
In Wales	Outside Wales	Course Title/Qualification	Course Format	Relevance	Relevance Rating (1-5) (*1-minor relevance, 5-significant relevance)*
Machynlleth, Centre for Alternative Technology		Renewable Energy and the Built Environment/MSc, PG Diploma, PG Certificate	1 week/month Sept-May residential	Theory of combustion covered in some detail as it relates to biomass	2
Cardiff University		Sustainable Energy and Environment /MSc, Diploma	F/T and P/T	Taught by staff delivering WEST LSPG theme	5
		Mechanical Engineering BEng, MSc	F/T	(Same as above) Heat Transfer – gas turbines, steam cycles	4
		Integrated Engineering BEng, MSc		Energy Studies – combustion and emissions and renewable energy systems	
University of Glamorgan (now the University of South Wales)		MSc Energy Systems Engineering / MSc	F/T and P/T	Overlap on combustion, heat transfer and industrial heating themes	3
Newport University (now the University of South Wales)		Energy Management Professional Development Certificate/ Diploma	Distance	Assuming some overlap with steam generation and using hot water	1
	University of Ulster	Renewable Energy and Energy Management PGD/MSc REEN (Renewable Energy)	Distance	Biomass sources, technologies and applications of energy from biomass for electricity generation, heat generation and as transport fuel	2
	Brunel University London	Building Services Engineering with Sustainable Energy MSc	Distance	Power cycles, CHP, combustion processes, boiler plant, thermal energy storage and environmental impacts of plant operation	3
	Loughborough University	Renewable Energy Systems Technology MSc	Distance	Covers the combustion of biomass, fuel composition and power efficiency in steam cycles etc.	3

Note. FT=Full-Time, PT= Part-Time.

3.1.2 Higher Education Institutions Within and Close to Wales

Given the LSPG focus on alternative fuels and efficiencies in combustion, with a special emphasis on gas turbine technology, a review of relevant HE courses available inside and outside Wales was conducted. It was found that none of the HE institutions in Wales are presently offering combustion related training courses directly to industry. Therefore, any combustion training offered by LSPG should not overlap with any other similar initiatives in Wales. In addition it was deemed valuable to ascertain the level of existing training provision delivered by HE institutions to traditional full time HE students in and around Wales.

In total, 42 different courses across 21 institutions were reviewed. Information regarding course content was gathered in a variety of ways. The content of courses being delivered within Cardiff School of Engineering was obtained directly through the 'School Office' itself. For other courses module descriptions could be found on websites in most instances, although many descriptions were too vague to be of practical use. In these cases the institutions were contacted directly to request further clarification. Only two institutions failed to respond about their respective courses (1 course each) but overlap was considered unlikely or minimal at best, so these courses have not been considered relevant.

As illustrated in Table 1, it was found that 9 courses currently being delivered in Wales possessed some relevant material, of these 6 were offered by Welsh institutions, with 5 delivered either within Cardiff University or within 15 miles of Cardiff City Centre. The University of South Wales (formerly known as Newport University) was the only Welsh Institution offering distance learning, however the relevance to LSPG was found to be minimal. Should anyone wish to pursue training in the field of LSPG, they may have to travel a considerable distance for traditional face-to-face training delivery. One could deduce that distance learning could be an important way to increase the numbers of employees willing and able to undergo training due to its convenience. It could also deliver the flexibility required for individuals anywhere in Wales who wish to pursue training and for whom full or part time study is not a realistic option without significant changes in their employment arrangements. Consideration will now be given to gas turbine specific courses.

3.1.3 Gas Turbine Specific Courses

There are gas turbine specific courses being delivered in many countries, but for relevance to the population of Wales only the courses delivered in English have been investigated (there are no courses delivered in Welsh).

Cranfield University is the primary provider of gas turbine related short courses in the UK, with approximately 20 gas turbine engineering continuing professional development (CPD) short courses advertised. During 2013, four training courses were offered: Combined Cycle Gas Turbines, Design and Performance of the Axial Turbine Component of Gas Turbine Engines, Gas Turbine Combustion and Gas Turbine Technology for Operations and Maintenance Engineers.

As an example, the Gas Turbine Combustion course covers the following topics:

a. Pre combustor diffusers – its role, design etc.
b. Combustion in the gas turbine – performance and design criteria, aerodynamics, fuel injection methods, retrofitting for multi-fuel capability, ignition flame stabilisation
c. Pollutants – carbon formation and gaseous pollutants
d. Fuels – variability and alternative fuels substitution
e. Computational modelling
f. Heat transfer

The course is primarily aimed at people actively engaged in the gas turbine industry or those about to enter the field with a degree in Science or Engineering (or equivalent).

The University of Leeds offers also CPD courses on combustion and gas turbine technology. Outside of the UK, the University of California, Irvine Combustion Laboratory offers a specific gas turbine course primarily for people already engaged in the research, design and development of gas turbine combustion. The course content is outlined below:

a. Fundamentals of combustion – flame temperature, speed and classification, flammability limits, stoichiometry, chemical dissociation and kinetics and evaporation of liquid sprays etc.
b. Gas turbine combustion – basic design features and types of chambers, performance criteria (e.g. emissions), performance and pressure loss, diffuser design, fuel preparation, ignition and more.

c. Low emissions gas turbine combustion – carbon formation, fuel properties, gaseous pollutants, low NOx design and regulation.

d. Experimental methods – model combustors, diagnostic tools, spray characterisation, high temperature and pressure experiments and computational analysis.

The syllabus of this course is very similar to the course delivered by Cranfield University, which suggests a possible core formula for the design of a gas turbine specific course for potential WEST training.

3.2 Establishing Industry Demand

Table 2 provides an overview of the nature of the organizations who responded.

Table 2. Industry respondents

Description
Power station
Gas/diesel engine provider
Energy consultancy
Biomass heating organisation
Power station
Forestry & timber company
A non-profit company operating an internet based network to support its members in the implementation of biopower technology fuels
Chemical manufacturer
Tissue manufacturer

It was deemed important to ascertain the level of interest in the proposed course topics across a range of educational levels (from undergraduate to industry professional/academic). The respondents were asked to rank their level of interest in each of the 13 proposed course topics on a scale of 1-4 (i.e. 1: of no interest, 2: of minor interest, 3: fairly interested, 4: definitely interested).

The topics 1-13 can be summarised as follows:

At undergraduate level:

a. Topic 1- The Nature of Fuels

b. Topic 2- Utilisation of Waste

c. Topic 3- Energy Conversion Processes – (i.e. combustion, gasification and pyrolysis)

d. Topic 4- Energy Conversion Technologies

e. Topic 5- Combustion Science

f. Topic 6- Improving Combustion Efficiency and Emissions

g. Topic 7- Combustion Risks and Hazards

h. Topic 8- A Bespoke Course concerning a company's specific energy conversion technologies and processes, developed in collaboration with the company

At graduate level:

i. Topic 9- Gas Turbine Emissions

j. Topic 10- Gas Turbine Flame Stability and Combustion Noise

k. Topic 11- Gas Turbines: Fuel Variability and Alternative Fuels

l. Topic 12- Combustion Performance in Gas Turbines and Thermodynamic Cycles

For industry professionals and academics:

m. Topic 13- Gas Turbine Theory, Design and Performance Course

Figure 2 displays the results from the questionnaire. The topic scores given by the respondents were added to give a total score for each topic (shown in brackets under the topic number). Each company's scores for individual topics are displayed in Figure 2. For example, for topic 1 the Tissue Manufacturer awarded a score of 4, whilst the Chemical Manufacturer assigned a score of 2.

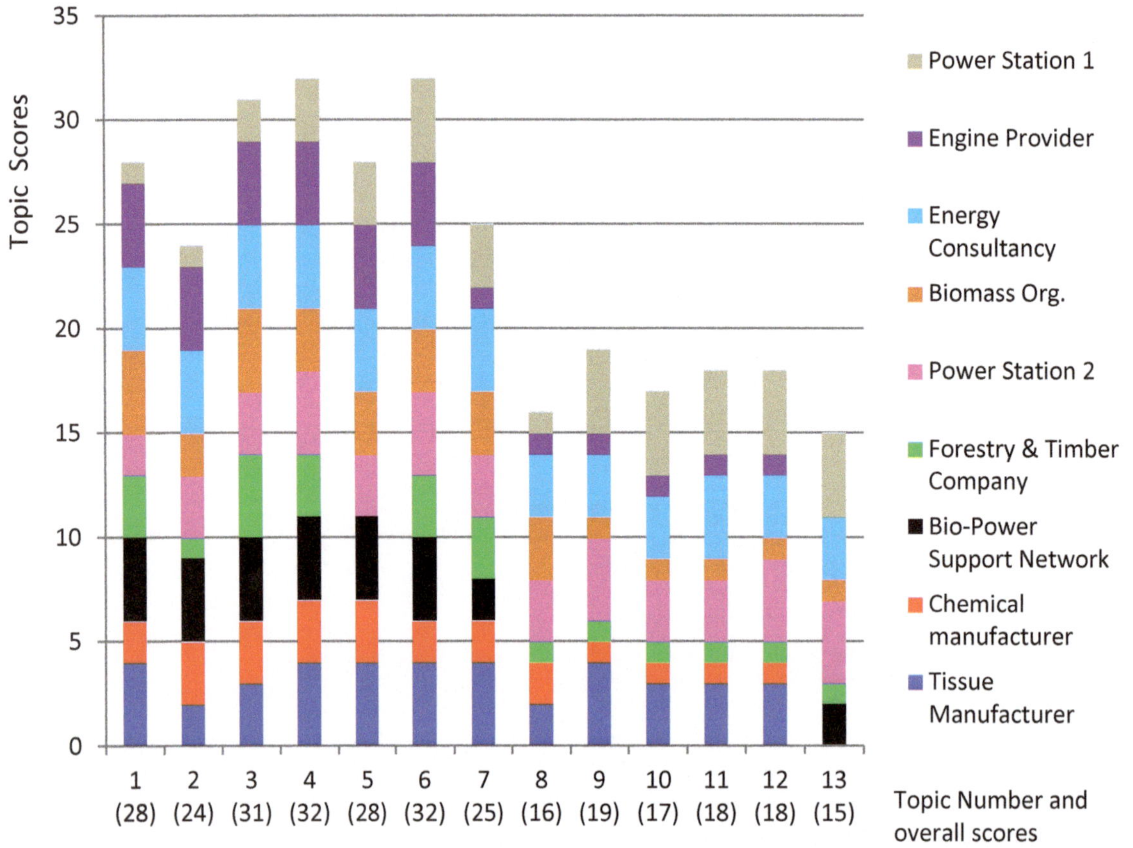

Figure 2. Proposed course topics feedback scores

As can be seen, topics 1-7 scored higher than the remaining topics. These topics are less specialised than topics 8-13 and so demonstrate an industry appetite for training at a fundamental, introductory level. It is not surprising that of the four companies showing a real interest in the more specialised gas turbine courses (i.e. scoring topics at 3 or more), two manage power stations. Those companies scoring the more specialised courses highly also scored the fundamental, introductory level courses highly, indicating the need for staff training at all levels within these companies. Training can be offered by LSPG at all educational levels however feedback indicates a more widespread demand for introductory level courses.

4. Discussion & Conclusion

4.1 Positive Industry Response

The importance of matching industry demand for training with teaching and research expertise is appreciated by Yadav (2012). Indeed training developed should marry up academic staff expertise, industry's established topics of interest and low carbon policy in the field of power, heat generation and utilisation. The survey of industry demand indicates a positive response to LSPG related training at an introductory level, which is well matched with the skills base of the GTRC. Indeed there is the potential to train the manufacturing and power generation sectors on a variety of combustion topics related to lowering carbon emissions in heat and power generation.

It is important to appreciate the dynamic nature of industry which means that needs evolve and hence training should be updated periodically to ensure the relevance of course materials (Yadav, 2012). The results could be strengthened with further industry surveying in order to increase the generalizability of findings, although 'questionnaire fatigue' presents a challenge to the survey approach. Efforts to engage with prospective survey targets directly in face-to-face meetings or further workshops could increase the response rate and develop a richer feeling for industry demand, although this would be a more time-consuming option.

Findings suggest that the lack of LSPG related training currently offered by Welsh institutions directly to industry means that training developed within WEST should not overlap with any existing projects and hence WEST training can be considered as meeting a genuine need within industry. The courses created should be delivered at a level which is accessible to individuals in a variety of workplace roles; efficient operation of a facility is an objective of workers involved in maintenance, operation and management. Indeed an understanding of efficient combustion processes, alternative fuels and new technologies is the key to lowering carbon emissions now and in the future, so training materials must be developed with the educational background of the potential audience in mind. However the WEST project stipulates training at level 4 and above of the CQFW, as such a balance must be achieved between delivering the desired level of training and ensuring that participants are able to achieve the learning outcomes. As Arslan and Kuş (2012) note, it is difficult to prepare training content for participants who may have varying levels of capabilities in terms of knowledge and skills.

Courses could be aimed at Wales' current workforce in the form of CPD training - to promote retention of skills and to build relationships between the University and industry participants. The importance of CPD training to engineers is well documented in Williamson et al. (2001), who state that the breadth of knowledge required by engineers and the rapid pace of technological change mean that CPD's are an imperative learning tool for the engineering industry. In addition the delivery format should be flexible enough to allow training to fit around a working life. Short courses or day release from work to attend the University or GTRC would be an option for individuals living close to the facilities. Distance learning (e.g. e-learning) would most likely be the preferred option for those individuals living further afield, as was stated in one of the questionnaire responses from a company based in Powys, North Wales. A combination of both e-learning and attendance for those in South Wales would also help to minimise time spent away from the workplace.

4.2 Creating Value for Welsh Industry

There is a very little empirical evidence on the demand for, development and delivery of low carbon training, specifically in the area of LSPG. This study could provide value for organisational bodies seeking to provide low carbon skills training. The findings suggest that industry in Wales is receptive to WEST LSPG training courses; initial consultation on the prospect of WEST LSPG educational courses has been wholly encouraging. The training to be developed fits well with Welsh Energy Policy, as such policy objectives and industry's perceived training needs can be considered as aligned. LSPG low carbon combustion courses can help deliver the knowledge necessary for the improved energy conversion efficiencies in the short term and can also introduce the workforce to the potential of alternative fuels. The courses could help to lay the groundwork for greater acceptance of future low carbon technological developments such as CCS.

The research has indicated the importance of geographical considerations in respect of the most effective mode of training. At present existing training provision in terms of HE establishments in Wales delivering direct to industry is limited, the training offered to full time HE students by establishments in and around Wales shows some relevance to LSPG, although not to a degree to which overlap or duplication may be an issue. Given the current concentration of existing training options in and around Cardiff City Centre, it would seem that distance learning, potentially in an e-learning format could be a valuable way to capture those participants who are unable to travel large distances (or who cannot spare the time to enrol as a part or full-time student) to attend traditional face-to-face training delivery. Considering a blended learning option with some element of e-learning along with traditional face-to-face delivery, the location of the GTRC in Port Talbot, South Wales, situated between the city of Swansea and the large town of Bridgend could also help to mitigate the travelling inconvenience for individuals in South Wales, should courses be delivered at the facility. In addition the criticality of delivering training at the appropriate level must be appreciated, current training is generally provided at masters degree level (level 5 of the CQFW) whilst industry is seeking training at a lower introductory level. This suggests that efforts to improve national green skills must be undertaken in collaboration with industry to ensure a coherent and appropriate approach. It must also be recognised that Energy Policy is vulnerable to economics as such a pragmatic approach to the development of training materials should be adopted (Hewlett, 2014). As RAE (2010:5) states, *"The future direction and success of the UK economy is largely predicated on the availability of the right skills."* It is further argued that, *"the right skills is a matter of strategic national importance and, as such, transcend party politics"* (2010:5). The need for a coordinated approach to Energy Policy and education is therefore vital in order to ensure that industry is confident in the direction of the market and likewise skills providers can develop relevant training. The formation of an Industrial Advisory Group (IAG) could be an important source of industry insight employing a collaborative approach to course development, ensuring that training delivered will truly be of value to Welsh industry in the transition to a low carbon economy.

Acknowledgments

The authors would like to thank Steve Morris and Dr Yura Sevcenco at the GTRC for their advice and guidance and Dr Dean Stroud, Dr Samantha Yandle, & Professor Andrew Thomas for their comments on earlier drafts.

References

Acworth, E. B. (2008). University–industry engagement: The formation of the Knowledge Integration Community (KIC) model at the Cambridge-MIT Institute. *Research Policy*, 1241-1254. http://dx.doi.org/10.1016/j.respol.2008.04.022

Alan, B., & Bell, E. (2011). *Business Research Methods.* Oxford: Oxford University Press.

Arlett, C., Lamb, F., Dales, R., Willis, L., & Hurdle, E.(2010). Engineering Education. *Journal of the Higher Education Academy Engineering Subject Centre, 5*(2), 18-25.

Arsland, R., Kus, A., (2012). A Modular Approach for Training Employees in the Automotive Service Sector. *Industry & Higher Education, 26*(5), 403-408. http://dx.doi.org/10.5367/ihe.2012.0116

Bertram, D., (2007). *Likert Scales.. are the Meaning of Life....*Canada. University of Calgary.

Bruneel, J., D'Esteb, P., & Salter, A. (2010). Investigating the factors that diminish the barriers to university-industry collaboration. *Research Policy*, 858-868. http://dx.doi.org/10.1016/j.respol.2010.03.006

Collis, & Hussey. (2009). *Business Research: A Practical Guide for Undergraduate & Postgraduate Students* (3rd ed.). Palgrae Macmillan.

Committee on Climate Change. (2013). *Reducing the UK's Carbon Footprint and Managing Competitiveness Risks.* United Kingdom: Committee on Climate Change.

Dasgupta, P., & David, P. (1994) in Bruneel, J., D'Esteb, P., & Saltera, A. (2010). Investigating the factors that diminish the barriers to university-industry collaboration. *Research Policy*, 858-868.

DECC. (2013). *UKCS Oil and Gas Production Projections.*

Edmondson, G., Valigra, L., Kenward, M., Hudson, R. L., & Belfield, H. (2012). *Making Industry-University Partnerships Work: Lessons from Successful Collaborations.* Science Business Innovation Board.

Fairweather, J. (1988). *Entrepreneurship and Higher Education: Lessons for Colleges, Universities & Industry.* Washington: ASHE-ERIC Higher Education Report No.6.

Gov. (2013). *The Ue of Low Carbon Technologies.* Retrieved 2013, from https://www.gov.uk/government/policies/increasing-the-use-of-low-carbon-technologies/supporting-pages/heat-networks

Hall, C., & Allan, F. (2014). Influencing Carbon Behaviours: What Pscyhological and Demographic Factors Contribute to Individual Differences in Home Energy Use Reduction and Transportation Mode Decisions. *Energy and Environment Research*, 1-14.

Helm, D. (2007). *The New Energy Paradigm.* Oxford: The Oxford University Press.

Hemingway, J., & Michaels, C. (2012). *Special feature – sub national electricity figures, Electricity generation and supply figures for Scotland, Wales, Northern Ireland and England, 2008 to 2011.* December 2012. DECC. Retrieved from https://www.gov.uk/government/uploads/system/uploads/attachment_data/file/65841/7345-elec-gen-2008-2011-et-article.pdf

Hewlett, S. (2014). *Sector Skills Training Needs Analysis.* LCRI. WEST

Jagger, N., Foxon, T., & Andy, G. (2013). Skills Constrains and the Low Carbon Transition. *Climate Policy, 13*(1), 43-47. http://dx.doi.org/10.1080/14693062.2012.709079

Kitson, M., & Hughes, A. (2009). University-Industry Knowledge Exchange: Demand Pull, Supply Push and the Public Space Role of Higher Education Institiuion: Full Research Report, ESRC End of Award Report, RES-171-25-0018. Swindon: ESRC.

Mead, N. A., Kathy, B., Jimmy, L., O'Mary, G., & Parish, C. (1999). Industry/university collaborations: different perspectives heighten mutual opportunities. *The Journal of Systems and Software, 49*, 155-162. http://dx.doi.org/10.1016/S0164-1212(99)00091-6

RAE. (2010). *Educating Engineers for the 21st Century.* The Royal Academy of Engineering

REA. (2012). *Renewable Energy: Made in Britain.* Renewable Energy Association.

Rosenberg, N. (1982). *Inside the Black Box: Technology & Economics.* Cambridge: Cambridge University Press.

Salter, A., Tartari, V., D'Este, P., & Neely, A. (2010). *Exploring UK Academic Attitudes to Collaborating with Industry and Entrepeneurship.* UK Innovation Research Council.

Stroud, D., & Fairbrother, P. (2008). The importance of workplace learning for trade unions: a study of the steel industry. *Studies in Continuing Education, 30*(3), 231-245. http://dx.doi.org/10.1080/01580370802439920

Vaidya, A., & Charkha, S. (2008). Barriers & Motivational aspects of University-Industry Linkage. *CURIE Journal, 1*(3), 58-63.

Viridor Ltd. (2013). *Transforming Waste – Cardiff EfW.* Retrieved April 23, 2013, from http://www.viridor.co.uk/our-developments/cardiff-efw

Wales Gov. (2010). *A Low Carbon Revolution.* Welsh Government.

Wales Gov. (2012). *Energy Wales: A Low Carbon Transition.* Welsh Government.

Williamson, S., Lamb, F., & Davis, L. (2001). *Education Issues are the Same Across the World.* International Conference on Engineering Education. Aug 6-10. Norway.

Yadav, D. (2012). University Aviation Degree Courses: A Study of Labour Market and Student Retention Challenges for Sustainable Course Development. *Industry & Higher Education, 26*(5), 393-401. http://dx.doi.org/10.5367/ihe.2012.0118

15

Oil Extraction From Oil Sludge and TPH Elimination of Solids/Water by Ozonation

Misri Gozan[1]

[1] Industrial Bioprocess Engineering Research Group, Chemical Engineering Department, Faculty of Engineering, Universitas Indonesia, Kampus UI Depok 16424, Indonesia

Correspondence: Misri Gozan, Industrial Bioprocess Engineering Research Group, Chemical Engineering Department, Faculty of Engineering, Universitas Indonesia, Kampus UI Depok 16424, Indonesia. E-mail: mgozan@che.ui.ac.id

Abstract

About 1.1 million barrels of oil sludge are generated yearly in Indonesia by many oil and refinery industrial activities. Oil is strongly bound with solid particles in the sludge. Oilin the sludge may range from benzene, phenols to polycyclic aromatic hydrocarbons which are classified as toxic, mutagenic and carcinogenic for human and pollute the environment. Oilextraction is favorable option because of environmental and economical reasons. Oil sludge from an oil fields inSumatra, Indonesia was extracted at low temperature (40–70 °C) resulting in Total Petroleum Hydrocarbon (TPH) around 1–2%. After addition of 10,000 ppm REKLIN G04 surfactant, TPH extraction was improved to 9%. After decantation, effluent water containing ± 15,000 ppm TPH was ozonated for 60 minutes resulting in TPH reduction to ± 5 ppm. After ozonation, TPH of discharge solids was only 7–40 ppm. Fingerprint by gas chromatography showed removal of light fraction of oil after ozonation.

Keywords: oil sludge, ozonation, surfactant, extraction, TPH, fingerprint

1. Introduction

Large amounts of oil sludge are generated during the cleaning of tanks, oil storage, maintenance of related facilities and processing operations prior to the sale of crude oil into the sea terminal. Disposed oil sludge normally contains toxic substances such as aromatic hydrocarbons (benzene, toluene, ethyl benzene and xylene), poly-aromatic hydrocarbons (Swoboda-Colberg, 1995).

In general, oil sludge is a complex water-in-oil (W/O) emulsion, typically including 30–50% of oil or total petroleum hydrocarbon (TPH), 30–50% of water and 10–12% of solids by mass (Ramaswamy et al., 2007). Total Petroleum Hydrocarbons (Ayotamuno et al., 2007) without adequate treatment can be a pollutant to the environment, especially on the ground.

Landfill, soil bioremediation and incineration are among common disposal methods for oil sludge. Disposal to landfill requires a lot of land that must be adequately insulated to prevent leaching of toxic compounds in the soil. Bioremediation applications with fertilizer and bacteria can lead to the accumulation of pollutants such as toxic metals. Oil sludge burning has limitations such as transportation and the release of toxic gas that evaporates into the environment. Toxic heavy metals can not be removed during the combustion process and will accumulate as solid particulates in the combustion zone (Dominguez et al., 2005). The objective of the research described in this paper is to find a more clean yet efficient method for treating oil sludge using ozonation and surfactant system.

Ozone is one of the alternative technologies that focus on achieving oxidation through chemical reactions. Ozone is very reactive and can oxidize compounds such as persistent mineral hydrocarbons. Ozone is readily generated when oxygen is exposured to electrical discharge. O atom ($E^{o}_{0/H2O}$ = 2.42 V) can be generated via dissociation of O_2 and boosts the rate of production of OH**.** (Bruggeman et al., 2009; Malik et al., 2001). Furthermore, O atom can directly react with contaminants and also takes part in the reactions with O_2 resulting in the formation of O_3, $E^{o}_{0/O2}$ = 2.07:07 V (Malik et al., 2001). Ozone as a strongly oxidizing allotropic form of oxygen reacts best when it can act as an electron transfer acceptor for the oxidation of metal ions, as an electrophile for the oxidation of phenol land other activated aromatics, and as a dipole addition reagent by addition to carbon–carbon multiple bonds. Ozone is unstable and decomposes through a cyclic chain mechanism to produce hydroxyl radicals in neutral and basic solutions (Glaze, 1987). Moreover, with the presence of H_2O_2 in plasma system, ozone can react with HO_2^- giving OH^- (Bruggeman et al., 2009; Malik et al., 2001; Joshi & Thagard, 2013).

In general, advanced oxidation processes (AOP) have typically been used for the decomposition of organic material, color and disinfectants. For example, aromatic hydrocarbon oxidation completely unraveled with a longer time than the other aliphatic hydrocarbons such as dodecane (Kornmuller et al., 1997). An oil emulsion can be characterized as hydrocarbon compounds. For the processing of oil-water emulsion, ozonation has not been studied so far. However, application of advanced oxidation may be a viable option for the degradation of the oil-water emulsion (Suthi, 2001). Various studies using surfactant or biosurfactant showed significant improvement of oil extraction solely (Yan et al., 2012) or in combination with other treatment such as ultrasonic (Zhang et al., 2012). Surfactants are amphiphilic molecules comprise distinct hydrophobic and hydrophilic moieties – a polar or ionic group connected to a long hydrocarbon tail (linear, branched or containing aromatic rings). In aqueous solutions and at critical micelle concentration (CMC), surfactants begin to form aggregates that are in dynamic equilibrium with the monomers in the bulk aqueous solution. Many factors affecting effectiveness of surfactant (Yazdi, 2011).

2. Method

2.1 Materials

Oil sludge was obtained from an oil pit field in eastern Sumatera, Indonesia. The surfactant (REKLING04) was a rhamnolipid base and obtained from PT Rekayasa Lingkungan, Jakarta-Indonesia. Chemicals such as Toluene, KI, H_2SO_4 and $Na_2S_2O_3$ are from Merck.

2.2 Procedure

Experiments were conducted in lab scaleto determine optimum conditions for oil extraction from oil sludge. Oil sludge, additional water (Sludge:water 1:1) and surfactant were put into 500 mL flasks. This mixture (500 mL) remained for settling 4 hours. Then, the lower stream was collected into a water-solid flask. The upper stream, together with surfactant, went to a second settling flask. In this tank, agitation and heating were applied for 120 minute, prior to sedimentation. Variation of temperature (40, 50, 60 and 70 °C) and surfactant concentration (0–15,000 ppm) were conducted and sampling were taken at minute 15, 30, 60 and 120. The lower stream of centrifuge comingled in the water-solid tank, while upper stream went to a centrifugation step for 10 minute. Part of centrifugal output was collected as recovered oil. The lower stream of centrifuge comingled in the water-solid flask. Water-solid stream was then ozonated for 1 hour. Sampling was carried out at minute 0, 15, 30, and 60. After setlement and filtration, water in the upper stream and solid were collected and analysed for TPH and Total Suspended Solids (TSS). The setup of oil extraction procedure is illustrated in Figure 1.

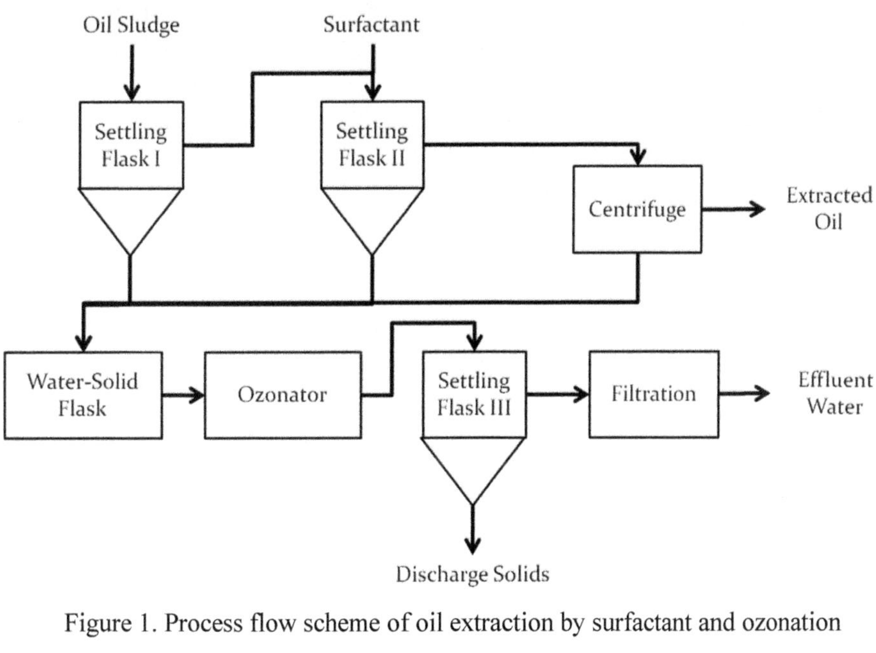

Figure 1. Process flow scheme of oil extraction by surfactant and ozonation

2.3 Analysis

Total Petroleum Hydrocarbon (TPH) analysis for discharged solids and effluent water were done by extraction with toluene, then filtration in filter paper with addition of sodium sulphate and measurement with Spectrophotometer HACH DR 3900. Fingerprint Analysis of Recovered Oil and Discharged Solids were conducted by Gas Chromatography-Flame Ionisation Detector 30 m × 0.25 mm film thickness 0.25 μm fused silica capillary columns, coated with methyl silicone (GC-FID, Shimadzu 14B series). Detector (FID) and injector temperatures were kept at 250 and 280 °C, respectively. The oven temperature was programmed from 60 to 280 °C at 4 °C/min with an initial hold time of 1 min and final hold time of 15 mins. Helium at a linear velocity of 2 mL/min was used as carrier gas.

Ozonator productivity testing was conducted by iodometric titration. Ozone is generated flowed into KI solution that oxidizes iodide ions into iodine solution. Iodine formed is then titrated with sodium thiosulfate solution so as to obtain the capacity of ozone produced.

3. Results

3.1 Heating

Oil sludge, water and surfactant were put into 500 mL flasks. Temperature, surfactant concentration and centrifugal speed were varied. Below centrifugal speed of 2700 rpm only less than 1% oil was recovered. Hence, for the rest of experiments, centrifugal speed of 2750 rpm was applied.

Oil sludge, water and surfactant were mixed in various temperature from 40 to 70 °C. Samples were taken at minute 15, 30, 60 and 120. After TPH analysis, oil recovered was depicted in Figure 2. In the beginning of mixing almost all samples showed around 1% of oil extraction. It was still in this percentage until 2 hours of experiments at sample with temperature 40 and 50 °C. At temperature 60 and 70 °C the oil extraction increased to around 2%. The peak oil extraction was achieved after 30 minute heating and mixing. The stable oil extraction of around 2% at temperature 60 °C lead to use these temperature as default for next experiments.

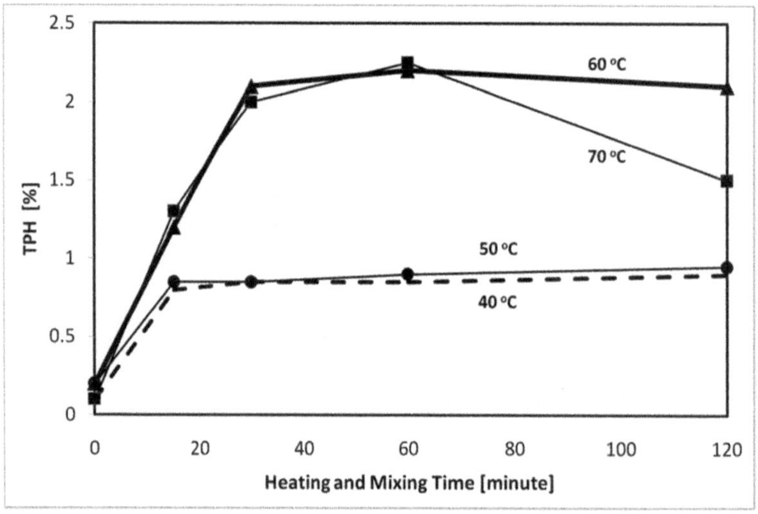

Figure 2. Effect of mixing temperature and heating time on oil extraction

3.2 Surfactant Concentration

Surfactant as deemulsifier was given to the mixture of oil sludge and waterat concentration of 200; 500; 1,000; 10,000 and 15,000 ppm. Samples were taken at minute 15, 30, 60 and 120. For every sampling, 30 minute of settling time were applied before upper sample were then undergone TPH analysis. Oil recovered was depicted in Figure 3. There was surfactant added for control samples. No significant different between 10,000 and 15,000 ppm surfactant was observed after 60 minute. There is no more increase in oil extraction observed after 60 minute. The highest oil extraction was ca. 9%.

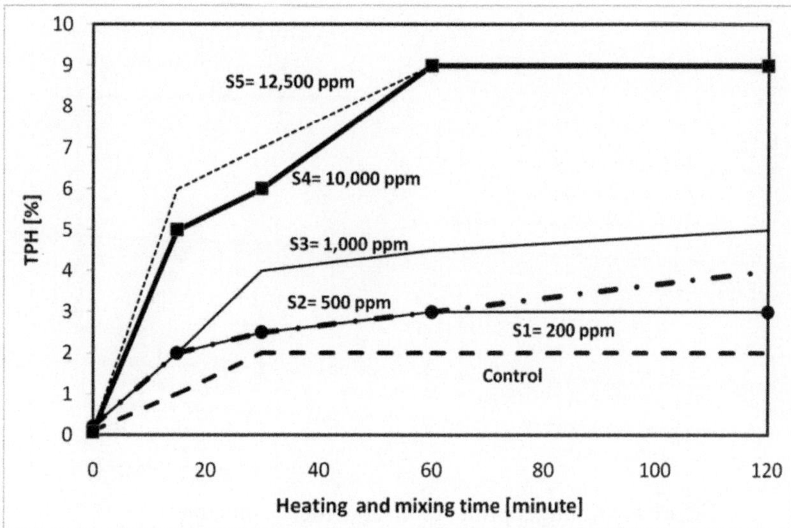

Figure 3. Surfactant effect on oil extraction. Heating was applied at 60 °C

3.3 Ozonation

Experiments on ozonation were aimed to reduce the TPH or oil content in the water effluent and discharge sands. After mixing with 10,000 ppm surfactant and heating at 60 °C, oil content in the upper part was decanted. Remaining water and solids were then stirred while 0.35 gram ozone/hour was applied. Samples were taken at minute 15, 30, 45 and 60 of ozonation. Figure 4 shows the TPH/oil content. It was clearly observed that after 45 minute the TPH content was almost not found.

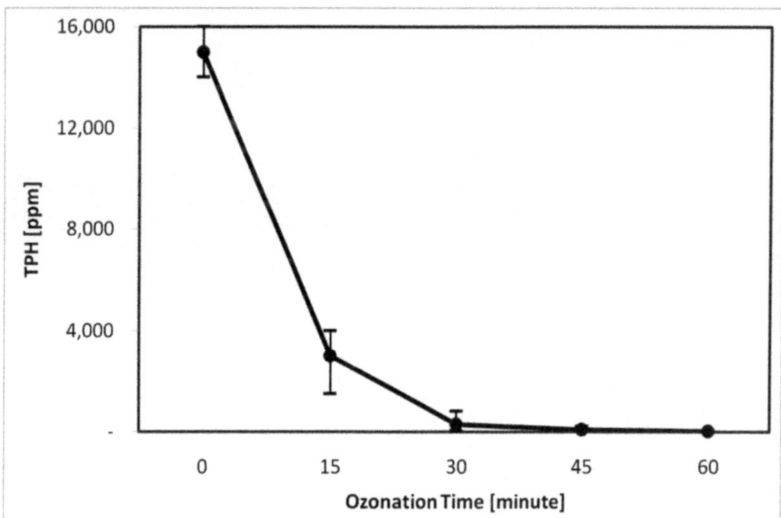

Figure 4. Oil content along with ozonation

To observe the effect of ozonation on hydrocarbon fraction, samples were analysed with GC as it is shown in Figure 5. The fraction of hydrocarbons was slightly changed. Before ozonation, fingerprint was predominated by lighter hydrocarbon. After 30 minute ozonation, heavier fractions were observed.

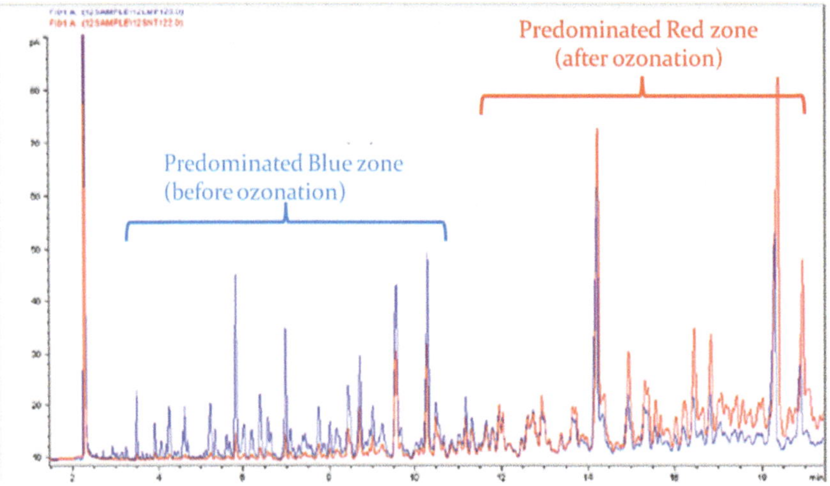

Figure 5. Oil finger print before and after ozonation

3.4 End Product Characterization

At the end, extracted oil, discharge solids and water effluent were caharacterized as depicted in Table 1.

Table 1. Characterization of end products: extracted oil, discharge solids and Effluent water

	BSW[1] in Extracted Oil [%]	**TPH in Discharge Solids** [ppm]	**TSS in Effluent Water** [ppm]	**Oil in Effluent Water** [ppm]
Experiments	0.5–0.9	7–40	10–57	4–5
Regulation	na	Na	200*	25*

[1] BSW = Basic Sediments and Water, ASTM D-96.

[2] KEP-42/MENLH/10/1996.

4. Discussions

4.1 Effect of Heating

This initial experiments were aimed to determine optimum conditions for separation of oil, water and sands. Heating the oil sludge mixture was expected to increase oil solubility, thus improve oil extraction. However, only slight increase, from 1 at temperature 40 °C to 2% TPH at the highest extraction. Increasing the temperature to 70 °C did not significantly increased the oil extraction. Instead, the oil extraction after 2 hours in the sample dropped to 1.5%. This might be due to evaporation of volatile hydrocarbons at slightly high temperature. The highest oil extraction was achieved at temperature 60 °C until 2 hour experiment.

4.2 Effect of Surfactant

Addition of surfactant and heating at 60 °C, evidently enhanced the oil extraction. The highest oil extraction was at TPH 9% by using 10,000 ppm REKLIN G04 surfactant. Oil extraction obviously also depends on the initial oil content of the sludge. The concentration of surfactant that should be given to the system would also affected by the oil content in the sludge. Addition of surfactant to 12,500 ppm did not show any significant effect. However, no more addition of surfactant was done to avoid reemulsion. CMC (Critical Micelle Concentration) is the critical threshold surfactant concentration in a solution. If the concentration is exceeded CMC it will lead to the formation of micelle reemulsion (Opawale & Burges, 1998). The surfactant used in this experiments was of Rhamnolipids, which are a class of glycolipid bio-surfactantsusually produced by specific bacterial strains (Mulligan, 2005). The CMC of Rhamnolipids has been reported between 10 and 230 ppm (Nitschke et al., 2005). However, our results showed that REKLIN G04 surfactant gave the highest extraction far beyond this concentration range. Nevertheless, addition of rhamnolipids into oil sludgemay affect the adsorption energy of hydrocarbons with solid particles and lower the energy required for desorption (Zhang et al., 2012), and thus the organic compounds could be easily removed at the ozone application.

4.3 Effect of Ozonation on Oil Content of Water Effluent and Discharge Solids

Immidiately after 0.35 gram ozone/hour application the TPH content was reduced significantly. After 45 minute the TPH content was almost not found. Ozone attacked lighter hydrocarbon fractions easier than the heavier ones. GC fingerprints confirmed that lighter fractions (0–16 minute, or atom C1 to C12) in the left part of Figure 5 were strongly degraded. Then after 16 minute retention time (heavier hydrocarbons, C12 to C39) we could see no significant degradation. Formation of hydroxyl radicals is the vital oxidants that dominate the plasma oxidations in water application of ozone. Their reactions with organic compounds may be differentiated into three different mechanisms: abstraction of hydrogen atom; electrophilic addition to unsaturated bond and electron transfer. In the case of saturated aliphatic hydrocarbons or alcohols, hydrogen abstraction is primary for organic abatement with yielding H_2O and an organic radical (Jiang et al., 2014). As for olefins or aromatic hydrocarbons, C-centered radical with a hydroxyl group at the a-C atom can be formed with OH radical addition to unsaturated double carbon–carbon bonds of organic compound. Besides, reduction of hydroxyl radicals to hydroxide anions by an organic substrate is of particular interest in the cases where hydrogen abstraction or electrophilic addition reactions may be disfavored by multiple halogen substitution or steric hindrance (Lukes, 2001). Ozone oxidation strength decreases with the number of branch and length of hydrocarbons. Higher intensity of ozone dose might be needed to eliminate the heavier fraction of hydrocarbons.

4.4 Environmental Compliance of Effluent Water and Discharge Sands

The main goal of oil extraction from oil sludge is to have as much as possible oil, while it is desirable to have less than 1% BSW in the extracted oil. In the other hand, due to stringent environmental regulation, effluent water from oil waste treatment must contain less than 25 mg/L Oil and Grease, and less than 200 mg/L TSS (KEP-42/MENLH/10/1996). All those expectation and regulation have been fulfilled by the method of extraction and ozonation.

5. Conclusion

Oil extraction of oil sludge was studied in these lab experiments. Using no surfactant, only low percentage of TPH was observed. The extracted oil increased significantly after application of heating (60 °C) and surfactant REKLIN G04. Elimination of TPH was achieved by using ozone resulting in TPH reduction to almost zero in both effluent water and discharge solids. Fingerprint by gas chromatography showed removal of light fraction of oil after ozonation.

Acknowledgements

Thanks to Irsham Villa for his helping hands in collecting some of the data.

References

Ayotamuno, M. J., Okparanma, R. N., Ogaji, S. O. T., & Probert, S. D. (2007). Bioremediation of a sludge containing hydrocarbons. *J. Appl. Energy, 84*(9), 936-943. http://dx.doi.org/10.1016/j.apenergy.2007.02.007

Bruggeman, P., & Leys, C. (2009). Non-thermal plasmas in and in contact with liquids. *Journal of Physics D: Applied Physics, 42*(5), 053001. http://dx.doi.org/10.1088/0022-3727/42/5/053001

Domínguez, A., Menéndez, J. A., Inguanzo, M., & Pis, J. J. (2005). Investigations into the characteristics of oils produced from microwave pyrolysis of sewage sludge. *Fuel Processing Technology, 86*(9), 1007-1020. http://dx.doi.org/10.1016/j.fuproc.2004.11.009

Glaze,W. H. (1987). Drinking-water treatment with ozone. *Environ. Sci. Technol., 21*, 224-230. http://dx.doi.org/10.1021/es00157a001

Jiang, B., Zheng, J., Qiu, S., Wu, M., Zhang, Q., Yan, Z., & Xue, Q. (2014). Review on electrical discharge plasma technology for wastewater remediation. *Chemical Engineering Journal, 236*, 348-368. http://dx.doi.org/10.1016/j.cej.2013.09.090

Joshi, R. P., & Thagard, S. M. (2013). Streamer-like electrical discharges in water: Part I. Fundamental mechanisms. *Plasma Chemistry and Plasma Processing, 33*(1), 1-15. http://dx.doi.org/10.1007/s11090-012-9425-5

Kornmuller, A., Cuno, M., & Wiesmann, U. (1997). Ozonation of Polycyclic Aromatic Hydrocarbons in Oil/water Emulsion. *Proceeding of the International Conference on Ozonation and Related Oxidation Processes in Water and Liquid Waste Treatment*. Berlin, Germany.

Lukeš, P. (2001). *Water Treatment by Pulsed Streamer Corona Discharge* (PhD Thesis). Institute of Chemical

Technology, Prague, Czech Republic.

Malik, M. A., Ghaffar, A., & Malik, S. A. (2001). Water purification by electrical discharges. *Plasma Sources Science and Technology, 10*(1), 82. http://dx.doi.org/10.1088/0963-0252/10/1/311

Mulligan, C. N. (2005). Environmental applications for biosurfactants. *Environmental pollution, 133*(2), 183-198. http://dx.doi.org/10.1016/j.envpol.2004.06.009

Nitschke, M., Costa, S. G., & Contiero, J. (2005). Rhamnolipid surfactants: an update on the general aspects of these remarkable biomolecules. *Biotechnology Progress, 21*(6), 1593-1600. http://dx.doi.org/10.1021/bp050239p

Opawale, F. O., & Burgess, D. J. (1998). Influence of Interfacial Properties of Lipophilic Surfaktant on Water-in-Oil Emulsion Stability. *Jurnal Colloid and interface science, 197*, 142-150. http://dx.doi.org/10.1006/jcis.1997.5222

Panpanit, S. (2001). *Oily wastewater treatment by coupling membrane filtration and ozonation* (Doctoral dissertation, Asian Institute of Technology).

Ramaswamy, D., Kar, D. D., & De, S. (2007). A study on extraction of oil from sludge containing oil using froth flotation. *J. Environ. Manage, 85*, 150-154. http://dx.doi.org/10.1016/j.jenvman.2006.08.009

Staehelin, J., & Hoigne, J. (1982). Decomposition of ozone in water: rate of initiation by hydroxide ions and hydrogen peroxide. *Environmental Science & Technology, 16*(10), 676-681. http://dx.doi.org/10.1021/es00104a009

Yan, P., Lu, M., Yang, Q., Zhang, H. L., Zhang, Z. Z., & Chen, R. (2012). Oil recovery from refinery oily sludge using a rhamnolipid biosurfactant-producing *Pseudomonas*. *Bioresource Technology, 116*, 24-28. http://dx.doi.org/10.1016/j.biortech.2012.04.024

Yazdi, A. S. (2011). Surfactant-based extraction methods. *Trends in Analytical Chemistry, 30*(6), 918-929. http://dx.doi.org/10.1016/j.trac.2011.02.010

Young, L. Y., & Caniglia, C. E. (1995). *Microbial transformation and degradation of toxic organic chemicals* (pp. 27-74). New York: Wiley-Liss.

Zhang, J., Li, J., Thring, R. W., Hu, X., & Song, X. (2012). Oil recovery from refinery oily sludge via ultrasound and freeze/thaw. *Journal of hazardous materials, 203*, 195-203. http://dx.doi.org/10.1016/j.jhazmat.2011.12.016

Zubaidy, E. A., & Abouelnasr, D. M. (2010). Fuel recovery from waste oily sludge using solvent extraction. *Process Safety and Environmental Protection, 88*(5), 318-326. http://dx.doi.org/10.1016/j.psep.2010.04.001

16

Transesterification of Soybean Oil with Ethanol Using Heterogeneous Catalysts Based on Hydrotalcites

Evelyn Alves Nunes Simonetti[1] & Gilberto Garcia Cortez[1]

[1] Escola de Engenharia de Lorena, University of São Paulo, Lorena, São Paulo, Brazil

Correspondence: Evelyn Alves Nunes Simonetti, Escola de Engenharia de Lorena, University of São Paulo, Lorena, São Paulo. E-mail: evelyn@dequi.eel.usp.br

Abstract

This work evaluated the potential of hydrotalcites based in Ca-Al on trasesterification of soybean oil with ethanol. The reaction was carried out at 78.5 °C with ethanol:oil molar ratio of 30 and 10 wt% of catalyst. The 2Ca-Al showed better performance than 3Ca-Al catalyst achieving conversion of 94%. The catalyst were characterized with XRD, BET and activity was evaluated on isopropanol decomposition reaction.

Keywords: biodiesel, ethanol, transesterification, hydrotalcite, soybean oil

1. Introduction

The search for renewable sources has opened a new field to fuels development and the biodiesel has stood out due to it same properties of convectional diesel and for the facility of being used directly in diesel engines or in mixtures. These properties are possible by the presence of mono-alkyl esters of long chain fatty acids on biodiesel structure. Another important fact is the biodiesel combustion, it has low emission of harmful gas such as sulfur oxide which is a great advantage to the environmental system (Gerogogianni et al., 2009; Helwani et al., 2009). Transesterification is the cheaper and easy route of biodiesel synthesis and it consists in a reaction between vegetable oil or animal fat with alcohol (methanol or ethanol) to form ester and glycerol. The stoichiometry is 3:1 (alcohol: triglycerides) and it is common to use an excess of alcohol to improve the yield (Enweremadu & Mbarawa, 2009). Transesterification is often applied as a homogeneous reaction using acid and basic catalysts, such as sulfuric and hydrochloric acids, KOH and NaOH (Meher et al., 2006; Demirbas, 2009). The advantages of these catalysts are low reaction time and high yields. Although there are many disadvantages such formation of soup, elevated ratio alcohol:oil, impossibility of reuse the catalyst and difficulty in separating the catalyst from the azeotropic system. To avoid these problems countless studies have been published to develop new heterogeneous catalysts - such like metal hydroxides, oxides, zeolites, supported catalysts and hydrotalcites (Lee & Saka, 2010; Semwal et al., 2011). Among these catalysts the hydrotalcites (LDHs) has opened a new source in heterogeneous catalysis due it innumerous composition and applications. The layered double hydroxides or hydrotalcites (LDHs) can be represented by the following general formula: $[M_{1-x}^{2+}M_x^{3+}(OH)_2]^{x+}A^{m-}_{x/m}nH_2O$. The structure of hydrotalcites is based upon layered double hydroxides with brucite like ($Mg(OH)_2$) hydroxide layers containing octahedrally coordinated M^{2+} and M^{3+} cations (Crepaldi & Valim, 1998). The thermal processing of HDL promotes the formation of metastable crystalline phases, such as the spinel phase and/or mixed oxides, forming acid and/or basic sites (Vaccari, 1998; Goh & Lim, 2008). Studies developed identified four different types of active sites identified in calcined HDL samples: isolated O^{2-} ions, pairs of $Mg^{2+}O^{2-}$ and OH^- groups and acid sites (Al^{3+}) (Di Cosimo et al., 1998). Many researches have been done on transesterification catalyzed by LDH. The most common hydrotalcite applied is the Mg-Al using methanol and vegetable oil as triglyceride source. It was achieved yields between 67-90.5% (Siano et al., 2006; Xie et al., 2006; Zeng et al., 2008). Another types of hydrotalcite can be used on transesterifcation such like MgCoAl-LDH. In this particular study was used canola oil and ethanol, the results showed elevated yields at high ethanol:oil ratio (Li et al., 2009).

The objective of this study is to evaluate the catalytic performance of hydrotalcite in the transesterification reaction using Ca-Al hydrotalcite obtained by continuous precipitation. It was evaluated the catalytic activity of the compounds with different molar ratio of Ca/Al.

2. Materials and Methods

2.1 Catalysis Synthesis

The hydrotalcite precursors were prepared from aqueous solutions containing $Ca(NO_3)_2.6H_2O$ and $Al(NO_3)_3.9H_2O$. The atomic ratio, y = Ca^{2+}/Al^{3+}, was equal to 2 and 3. An aqueous solution of 1.5 mol/L of K_2CO_3 was used as precipitating agent. These two solutions were maintained at 55 °C and mixed in a reactor keeping pH of 10.0±0.4 during all process. The precipitate was kept at 40 °C for 24 hours under slow agitation. Then, it was separated by vacuum filtration and washed with deionized water up to neutral pH. The catalysts were calcined at 450 ºC and named as 2Ca-Al calcined and 3Ca-Al calcined.

2.2 Surface Area and Pore Volume

Samples were treated at 300 °C for 3 hours in order to remove chemicals adsorbed. The surface area and pore volume were analyzed in an equipment BET, model Belsorp II, using nitrogen adsorption at 196 °C and relative pressures (P/P_0) ranging from 0.05 to 0.35. The pore distribution was determined using the BJH method from the adsorption/desorption isotherms and the specific area by BET method.

2.3 X-ray Diffraction (XRD)

X-ray diffraction were performed using the powder method an SEISERT equipment, model 1001, radiation CuKα (λ= 1.54178 Å) and nickel filter. Crystalline phases were identified using JCPDS (Joint Committee of Powder Diffraction Standards, 1983).

2.4 Isopropanol Decomposition Reaction

The catalytic performance was evaluated through the decomposition of isopropanol, using a fixed bed reactor (quartz), with a continuous flow of reagents at atmospheric pressure. The activation of the catalyst was performed in situ at 200 °C for 2 h at oxygen (AGA, 99.9999%) atmosphere. The reactor was fed with isopropanol (99.7%, Merck) using a pump Thermo Separation Products, Model Spectra P100, at 0.05 mL of flow, diluted in a mixture of He and O_2 (molar ratio 3/1) at a flow rate of 74 $mL{\cdot}min^{-1}$. The reaction was carried out at a range of 170 °C to 310 °C, using catalyst mass of 100 mg and a residence time factor (W/F_{A0}) equal to 2.4 $g.h/mol_{Isop}$. Gases released by the catalyzed reaction were analyzed on line by a chromatograph Varian GC-3350, equipped with a thermal conductivity detector (TCD) and provided with a column packed Carbowax 20M, operating with a heating temperature between 30 and 150 °C. Helium was used as carrier gas in the column.

The conversion (X_A), selectivity (S_p) and specific activity (mols of converted products) were calculated using Equations 1-3 and they were based on the input and output concentration and assuming a differential reactor:

$$X_A(\%) = \left(\frac{n_I}{n_{Io}}\right)100 \tag{1}$$

$$Sp(\%) = \left(\frac{n_P}{n_I}\right)\left(\frac{N_{CP}}{N_{CI}}\right)100 \tag{2}$$

$$\text{SpecificActivity} = \frac{F_{Ao}X_AS_p}{W_AS_g} \tag{3}$$

Where:

n_I : mols of isopropanol consumed;

n_{Io}: mols of isopropanol fed;

N_{CP}: mols of carbon atoms formed;

N_{CI}: mols of carbon atoms in isopropanol;

F_{Ao}: mols of isopropanol fed per minute;

W_A: catalyst mass;

S_g: catalyst specific area (m^2/g).

2.5 Transesterification with Ethanol

The catalysts were activated in an oven at 200 ºC for 24 h and the transesterification reactions were conducted in a spherical reactor (500 mL) equipped with a mechanical stirring. The amount of catalyst used in each experiment was 10% relative in mass of soybean oil. The experiments were performed in fixed conditions of temperature (78.5 °C) and molar ratio of 1 to 30 (oil/alcohol). In many works it is common to use a molar ratio up to 15:1 (alcohol:oil). Although, this proportion is more usually in reactions with methanol. In this work we use ethanol as alcohol font which is less reactive than methanol, so it was required a higher molar ratio. After the reaction the catalyst was separated by vacuum filtration. Then, distilled water was added to filtrate and homogenate. The upper phase was separated and washed twice with distilled water. Then, the sample was centrifuged and trace of methanol and ethanol were removed by distillation. The biodiesel was finally dried with anhydrous sodium sulfate and stored at 4 °C for further analysis.

2.6 Proton Nuclear Magnetic Resonance (^1H NMR)

The biodiesel samples were diluted in deuterated chloroform (CDCl 3, Aldrich) and analyzed in a spectrometer model Varian MERCURY 300 MHz operating in the region from 4.4 to 4.0 ppm. The absorption signals of chloroform were used as internal reference for the standard scale. The values of chemical shifts are expressed in units (ppm), coupling constants (J) and Hertz (Hz).

The conversion in ethyl ester was evaluated based on the signals present in the region of 4.05 to 4.35 ppm of ^1H NMR spectrum. In the ^1H NMR spectrum of soybean oil, this region represents the methylene hydrogens of triglyceril (Garcia, 2006). According to Equation 4 the resonance peak of the etoxyl hydrogen atoms of the ethyl esters split into a quartet. The area under the first and fourth peak is $A_t/8$ and that one under the second and third is $3A_t/8$, where A_t is the sum of areas under the four peaks. The fourth component (A_{c4}) of the quartet of ethyl esters (range 4.07 to 4.08 ppm) is a single peak that does not suffer overlap from any other peak, so that component was used to calculate the conversion of ethanolysis vegetable oils.

$$EE(\%) = \left(\frac{8A_{c4}}{A_{dd+ee}} \right) 100 \tag{4}$$

Where:

A_{c4} = area of the component fourth peak;

A_{dd+ee} = area of all signals between 4.35 and 4.05 ppm;

EE(%) = percentage of ethyl esters in vegetable oils.

3. Results and Discussion

3.1 X-ray Diffraction (XRD)

Figure 1 shows the X-ray diffraction (XRD) profiles of the 2Ca-Al not calcined, 2Ca-Al calcined, 3Ca-Al not calcined and 3Ca-Al calcined samples. It can be observed that the samples 2Ca-Al not calcined and 3Ca-Al not calcined showed small peaks at 15°, 18°, 20°, 23° which are associated to layered structure. These peaks exhibit low intensity when compared to the conventional hydrotalcite Mg-Al. There are also the peaks related to the formation of $CaCO_3$ crystals (calcite) in 29°, 47° and 48° (JCPDS 72-1652). After the calcination process the peaks related to layered structure disappear in 2Ca-Al and 3Ca-Al calcined samples. The elevated temperature of drying and calcination increases the rate of water and carbon dioxide released. This phenomenon damages the solid structure leading to a loss of the lamellar arrangement. Besides, the preparation method not ensure the uniform mixing of the Ca^{2+} and Al^{3+} and the Ca^{2+} cations can agglomerate, favoring the formation of $CaCO_3$ crystals (Frost et al., 2009; Yu et al., 2011; Gao, 2010).

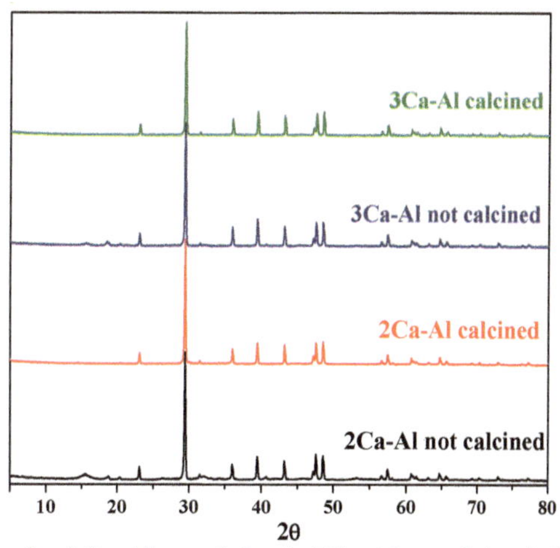

Figure 1. XRD patterns of samples 2Ca-Al uncalcined, 3Ca-Al uncalcined, 2Ca-Al calcined at 450 °C and 3Ca-Al calcined at 450 °C

3.2 Surface Area and Pore Volumes/Adsorption-Desorption

Figure 2 shows the nitrogen adsorption/desorption isotherms measured at -196 °C on 2Ca-Al calcined and 3Ca-Al calcined. Table 1 shows the characteristic parameters of the porosity of these catalysts, such as specific surface areas (S_{BET}) and total pore volumes (V_p). The isotherms can be classified as type III (IUPAC classification – International Union of Pure and Applied Chemistry). The shape of these isotherms is characteristic of non-porous or macroporous particles which interact very weakly with the adsorbent molecules. It also related to system with adsorption heat virtually nil and related to small gas volume - this can be observed in the sintered samples with surface area less than 5 m^2/g (Sing et al., 1982; Santilli & Pulcinelli, 1993). The hysteresis type is H3 and there are mainly two cases in which this type can be formed: very wide capillaries having narrow openings and an interstice between the parallel plates. It is also found in adsorption isotherms of many crystalline metal oxide aggregates of sheet-like shape. The slit width increase during adsorption and decreases irreversibly during desorption (Kaneko, 1994). Figure 2 also shows that the sample 2Ca-Al decreases the amount adsorbed as well the values of surface area and specific pore volume when compared to 3Ca-Al as showed in Table 1. Despite the smaller specific surface area the sample 2Ca-Al has a larger average pore diameter in accordance with Table 1. This may suggest that the 2Ca-Al catalyst has pores wider than those in 3Ca-Al catalyst and also has a smaller number of pores (Teixeira et al., 2001).

Table 1. Values of specific surface area (S_{BET}) and specific pore volume (V_p) of the samples 2Ca-Al calcined and 3Ca-Al calcined

Catalyst	S_{BET} (m^2/g) (BET)	V_p (cm^3/g) (BJH)	Pore Diameter (nm)
2Ca-Al calcined	6.60	0.07	45.10
3Ca-Al calcined	13.7	0.11	32.18

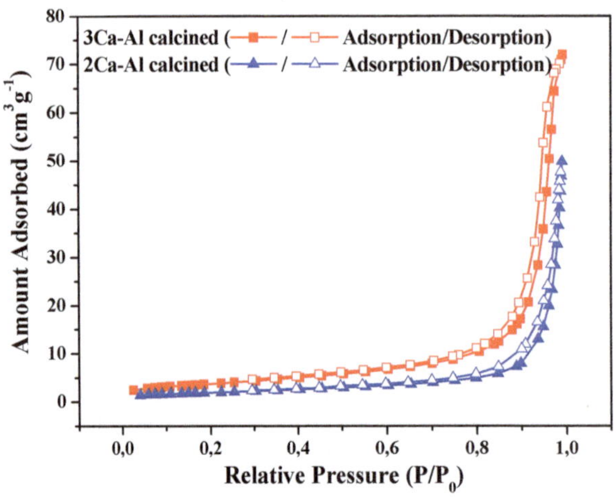

Figure 2. Nitrogen adsorption/desorption isotherms at -196 °C of samples 2Ca-Al calcined and 3Ca-Al calcined

3.3 Pore Size Distribution

Figure 3 shows the pore size distribution (estimated by N_2 adsorption) of the 2Ca-Al calcined and 3Ca-Al calcined catalysts. It can be note that the porous structure of 2Ca-Al calcined and 3Ca-Al calcined are basically composed of pore in the mesoporos region (2 nm < w <50 nm) with multimodal distribution indicating heterogeneity in the samples. In this region can be observed the presence of two peaks in about 10 nm for the 2Ca-Al calcined and 50 nm for the 3Ca-Al calcined sample.

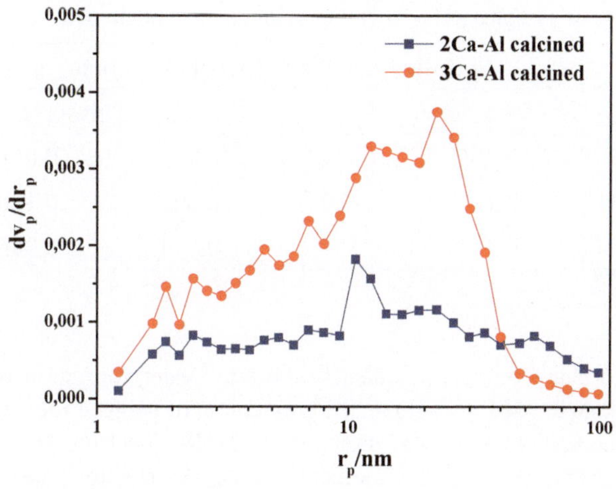

Figure 3. Pore size distribution of the samples 2Ca-Al calcined and 3Ca-Al calcined

3.4 Isopropanol Decomposition Reaction

Table 2 shows the results of specific catalytic activity of the catalysts evaluated in isopropanol decomposition reaction at 250 °C, 280 °C and 300 °C. The main products were propylene and diisopropyl ether in both catalysts. The 2Ca-Al calcined sample showed higher specific catalytic activity at different temperatures when compared to the 3Ca-Al calcined for both products, propylene and diisopropyl ether according to Table 2. Isopropanol conversion is often applied as catalytic test to obtain the effective catalytic properties of acid and base sites. Thus, the formation of propylene (from dehydration of isopropanol) or acetone (from dehydrogenation) provides information about the presence of acid and basic (or redox) centers, respectively (Gervasini et al., 1997). It can be observed that all samples showed the formation of propylene and diisopropyl ether. These products are related to dehydration of isopropanol using acid sites of the catalyst (Gervasini et al., 1997). The main reason to the acid sites are associated with great amount of aluminum ions represented to the high density of $Al^{3+}O^{2-}$ groups on the surface. However the basic groups are provides by MgO groups on the surface of the conventional hydrotalcites (Diez et al., 2003; Aramendia et al., 1996). The absence of basic represented by $Ca^{2+}O^{2-}$ groups is related to the bond between the calcium and carbonate. Both mechanism of propylene and diisopropyl formation are correlated to the abstraction of the alcohol molecule on catalysts surface by the acid site. This abstraction promotes the formation of cabanions and then olefins and ether (Robles-Medina et al., 2009). At 250 °C the diisopropyl formation was more intense with 3Ca-Al catalyst however at 280 °C and 300 °C the formation was much more intense with 2Ca-Al. The mechanism of ether and propylene formation can be represente below, were A represents an acid site and B a basic site.

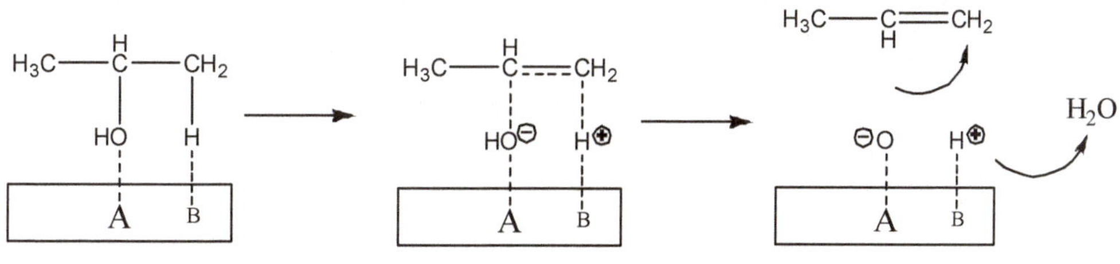

Figura 4. Mechanism of propylene formation on decomposition of isopropanol

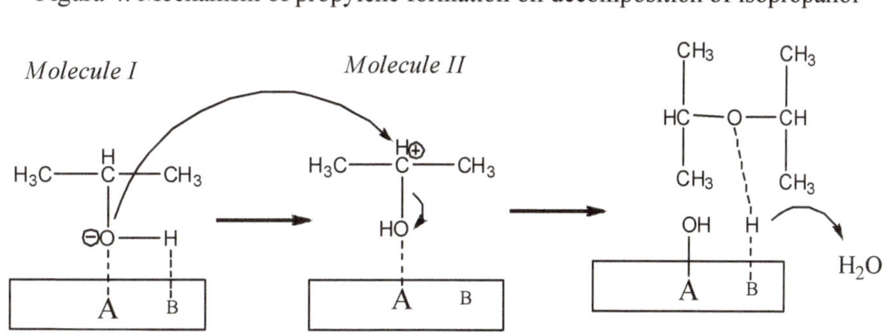

Figura 5. Mechanism of ether formation on decomposition of isopropanol

Table 2. Specific catalytic activity of samples 2Ca-Al calcined and 3Ca-Al calcined

Catalyst	Specific Catalytic Activity (10^{-4} mol m^{-2} min^{-1})					
	Propylene			Diisopropyl Ether		
	250 °C	280 °C	300 °C	250 °C	280 °C	300 °C
2Ca-Al calcined	0.110	0.440	1.400	0.002	0.140	0.660
3Ca-Al calcined	0.044	0.234	0.564	0.044	0.085	0.239

3.5 Transesterification Reaction

^{1}H NMR technique was used to characterize and quantify the ethyl ester formation by the transesterification reaction of soybean oil. The technique used to evaluate the conversion of trilgycerides in ethyl esters is based on the signals presents in the region of 4.05 to 4.35 ppm of ^{1}H NMR spectrum (Garcia, 2006). This region represents the methylene hydrogens of triglycerides molecules of soybean. After transesterification reaction the signals of etoxyl hydrogen of ester molecules takes place. The Figure 4 shows the spectra of the reactions catalyzed by 2Ca-Al, 3Ca-Al calcined and soybean oil. It can be observed that the transesterification catalyzed by 2Ca-Al calcined showed the typical spectra etoxyl hydrogen of ethyl ester molecule. When compared with the soybean spectra it can be observed that there is no overlap with the signals of the triglyceride molecules indicating high conversion in ethyl esters. Table 3 shows the integration values of signals between 4.06 and 4.35 ppm where conversion with 2Ca-Al calcined reached 94%. The spectra of the reaction catalyzed by 3Ca-Al calcined showed an overlap of signals between the methylene hydrogens of triglycerides and the signals of etoxyl hydrogen of ethyl ester molecule. It can be observed that the 3Ca- Al calcined spectra approaches to the soybean spectra indicating low conversion. The table 3 shows that the reaction catalyzed by 3Ca-Al reached only 31% of conversion.

Table 3. Integration peaks and the conversion percentage determined by ^{1}H NMR

Catalyst	Area A_{c4}	Area A_{dd+ee}	Conversion (%)
2Ca-Al calcined	1.06	7.58	94
3Ca-Al calcined	0.04	0.99	31

The addition of calcium in the 3Ca-Al catalyst was expected to enhance the conversion of ethanolysis due to the creation of basic sites which are more active than acid sites, but the effect was the opposite. The carbonate present in the samples displays no particular catalytic activity on the reaction. This function belongs to the mixed oxides after calcination (Xie et al., 2006). The Al^{3+} cations occupy the octahedral sites in hydrotalcites before calcination and the structure rearranges at high temperature in a tetrahedral form combined with Ca^{2+} cations. A positive charge is generated by this rearrangement and it is compensated by the formation of two types of defects: cationic vacancies and/or inclusion of intersticial oxygen which promotes the formation of sites like $Al^{3+}O^{2-}$ or $Ca^{2+}O^{2-}$ (Xie et al., 2006). In this study the main site formed was related to $Al^{3+}O^{2-}$ as evidenced in isopropanol decomposition reaction and the calcium content promotes the formation of carbonate. Adding more calcium in structure enhances the formation of carbonate and decreases the aluminum sites. With less active site available the access to the bulky triglyceride molecules decreases and then the conversion in ethyl esters is lower. Another important factor are the wider pores present in the 2Ca-Al catalyst which provides more contact with the triglyceride molecules and then increase the conversion in ethyl esters. Despite the higher surface area of 3Ca-Al catalysts it does not contributed to the elevation of conversion. The amount of carbonate formed occupies a large extension of catalyst inhibiting the action of the active sites and blocking contact with the triglyceride molecules.

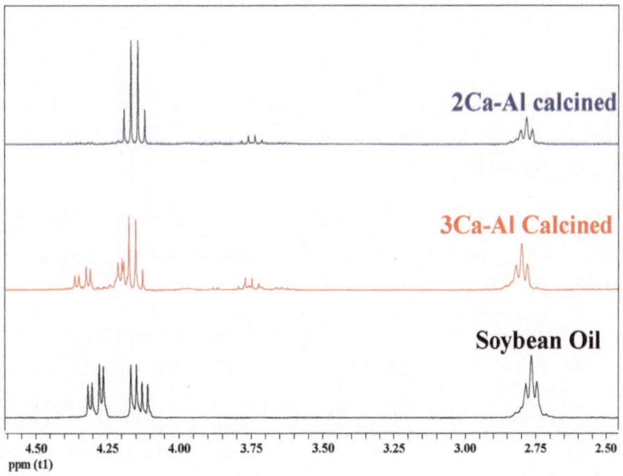

Figure 6. ^{1}H NMR spectra of soybean oil and ethyl ester catalyzed

4. Conclusion

In this work was proposed the use of heterogeneous catalysts using hydrotalcites in soybean oil ethanolysis. The Ca-Al LDHs were prepared by the co-precipitation method using Ca/Al ratio of 2 and 3. The XRD spectrum showed the appearance of peaks related to $CaCO_3$ crystals in all samples – that can be the elevated temperature of drying and calcination used in this work. The isotherms were classified as type III characteristic of non-porous or macroporous particles which interact very weakly with the adsorbent molecules. The surface area of the catalyst increased with the addition of calcium content as also the pore volume. The 2Ca-Al showed higher specific catalytic activity than 3Ca-Al. The acid sites formed were probably related to the aluminum content. The calcium addition inhibits these acid sites decreasing the production of propylene and ether in the isopropanol decomposition reaction. The transesterification of soybean oil reached the conversion of 94% with 2Ca-Al catalyst and 31% with 3Ca-Al. The main reason for the difference in reaction yields is related to higher amount of active sites in the catalyst 2Ca-Al. The larger surface area of the catalyst 3Ca-Al did not transform in higher yields.

Acknowledgements

The authors gratefully acknowledge the financial assistance to CAPES.

References

Aramendia, M. A., Borau, V., Jimenez, C., Marinas, J. M., Porras, A., & Urbano, F. J. (1996). Magnesium Oxides as Basic Catalysts for Organic Processes: Study of the Dehydrogenation–Dehydration of 2-Propanol. *J. Catal., 161*, 829-838. http://dx.doi.org/10.1006/jcat.1996.0246

Crepaldi, E. L., & Valim, J. B. (1998). Hidróxidos Duplos Lamelares: Síntese, Estrutura, Propriedades e Aplicações. *Quim. Nova, 21*, 300-311. http://dx.doi.org/10.1590/S0100-40421998000300011

Demirbas, A. (2009). Progress and recent trends in biodiesel fuels. Energ. Convers. *Manage, 50*, 14-34. http://dx.doi.org/10.1016/j.enconman.2008.09.001

Di Cosimo, J. I., Diez, V. K., Xu, M., Iglesia, E., & Apesteguia, C. R. (1998). Structure and Surface and Catalytic Properties of Mg-Al Basic Oxides. *J. Catal., 178*, 499-510. http://dx.doi.org/10.1006/jcat.1998.2161

Díez, V. K., Apesteguía, C. R., & Di Cosimo, J. I. (2003). Effect of the chemical composition on the catalytic performance of MgyAlOx catalysts for alcohol elimination reactions. *J. Catal., 215*, 220-223. http://dx.doi.org/ 10.1016/S0021-9517(03)00010-1

Enweremadu, C. C., & Mbarawa, M. M. (2009). Technical aspects of production and analysis of biodiesel from used cooking oil-A review. *Renew Sust. Energ. Rev., 13*, 2205-2224. http://dx.doi.org/10.1016/j.rser.2009.06.007

Frost, R. L., Palmer, S. J., & Spratt, H. J. (2009). Thermal decomposition of hydrotalcites with variable cationic ratios. *J. Therm. Anal. Calorim., 95*, 123-129. http://dx.doi.org/10.1007/s10973-008-8992-4

Gao, L., Teng, G., Xiao, G., & Wei, R. (2010). Biodiesel from palm oil via loading KF/Ca-Al hydrotalcite catalyst. *Biomass Bioenerg., 34*, 1283-1288. http://dx.doi.org/10.1016/j.biombioe.2010.03.023

Garcia, C. M. (2006). *Vegetable oil transesterification.* (M. S. Thesis). Universidade Estadual de Campinas, Campinas, SP.

Gerogogianni, K. G., Katsoulidis, A. P., Pomonos, P. J., & Kontominas, M. G. (2009). Transesterification of soybean frying oil to biodiesel using heterogeneous catalysts. *Fuel Process Technol., 90*, 671-676. http://dx.doi.org/ 10.1016/j.fuproc.2008.12.004

Gervasini, A., Fenyvesi, J., & Auroux, A. A. (1997). Study of the acidic character of modified metal oxide surfaces using the test of isopropanol decomposition. *Catal Lett., 43*, 219-228. http://dx.doi.org/10.1023/A:1018979731407

Goh, K. H., Lim, T. T., & Dong, Z. (2008). Application of layered double hydroxides for removal of oxyanions: A review. *Water Resour., 42*, 1343-1368. http://dx.doi.org/ 10.1016/j.watres.2007.10.043

Helwani, Z., Othman, M. R., Azis, N., Fernando, W. J. N., & Kim, J. (2009). Technologies for production of biodiesel focusing on green catalytic techniques: A review. *Fuel Process Technol., 90*, 1502-1514. http://dx.doi.org/10.1016/j.fuproc.2009.07.016

Joint Committee on Power Diffraction Standards. (1983). *Powder Diffraction File. Swarthmore, PA International Center of Diffraction Data.*

Kaneko, K. (1994). Determination of pore size and pore distribution 1. Adsorbent and catalysts. *J. Membrane Sci., 96*, 59-58. http://dx.doi.org/10.1016/0376-7388(94)00126-X

Lee, J. S., & Saka, S. (2010). Biodiesel production by heterogeneous catalysts and supercritical technologies. *Bioresourse Technol., 101*, 7191-7200. http://dx.doi.org/10.1016/j.biortech.2010.04.071

Li, E., Xu, Z. P., & Rudolph, V. (2009). MgCoAl-LDH derived heterogeneous catalysts for the ethanol transesterification of canola oil to biodiesel. *Appl. Catal. B-Environ., 88*, 42-49. http://dx.doi.org/10.1016/j.apcatb.2008.09.022

Meher, L. C., Vidya Sagar, D., & Naik, S. N. (2006). Technical aspects of biodiesel production by transesterification-A review. *Renew Sust. Energ. Rev., 10*, 248-268. http://dx.doi.org/10.1016/j.rser.2004.09.002

Robles-Medina, A., González-Moreno, P. A., Esteban-Cerdán, L., & Molina-Grima, E. (2009). Biocatalysis: Towards ever greener biodiesel production. *Biotechnol. Adv., 27*, 398-408. http://dx.doi.org/10.1016/j.biotechadv.2008.10.008

Santilli, C. V., & Pulcinelli, S. H. (1993). Analise da Textura de Materiais Ceramicos A Partir das Isotermas de Adsorcao de Gases. *Cerâmica, 39*, 11-16.

Semwal, S., Arora, A. K., Badoni, R. P., & Tuli, D. K. (2011). Biodiesel production using heterogeneous catalysts. *Bioresourse Technol, 102*, 2151-2161. http://dx.doi.org/10.1016/j.biortech.2010.10.080

Siano, D., Nastasi, M., Santacesaria, E., Di Serio, M., Tesser, R., Minutillo, G., ... Tenore, T. (2006). Process for producing esters from vegetable oils and animal fats by using heterogeneous catalysts. WO 050925 A1.

Sing, K. S. W., Everett, D. H., Haul, R. A. W., Moscou, L., Pierotti, R. A., Rouquérol, J., & Siemieniewska, T. (1982). Reporting physisorption data for gas/solid systems with special reference to the determination of surface area and porosity. *Pure Appl. Chem., 54*, 2201-2218. http://dx.doi.org/10.1351/pac198254112201

Teixeira, V. G., Coutinho, F. M. B., & Gomes, A. S. (2001). Principais Métodos de Caracterização da Porosidade de Resinas à base de Divinilbenzeno. *Quim. Nova, 24*, 808-818. http://dx.doi.org/10.1590/S0100-40422001000600019

Vaccari, A. (1998). Preparation and catalytic properties of cationic and anionic clays. *Catal. Today, 41*, 53-71. http://dx.doi.org/10.1016/S0920-5861(98)00038-8

Xie, W., Peng, H., & Chen, L. (2006). Calcined Mg–Al hydrotalcites as solid base catalysts for methanolysis of soybean oil. *J. Mol. Catal. A-Chem., 246*, 24-32. http://dx.doi.org/10.1016/j.molcata.2005.10.008

Yu, C., Wang, C., Hsu, M., & Chyou, Y. (2011). Development of a novel Ca/Al carbonates for medium-high temperature CO2 capture. *Energy Procedia, 4*, 787-794. http://dx.doi.org/10.1016/j.egypro.2011.01.120

Zeng, H. Y., Feng, Z., Deng, X., & Li, Y. Q. (2008). Activation of Mg-Al hydrotalcite catalysts for transesterification of rape oil. *Fuel, 87*, 3071-3076. http://dx.doi.org/10.1016/j.fuel.2008.04.001

17

Improving Energy Efficiency and Environmental Sustainability of Commercial Insulation

Celine Manoosingh[1], James Mihelcic[2] & Manjriker Gunaratne[2]

[1] Georgia Southern University

[2] University of South Florida

Correspondence: Celine Manoosingh, Georgia Southern University. E-mail: cmanoosingh@gmail.com

Abstract

With increasingly stringent energy standards set in place by the Department of Energy, energy efficiency is becoming a paramount concern to manufacturers of appliances. Additionally, the production and disposal of the voluminous amount of polyurethane foam commonly utilized as insulation in refrigeration units poses a significant environmental challenge. In this context, this study investigated an alternative insulation for use in commercial refrigerator/freezer units. A prototype exploring the use of evacuated packets of pyrogenic silica substituting for conventional insulation was assessed. Assessment criteria included experimental comparison of heat transfer characteristics and the energy efficiency of the new insulation as well as its life cycle as it is related to environmental sustainability. Results indicate that in the new insulation design applied to the unit's cover, heat flux decreased by an average of 36%, and energy efficiency improved by 5.1% over a 24 hour period. The new insulation design also resulted in improved environmental sustainability, resulting in a savings of 0.257 metric tons of CO_2e over 20 years for a single unit. Results provide an alternative insulation design for use in commercial refrigerator and freezers, and a framework by which to assess the efficiency and environmental performance of similar products.

Keywords: life cycle assessment, insulation, refrigerator, greenhouse gas emissions, carbon dioxide, built environment, LEED

1. Introduction

In the United States, resident and commercial buildings together use more energy and emit more carbon dioxide than either the industrial or transportation sector. In fact, infrastructure in the United States now accounts for 41% of all energy needs, two thirds of electricity consumption, and one eighth of all water use (U.S EIA, 2011). Commercial refrigeration systems also account for approximately 8% of all commercial building energy use (DoE, 2011).

Since the introduction of mandated energy efficiency standards (e.g., in the United States, Energy Star) the goal of every refrigerator commercial and residential manufacturer has been to improve the design and processing of the refrigerator unit to improve efficiency and sustainability. This is especially important because the average refrigerator is comprised of over twelve different types of material, including voluminous amounts of polyurethane foam (AHRI, 2011). Energy improvements in the United States are typically measured in power savings, as expressed by KWh/day. The Environmental Protection Agency (EPA) also requires that the end-of-life of a unit include safe disposal and with minimal environmental impact. Unfortunately, the volume of insulation material, as well as variances in material composition make efficient reuse and recycling difficult. In fact, the process of completely recycling a commercial refrigerator is in fact both tedious and inefficient. In particular, the disposal of polyurethane foam used for insulation presents a significant environmental challenge. Polyurethane foam serves a dual purpose in commercial refrigerators. In addition to functioning as the primary insulator of the unit, foam is injected throughout the body of the unit to provide structural support. It has been traditionally used as the insulator of choice due to its relatively high 'R' value, or the ability of a material to resist heat. When polyurethane foam is incinerated, it emits carbon monoxide, nitrogen oxides, isocyanine, acrylonitrile, and hydrogen cyanide in levels exceeding those allowed by the EPA (EPA, 2011).

Significant strides have been made in improving the energy efficiency of residential refrigeration, but few are reported in the field of commercial refrigeration. While no literature was identified for this study on efforts to improve energy efficiency in non-residential refrigeration units, a substantive review of advances in alternative technology to improve the energy efficiency of residential unit is provided by Bansal et al. (2011). That study reviewed emerging technologies that have the potential to improve energy metrics in residential refrigerators. The review concluded that significant breakthroughs are needed for alternative mechanical systems to compete with traditional vapor compression technology. A key similarity in many publications evaluating energy efficiency improvements in refrigerators (e.g., Waide, 2002; Boardman 2004; Mahlia, 2010) is that a fundamental change is needed in the industry. In fact, many of the options explored by major residential refrigerator manufacturers involve the reduction of the amount of polyurethane foam used in the insulation.

Wu et al. (1998) investigated the heat transfer characteristics of polyurethane foam and showed that evacuating the air from the foam cells can reduce the thermal conductivity of the polyurethane foam in a refrigerator by up to 75%. While the idea of alternative types of insulation have been explored by many manufacturers producing residential refrigeration units, advanced insulation technologies to reduce the amount of heat gained by the refrigerator storage cabinet is limited by manufacturers' desire to achieve maximum internal volume for product storage. Additional limitations are cost of the insulation material, available manufacturing techniques and labor, and material degradation over time (DoE, 2010).

Vacuum paneling consists of a rigid core with cells of evacuated air and is a developing technology to increase energy efficiency. This type of insulation in buildings and other devices has gradually increased in recent years, as manufacturers have developed more economic panels. A redesigned vacuum panel reduces the volume of air molecules in an evacuated space, thus reducing the rate of heat transfer. There are two main approaches to vacuum technology: vacuum powder and vacuum paneling. Vacuum paneling can consist of powder- and fiber-filled panels, compact vacuum insulation (with stainless steel walls), or aerogels. Other insulation techniques explored in previous research are the use of baffle-type argo-filled panels, proven to substantially decrease thermal conductivity (Kudoh, 2006). When vacuum paneling was employed in residential refrigeration units, a 20.4% energy savings was achieved (EPA, 2009). Vacuum paneling prototypes were also tested by Eberhardt (2007) and Weaver et al. (1996) and they observed respective improvements in energy efficiency of 15% to 25%. An advanced version of polyurethane foam was also developed by Parenti (2007) that improved the efficiency of the foam insulation. It combines the idea of the vacuum panel with traditional injection molding. The insulation method employed in that study was found to increase energy efficiency by 35%.

Bouquerel, Duforestel, Baillis and Rusaouen, (2012) investigated heat transfer through a VIP in a building through empirical models. While the applications are very different in comparison to refrigerators, the heat transfer can be determined through similar models. There has been some debate as to the effectiveness of the filling materials used in VIPs. Kwon, Jang, Jung and Song (2009) examined the use of powder, foam, and fibers as filling materials for VIPSs. The results show that that the solid conductivities of the fiber and more solid types of filling for the VIP insulation are lower than those of the powder and foam due to the relatively long thermal path. It can concluded from examination of the literature that the higher density of the filling material of the VIP, the more effective the insulation is for the case of commercial refrigeration. While vacuum insulation panels are considered one of the foremost technologies for insulation in buildings, the technology is less than ideal for use in refrigerators due to the cost as well as the ability of vacuum insulation panels to be cut to fit specialized models. Another emerging technology in the insulation field is industrial grade silica beads and aerogels. Silica is well known for its thermal resistance. In a study by Fricke, Hummer and Scheuerpflug (1995), vitreous silica was compared to aerogels. The result showed an increase in radiative transfer through aerogels, but a significant decrease in thermal conduction with the use of aerogels. Vitreous silica was also proven to be an excellent inhibitor to heat transfer. While aerogels are not as yet feasible for use in commercial refrigeration systems due to its high cost, silica in the form of non-porous beads and fibers are still a viable option to consider.

Further, a study by Zhao (2006) investigated the radiative heat transfer properties of a silica aerogel composite. The results show that the effectiveness of this type of material is greatly increased when the inclination, diameter and length are configured based on experimental results. The optimum parameters combined with the fibrous silica utilized in the study produce an effective insulation. The idea of combining silica as the filling material for a vacuum panel was explored by Fricke et al. (1995). The authors assert that even though VIPs have a thermal resistance ten times that of foam-based insulation of the same thickness, the filling material should contribute to the heat- bearing load , and will then reduce some of the associated draw-backs related to VIPs—i.e.- the loss of vacuum over time. The study suggests fumed silica as a filling material, where the suggested silica material has a thermal conductivity of about 0.004 $W\,m^{-1}\,K^{-1}$ at room temperature. Another material consideration was employed by Yu, Li and Zhu (2009) where thermal performance of insulation was investigated through the walls of a low temperature cold box. The results of their testing indicate that a reflective heat shield significantly reduced heat losses by up to 18%. The study also showed a reduction in heat loss when aluminum foil was used as an alternative to the heat shield.

Accordingly, based on the literature review and the need to improve energy efficiency and environmental sustainability of refrigeration, the primary objective of this study is to explore commercial refrigeration design measures that will reduce the volume of polyurethane foam utilized in commercial-grade refrigeration units, while improving or sustaining key metrics such as the maximum allowable energy consumption, internal volume, and integrated average temperature. This is achieved through the development and laboratory testing of an alternative insulation prototype, evaluated by measurements of energy consumption and heat flux in a redesigned commercial refrigerator/freezer unit. The research also evaluates the redesigned insulation in terms of its environmental sustainability through the use of life cycle assessment. With increasingly stringent and mandatory efficiency regulations being employed by Energy Star, the California Energy Commission, and the United States Department of Energy, the need for redesign of key refrigerator components is vital. In fact, research shows that if all commercial and residential solid door refrigerators and freezers were just replaced with Energy Star models, greenhouse gas emissions would be reduced by 6 billion pounds (Energy Star, 2011). Hence, the market success of refrigerator manufacturers relies on innovative and efficient unit design.

2. Methods

2.1 Power Consumption and Heat Flux Testing

A commercial 5 $ft.^3$ Whirlpool refrigerator chest (Model EH070FXRQ), shown in Figure 1 (left), was outfitted with measurement equipment for testing and evaluation purposes. Heat flux sensors, Omega- HSF-4 (Stamford, Connecticut) and thermocouples, Omega- Type-T (Stamford, Connecticut) were installed in the internal refrigerator/freezer cabinet, in locations described in the right of Figure 1, to measure heat transfer and to monitor internal temperature respectively. Sensors 1 and 2 were placed in the cover of each unit, where the insulation design was varied depending on the unit; while sensors 3 and 4 were placed across the polyurethane insulation of the walls, that remained unchanged, to act as the control. A power meter, P3 Kill-A-Watt (New York, New York) was utilized to determine the total energy used by the unit over the testing period.

The refrigerator/freezer was placed, unmodified, in a testing chamber to a controlled ambient temperature of 75°F, relative humidity of 45% and air pressure of 101MPa, as prescribed by the ASHRAE 72-2005 testing standard (ASHRAE, 2005). This unchanged, traditionally insulated unit is referred in this study as 'Unit A'. Heat flux and power consumption were measured and recorded over a 24 hour period for Unit A. Details of the test are provided later in this manuscript.

After the testing of Unit A was complete, the cover of the refrigerator/freezer chest was removed; the polyurethane insulation located in the cover was removed and replaced with an alternative insulation configuration consisting of an insulation cartridge in which packets of evacuated silica are inserted into a honey-comb like recycled plastic structure. A reflective heat shield made of recycled polyester fiber encapsulated the cartridge, and a small amount of polyurethane was applied to seal the structure. This unit is referred to as 'Unit B'. Unit B was then placed in the testing chamber and was made to undergo the same heat flux and power consumption testing over a 24 hour period as Unit A.

In both the traditional unit (Unit A) and the modified unit (Unit B), the mechanism by which cooling occurs remained the same. A vapor-compression system was utilized, wherein the refrigerant travels along tubing through the compressor. From there, the refrigerant moves through smaller tubes to the evaporator. The movement of Freon from the compressor to the evaporator causes a significant drop in the pressure of the liquid, allowing it to absorb warm air from the internal compartment, changing from liquid to heated gas which then moves through the coils of the condenser and is released into the cooler ambient air. The cycle then refreshes itself when Freon loses heat and returns to a liquid state.

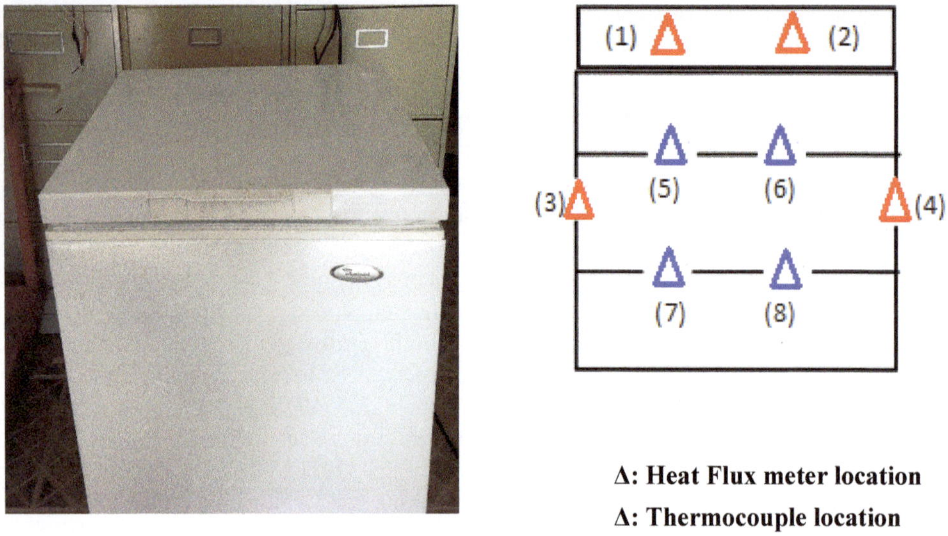

Figure 1. The testing unit was a 5 ft.^{3} commercial refrigerator/freezer (left) outfitted with a network of monitors (right) used to measure heat flux through insulation and internal temperature of the unit.

Polyurethane insulation was also used in the cover and walls of the traditional commercial refrigerator/freezer. Figure 2 illustrates the various materials that typically comprise the cover and walls of these units. In the system studied here, two separate metal trays measuring 20.7 in. × 23.8 in were coated with polystyrene to form the exterior of the unit. The process by which it was manufactured involved molding steel sheeting to form the inner and outer surfaces of the walls and chest cover. The inner plastic sheeting was molded separately and attached to the metal tray by plastic fasteners. Polyurethane was injected into the 3.35 in. cavity between the outer steel and inner plastic liner. The injected polyurethane took the form of foam and is inserted at a pressure of 4.5lbs/in^2. After a 36 hour setting time, this foam hardened between the liners and the steel strips and provided insulation and structural support to the unit.

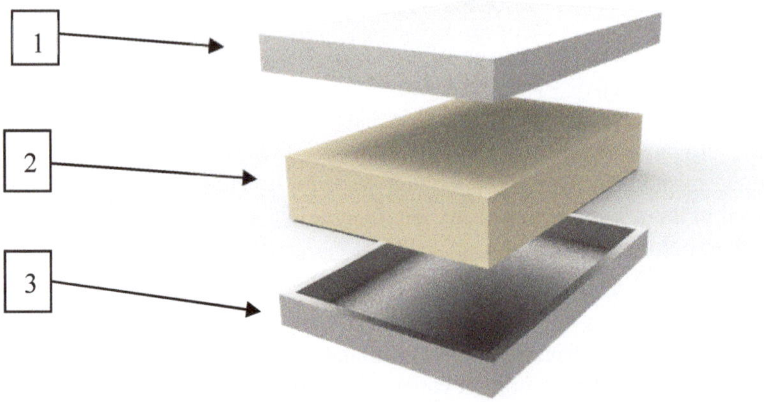

Figure 2. Traditional refrigerator cover and walls incorporates insulation that is comprised of liquid polyurethane foam that is injected in between two metal trays and allowed to harden. (1) external metal tray with an acrylic coating; (2) hardened polyurethane, the primary insulator; (3) opposite metal tray also coated with acrylic.

Figure 3 illustrates the idealized alternative insulation design, employed in the cover of Unit B. A specific assembly sequence to optimize energy efficiency as well as material recovery at the end of life was developed. The redesigned insulation utilized two separate production sequences before the final assembly of the refrigerator/freezer chest. The first part of the production sequence was the fabrication of the insulation cartridge. The insulation system employed a honey-comb cartridge; manufactured using vacuum formed recycled plastic obtained from the waste stream of the disassembly and recovery facility to form hexagonal cavities of size 5" with an outside 90 degree flange, 1" wide with predrilled holes. The recycled plastic 'honeycomb' mold was used for its ability to provide structural support to the unit.

Cavities within the honeycomb sheet were subsequently sprayed with heat release adhesive and fitted with 3 $in.^2$ hexagonal plastic packets of industrial grade fumed silica beads. Specialized industrial grade fumed silica was utilized due to its ability to absorb moisture, prevent conductive and convective heat losses, and regenerate after use. In the production of the silica packets, each plastic packet was subjected to the generation of a vacuum to a level of 3.4 kPa. The generation of a vacuum ensured that heat loss from the cold cabinet to the inside air was minimized. The honeycomb cartridge was manufactured with proportionate dimensions so that could be inserted into the external and internal refrigerator metal trays, the metal skin was also manufactured with an inside flange with predrilled holes proportionate to the cartridge.

The second part of the production sequence began with outfitting the metal skin of both the refrigerator/freezer walls and chest. Heat release adhesive was sprayed into space between the metal tray and a reflective heat shield was inserted into the tray for further insulation. The insulation cartridge manufactured in part 1 of the production sequence was fitted into the space between the metal tray and a small amount of polyurethane foam was injected into the area between the cartridge and the external metal. The cartridge was sealed into the metal tray with plastic panel retainer clips. The walls and the cover of the refrigerator/freezer were assembled through traditional means: e.g., welding at the points of contact. After the assembly was fully complete, the white plastic acrylic was fastened to the outside of the unit to maintain traditional aesthetics.

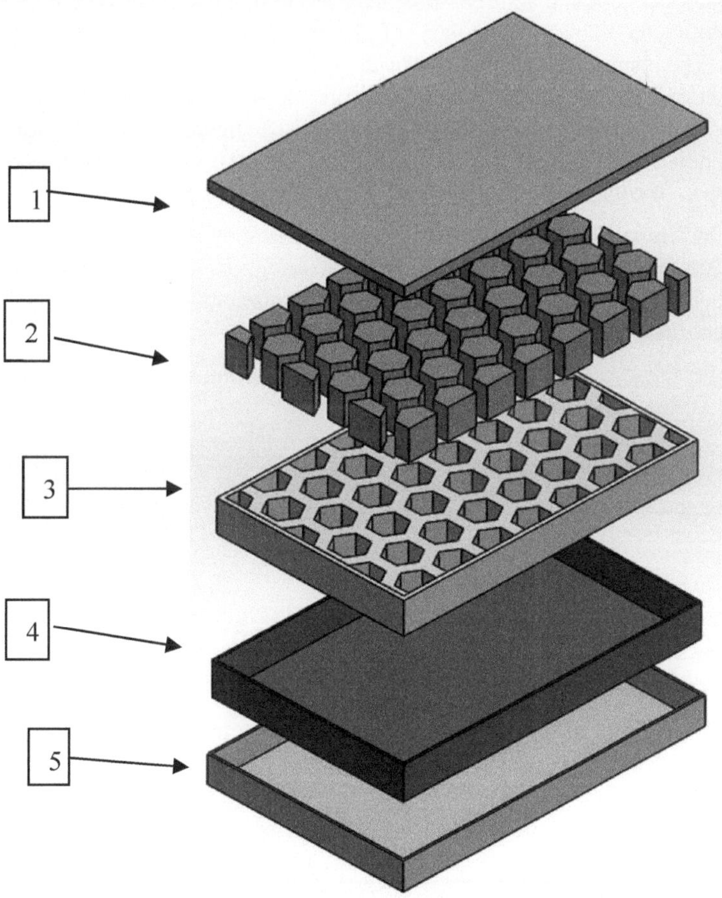

Figure 3. Commercial freezer chest cover. (1) external metal cover that is covered with a white acrylic sheet; (2) packets of evacuated fume silica are inserted into the recycled plastic cartridge (3); (4) reflective heat shield that encompasses the cartridge; (5) external metal tray that is also covered with a sheet of white acrylic for standard aesthetic purposes.

The refrigerator/freezer unit utilized in the testing measured 20.7 in. × 23.8 in. × 34.45 in., had a net weight of 117 lbs., and was outfitted with an 115V AC reciprocating hermetically-sealed compressor, automatic electrical defrost heater, and was charged with 140 grams of R134a refrigerant as a working fluid. The compressor was a single fixed-speed type, with thermostat based on and off control. The testing chamber contained the refrigerator unit, laptop computer, and other test equipment. The control room was outfitted with a thermostat, humidity gage, and room pressure monitor. The humidity, air pressure, and temperature in the room were brought to steady state conditions as prescribed by the ASHRAE standard 72-2005 which mandated a humidity level of 45%, air pressure of 101MPa, and a constant ambient temperature of 75° F.

Power consumption of each unit was also measured simultaneously. The temperature and heat flux measurements obtained from the measurements were recorded digitally at the top of every hour of the test. The testing began after 7 hours after the unit temperature had been stabilized under steady state conditions. The heat flux data across sensors 1 and 2 (labeled in Figure 1) were collected beginning at hour (7).

2.2 Methodology for Life Cycle Assessment

This component of the research examines the life cycle consequences associated with the two different types of insulation utilized in commercial refrigeration study. This study utilized SimaPro life-cycle inventory (LCI) software (Amsterdam, The Netherlands). SimaPro integrates ISO 14040: 2006 standards (International Organization for Standardization, 2006) for environmental management and standardized LCI formats to analyze model data. This version has built-in databases from various sources containing energy and emission data characteristics of those found in the United States and abroad. In addition to manufacturer data discussed in the previous section, these databases were utilized to quantify emissions from unit processes.

The manufacturing of a commercial refrigerator/freezer typically takes place in one central facility where a number of mechanical components, manufactured separately are assembled. The compressor, evaporator and condenser are usually manufactured by a subsidiary of the main company or a separate entity, and then assembled in the final unit. The insulation for the unit, as well as the metal trays coated with acrylic, are typically added at the main manufacturing and assembling facility before the final assembly takes place. The functional unit for this life cycle assessment was a generation of an internal temperature of -6°F for 200,000 hours of operation, under average ambient conditions of 75°F and 40% humidity. This is equivalent to roughly 20 years of operation, the typical life span of commercial refrigerator/freezer unit of this type.

The system boundary of the life cycle component of the research is shown in Figure 4. It includes the inputs from raw material acquisition, material processing, manufacturing, use, decommissioning, treatment and disposal and recycling, where material is salvaged from the decommissioning and recycling of the original unit for one cycle of remanufacturing.

In order to create a realistic life cycle assessment, the following assumpions about the manufacture, use and disposal scenarios for the two types of units are made:

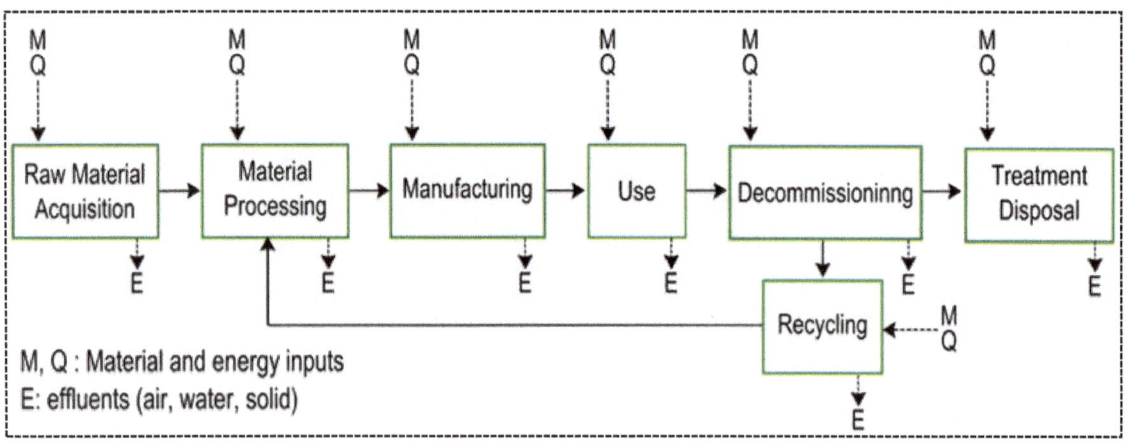

Figure 4. LCA system boundary of this research. The LCA encompasses material and energy inputs from raw material acquisition, material processing, manufacturing, use, decommissioning, recycling, treatment and disposal, and one cycle of remanufacturing.

Between January and July 2012, the primary data for the reporting year 2011 were collected from three representative manufacturers contacted for the study. The manufacturers were surveyed using a self-administered questionnaire based on study objectives, and in accordance with ISO protocol. The participating companies were ensured confidentiality, and all three participated in the study. The returned surveys indicated a combined production of 3.64 million 5 $ft.^3$ units was manufactured in 2011 by these factories combined, accounting for 58%

of the total production of these types of units internationally. Incomplete data sets were completed by conducting follow up telephone calls and interviews with representatives from each manufacturing plant.

Product yields reported in the survey showed how the input materials were manufactured into products, co-products and waste. It was observed by manufacturers that approximately 65% of the input materials were directly utilized in unit production. The other 35% was scrapped and quite frequently ended up being disposed of in a landfill. The manufacturer surveys also indicated that all deliveries of raw materials to the manufacturer facility were by truck. It was found by this research that manufacturers typically selected raw material suppliers based on geographic proximity to decrease costs associated with transportation. Environmental burdens associated with this transportation are included in the cumulative system boundary assessment. Energy and material utilized as packaging material are not included within the bounds of this study.

It should be noted that each factory outsourced the manufacturing of some components of the refrigerator/freezer, but were asked to include this data in the data reporting. For the purposes of this study, independent companies that supply components or whole parts to the parent factories were included in the data reported by those respective parent companies. The factories were also asked to document what data were unavailable with regard to reporting this type of information; incomplete data was assumed based on existing relevant sources from built-in inventory in SimaPro. The data sets available in SimaPro were chosen based on their geographic relevance to this study. The complete list of sources utilized is listed in Table 1. The manufacturer surveys proved to be quite useful in determining the manufacturing techniques and processes.

Table 1. LCA unit processes and corresponding data sources

LCA Unit Process	Data Source
Fuel, Energy and Raw Material Extraction	
Input materials	Manufacturer survey
Raw material extraction processing	Ecoinvent Database: PRé Consultants and Sylvatica, 2001
Energy/fuel utilized in processing	Ecoinvent Database: PRé Consultants and Sylvatica, 2001
Emissions	Ecoinvent Database: PRé Consultants and Sylvatica, 2001
Refrigerator Component Manufacturing	
Fuel: assembly and processing	Manufacturer survey
Emissions	Ecoinvent Database: PRé Consultants and Sylvatica, 2001
Transportation	
Fuel	Manufacturer survey
Emissions	USLCI database: PRé Consultants and Sylvatica, 2001
Installation	
Tooling emissions	USLCI database: PRé Consultants and Sylvatica, 2001
Operation	
Life span energy use	Manufacturer survey
Energy and material use: reasonable expectation of repairs	Manufacturer survey
Emissions from tooling	USLCI database: PRé Consultants and Sylvatica, 2001
Disassembly and Recycling	
Input materials and general process	Manufacturer survey
Energy use for disassembly	USLCI database: PRé Consultants and Sylvatica, 2001
Incineration emissions	USLCI database: PRé Consultants and Sylvatica, 2001
Landfill consequences	USLCI database: PRé Consultants and Sylvatica, 2001
Reuse processing	USLCI database: PRé Consultants and Sylvatica, 2001

To ensure quality of data, the three manufacturers surveyed in the research were also chosen from three different countries and regions used similar, but not identical machinery for extraction, manufacturing and assembling purposes. While many of these processes were unique, several processes for specific unit processes overlapped. Thus, in order to normalize the data for material, energy and emissions for each unit process, the overlapping data were averaged, and the mean of the data used as the primary input for the life cycle assessment. The collected data were used to construct the basic life cycle inventory for the traditional model. "Unit A" employs traditional injected polyurethane as the insulation and "Unit B" utilizes a non-traditional insulation applied to the cover and walls.

Eco-indicator 99 (H) was chosen as the primary environmental impact assessment method. It was selected because of its relevance to chief environmental sustainability concerns associated with the production of different components of the commercial refrigerator. Indicators such as global warming potential, depletion of natural resources, and overall energy use were all determined to be critical metrics in determining whether a more sustainable design than Unit A is achieved.

3. Results and Discussion

3.1 Energy Consumption and Heat Flux Testing

Following an initial temperature anomaly at the commencement of testing (during hour 2) the ambient temperature remained at 75° F with an average deviation of 0.1°F. There was a significant decrease in the internal temperatures of the units due to a 7 hour stabilization period where the internal cabinet air went from ambient room temperature to a standard operating temperature of -11°F. Accordingly, heat flux and power consumption monitoring commenced at hour 7.

Figure 5 shows the results of the heat flux measured at sensor 1, the sensor closest to the west wall of the unit (refer to Figure 1). The heat flux for the polyurethane through this sensor fluctuated from 5.3W/m^2 at hour 18 to 6.4W/m^2 at hour 15 to. By comparison, the heat flux through this sensor for Unit B varied from a minimum heat flux of 3.5 W/m^2 at hour 8 to a peak of 3.9 W/m^2 at hour 10. This suggests that the insulation utilized in Unit B had a more consistent thermal resistance.

Figure 5. Time variation of heat flux through the cover of testing units A and B recorded at ensor 1

The time variation of the heat flux measured by sensor 2 is shown in Figure 6. Both sensors display similar trends in heat flux fluctuation with that of Unit A being more significant. Figure 6 shows that the lowest level of heat flux observed by sensor 2 in unit A was 5.5 W/m^2 occurring at hour 8. The minimum heat flux for sensor 2 was 3.4 W/m^2 at hour 10 and the maximum was 6.7 W/m^2 at hour 19. For Unit B, There was a significant increase in heat flux through the insulation of Unit A between hour 18 and 19, where an increase of 1.1 W/m^2 was observed.

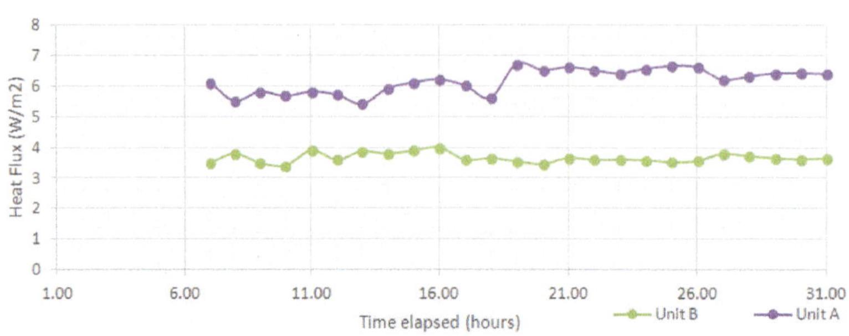

Figure 6. Heat flux through the cover of testing unit A and B recorded at sensor 2

When comparing Unit A to Unit B, it was observed that there was an average decrease in heat flux of approximately 40% across sensor 1 and sensor 2, where the insulation varied from polyurethane in Unit A, to the modified insulation in Unit B. Additionally, there is less fluctuation in recorded values of heat flux in Unit B that is characteristic of a more uniform heat resistance configuration. This suggests that the honeycomb structure in conjunction with the primary insulator of vacuumed silica packets is an effective pairing for the purposes of thermal efficiency.

Because the ambient conditions and the internal temperature of the refrigerator/freezer remained more or less constant during the testing, the fluctuations in heat flux across each type of tested insulation are indicative of the effectiveness of the insulation materials and configuration. Since the insulation material had not changed across sensors 3 and 4, this is further supported by the heat flux data collected by the sensors located on the unit walls (Figures 7 and 8). The conventional insulation was not replaced in the walls of Unit B; this type of polyurethane insulation was utilized as the primary insulator for both Unit A and Unit B (except for the cover of Unit B). As shown, the heat flux remains relatively constant through Unit A and Unit B, and indicates that the heat flux was heavily dependent on the insulation material utilized in testing.

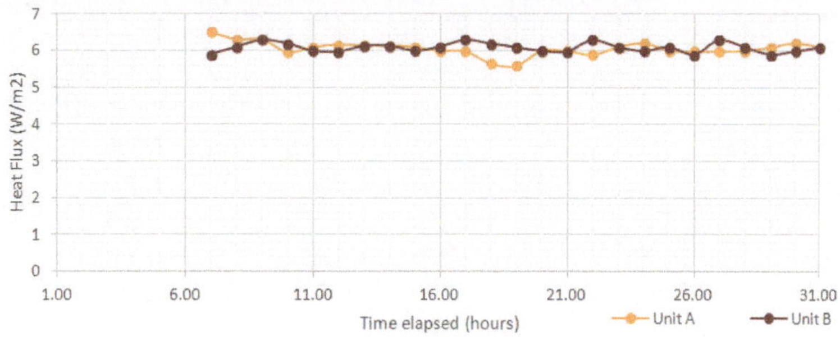

Figure 7. Heat flux through the cover of testing unit A and B recorded at sensor 3. The heat flux through the east wall (sensor 3) of the refrigerator/freezer unit remained relatively constant for Unit A and Unit B. Both units employed polyurethane insulation as the primary insulator.

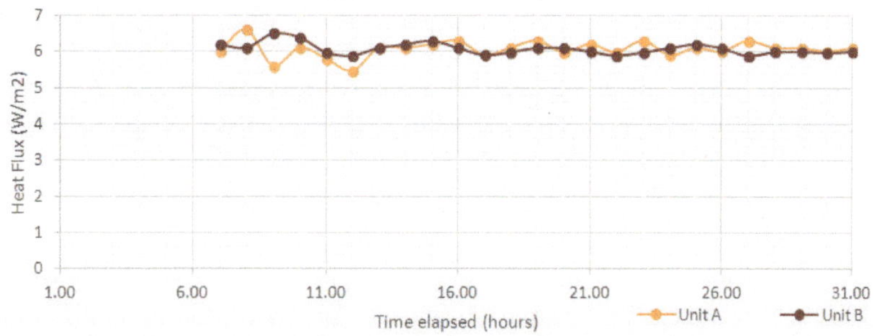

Figure 8. Time variation of heat flux through the cover of testing unit A and B recorded at sensor 4

Results of the power consumption measurements for both units are provided in Figure 9. The power meter 7 hours after the internal temperature had stabilized. Figure 9 shows the power consumption was consistently higher for Unit A. In fact, cumulative power consumption after 24 hours of testing was recorded to be 0.92 kWh for Unit A and 0.869 kWh for Unit B. The results indicate there is a marked increase in thermal and energy efficiency achieved by the use of the modified insulation in Unit B (when placed only in the unit's cover). On average, the alternative insulation performed 25-40% better with direct heat transfer. Consistent with this trend, there was an observed approximate 5% reduction in overall energy usage per day. This is significant when considering the annual power consumption of a commercial refrigerator/freezer and the power consumption over the life time of the unit.

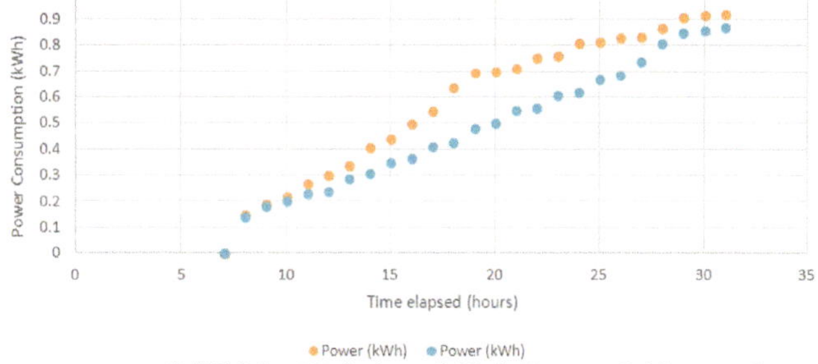

Figure 9. Cumulative power usage (kWh) by Unit A and Unit B over 24 hours. Power usage was recorded on an hourly basis as each separate test was conducted

3.2 Life Cycle Assessment

Figure 10 shows the characterization of Unit A and Unit B with respect to each key environmental sustainability impact category by the Eco-Indicator 99 (H) assessment methodology. Eco-Indicator 99 (H) utilizes a damage-oriented approach to quantify the impact of each system on emissions, land use, and resource depletion. In each of these three categories, there are indicators representative of the key metrics. For example, in the area of emissions some the impact indicators are: carcinogens (i.e., effects due to emissions of carcinogenic substances to air, water, and soil, expressed in Disability adjusted Life Years (DALY)/kg emission); and ozone layer depletion (ozone layer damage, expressed in DALY/kg emission, due to increased UV radiation as a result of emission of ozone depleting substances to air). Eco-Indicator 99 utilizes a point system to determine the overall damage impact. The higher the point value, the larger the damage associated with that impact category. Figure 10 also shows the relatively lower impact of Unit B compared to Unit A in all impact categories.

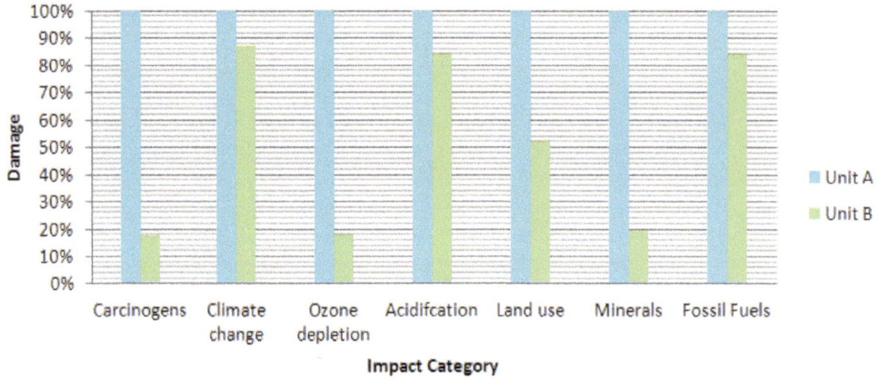

Figure 10. Characterization of Unit B relative to Unit A in each key environmental impact category

Figure 10 shows that the most significant relative decreases were noted in the areas of carcinogens, ozone depletion potential, and mineral usage, which experienced a relative decrease of over 80% when the modified insulation design was utilized in the refrigerator/freezer unit. The overall use of fossil fuels, a key metric in determining overall energy efficiency, showed a relative decrease of 16%, which can be attributed to both the reduction in energy utilized in the use phase, and the energy saved from the production of additional insulation.

Figure 11 shows the normalized data from the analysis. For LCA normalization, the data obtained from the impact assessment is divided by the equivalent regional data to provide a better indication of which impact categories contributed the most to overall environmental impact. As shown, the normalized data revealed the most impactful category to be the use of fossil fuels.

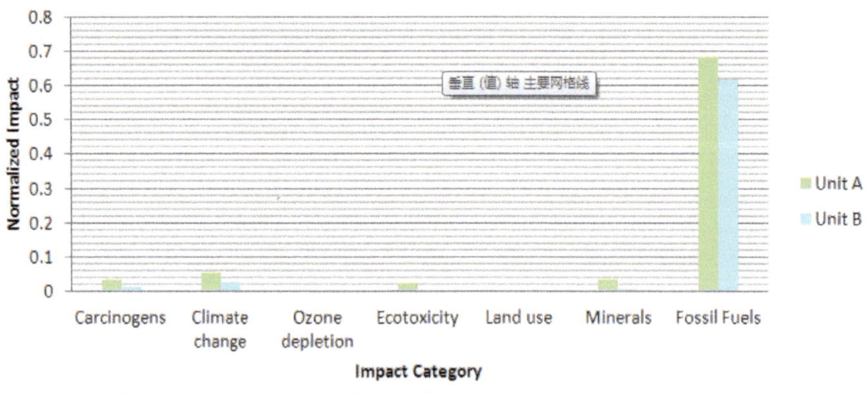

Figure 11. Normalization of each type of insulation in key environmental impact category based on equivalentEuropean data. Each impact indicator is shown in terms of its contribution to the overall life cycle impact

Other results obtained from the LCA (not provided) demonstrated that the value of the energy used during the total life cycle was estimated to be 5,862 MJ for Unit A and 4,698 MJ for Unit B, an overall decrease of about 20%. The weighting scheme associated with Eco-Indicator 99 quantified the fossil fuel use for Unit A as 205 Pts., with Unit B carrying a lower overall impact of 192 Pts.

The production of both units was energy intensive, utilizing 27% (55 Pts.) of the total energy consumption for Unit A, and 38% (73 Pts.) of the total energy consumption for Unit B. In particular, process contribution metrics from SimaPro indicated that the processes involved in the fabrication of the metal frame of the unit, as well as the forming of the plastic sheeting covering the unit expends a large amount of energy and greenhouse gas emissions. A similar trend also resulted from the fabrication of the mechanical components of both units, contributing 42 Pts. out of an estimated 55 for the manufacturing of Unit A, and 42 Pts. out of an estimated 73 for Unit B. Production of the insulation of Unit B was more energy intensive due to the energy associated with the production of silica, which contributed 24 Pts. However, a decrease in overall energy consumption was observed when an energy analysis was conducted over the life cycle of the unit, where the total points attributed to energy consumption for Unit A was 205 and Unit B carried a point value of 192, indicating that the impact points gained through the manufacturing phase was later offset by the lower point value in the use and disassembly phases.

A critical issue that was noted in the manufacturer survey was the management of waste and reuse of scrap metal from the manufacturing of mechanical components. Reusing metal discarded from previous production runs is important to conserving natural resources and ultimately mitigating the overall impact. Of the three manufacturers surveyed, only one reused scrap metal in the manufacturing. Assuming a U.S national energy mix, recycling one metric ton of steel would save approximately 1.1 metric tons of iron ore which results in savings of 2.79 metric tons of CO_2e (EPA, 2013).

The use phase contributed more than 40% of the overall points toward the overall environmental impact. Thus, the operation phase presents a significant opportunity for energy efficiency improvement. In addition, over the complete life cycle of both units, the overall energy use from a unit employing a fully fabricated modified insulation system is expected to be approximately 15% less than that of the traditional unit, as supported by the literature, as well as the experimental metrics.

Results of the disposal and disassembly phase analysis by the Eco-Indicator 99 (H) assessment method showed that Unit A utilized 36% greater landfill space (17.6 kg of Unit A was landfilled) due mainly to the voluminous amount of foam associated with the disposal. Alternatively, in Unit B, 86% of the insulation is recoverable for reuse, and hence only 11.3 kg of Unit B was landfilled. The overall energy use for disassembly was slightly greater for Unit B, attributed mainly to the additional energy needed to recover the modified insulation. However, this is offset by the significant amount of material salvaged, and the energy saved from not having to remanufacture the same insulation materials for a second production run.

Experimental and environmental assessment of the unit revealed that significant opportunities exist to reduce the overall environmental and energy impact of the commercial refrigerator unit are in the use phase, which can be achieved by reducing the energy consumption of these units on a daily basis. The commercial refrigerator/freezer unit investigated in this study that employed an alternate insulation provides an example of modifications that could be made to refrigerator design in order to increase the energy efficiency during the use phase. The experimental test conducted in this study showed that an estimated energy consumption improvement of 5.1% over 24 hours was experienced by one unit employing the modified insulation in the cover of the unit alone, amounting to a significant energy savings of 373 kWh over the lifetime of the refrigerator/freezer. This equates to a savings of roughly 0.257 metric tons of CO_2e emissions for a singular unit over its 20 year life span (EPA, 2014). For a unit fully outfitted with the alternative insulation described in this work, energy savings of 1,119 kWh and 0.771 metric tons of CO_2e emissions are expected.

References

Air-Conditioning, Heating and Refrigeration Institute (AHRI). (2011). Polyurethane and olyisocyanurate foams. Washington D. C.

American National Standards Institute (ANSI/ASHRAE). (2005). Method of testing commercial refrigerators and freezers ANSI/ASHRAE Standard 72-2. Atlanta, GA.

Bansal, P. K, Vineyard, E. A, & Abdelaziz, O. A. (2011). Review and status of not-in-kind refrigeration technologies for household space conditioning, water heating and food refrigeration. International Journal of Sustainable Build Environment, 1(1), 85-101. http://dx.doi.org/10.1016/j.ijsbe.2012.07.003

Boardman, B. (2004). Achieving energy efficiency through product policy: the UK experience. Environmental Science and Policy, 7(3), 165-176. http://dx.doi.org/10.1016/j.envsci.2004.03.002

Bouquerel, M., Duforestel, T., Baillis, D., Rusaouen, R. (2012). Heat transfer modeling in vacuum insulation panels containing nanoporous silicas—A review. Energy and Buildings, 54(2), 320-336.

http://dx.doi.org/10.1016/j.enbuild.2012.07.034

Department of Energy (DoE). (2010). U.S residential and commercial buildings total primary energy consumption. Buildings Energy Data Book. Washington D.C.

Department of Energy (DoE). (2011). Commercial energy end-use splits. Buildings Energy Data Book. Washington D.C.

Eberhardt, H. F. (2007). Vacuum Insulation Panels (VIP) in energy efficient cooling appliances for improving the environment of our children's world. Proceedings of the 8th International Vacuum Insulation Symposium, March 2007.

Energy Star (EPA). (2011). Buy products that make a difference. Washington D.C. Retrieved from http://www.energystar.gov/certified-products/detail/commercial_refrigerators_freezer

Environmental Protection Agency (EPA). (2009). Substitutes in rigid polyurethane. Washington D.C. Retrieved from http://www.epa.gov/ozone/snap/foams/lists/appliance.html

Environmental Protection Agency (EPA). (2011). Global Warming Potentials of ODS Substitutes, Ozone layer protection. Washington D.C.. Retrieved from http://www.epa.gov/ozone/geninfo/gwps.html

Environmental Protection Agency (EPA). (2013). *Waste Reduction Model (WARM).* Washington D.C. Retrieved from http://www.epa.gov/climatechange/wycd/waste/calculators/Warm_Form.html

Environmental Protection Agency (EPA). (2014), Greenhouse Gas Equivalencies Calculator. Washington D.C. Retrieved from http://www.epa.gov/cleanenergy/energy-resources/calculator.html

Fricke, J., Hummer, E., & Scheuerpflug, P. (1995). A Study of Thermal Properties of Silica Aerogels. Proceedings of 2nd International Symposium of Aerogel, April 2005.

Kudoh, A., Ohira, H., Nakamura, K., & Araki, K. (2006). Simple method to support optimization of layouts for vacuum insulation panels (VIPs) for refrigerator. Heat Transfer Asian Research, 35(6), 58-63.

Kwon, J. S., Jang, C. H., Jung, H., & Song, T. H. (2009). Effective thermal conductivity of various filling materials for vacuum insulation panels. International Journal of Heat and Mass Transfer, 52(23–24), 5525–5532. http://dx.doi.org/10.1016/j.ijheatmasstransfer.2009.06.029

Mahlia, T. M., & Saidur, I. (2010). A review on test procedure, energy efficiency standards and energy labels for room air conditioners and refrigerator–freezers. Renewable and Sustainable Energy Reviews, 14(7), 1888-1900. http://dx.doi.org/10.1016/j.rser.2010.03.037

Parenti, V. (2007, June). Novel rigid polyurethane formulations offering combined high energy efficiency and fast productivity for outstanding performance in the domestic appliance industry. Paper presented at Polyurethane Technical Conference, Orlando, FL.

U.S Energy Information Administration (USEIA). (2011). Annual Energy Review 2011. Washington D.C.

Waide, P., Lebot, B., & Hinnells, M. (2002). Appliance energy standards in Europe. Energy and Buildings, 45(67), 56-80.

Weaver, P. M., Ashby, M. F., Burgess, S., & Shibaike, N. M. (1996). Selection of materials to reduce environmental impact: A case study on refrigerator insulation. Materials & Design, 17(1), 11-17. http://dx.doi.org/10.1016/0261-3069(96)00024-6

World Energy Council (WEC). (2013). Energy Efficiency Policies around the World: Review and Evaluation. London, U.K.

Yu, F., Li, Y., & Zhu, Y. (2009). Numerical and experimental investigation of the thermal insulation performance of low temperature cold box. International Communication in Heat and Mass Transfer, 36(9), p.908-911. http://dx.doi.org/10.1016/j.icheatmasstransfer.2009.05.008

Zhao, Y. (2006). Research on synthesis process of silica aerogel/polystyrene core/shell composite materials.

18

Groundwater Resources Assessment in Tain River Basin of the Black Volta of West Africa

D. F. Dramani[1], G. K. Anornu[1], A. T. Kabo-bah[2] & S. Oduro-Kwarteng[1]

[1] Department of Civil Engineering, Kwame Nkrumah University of Science and Technology, PMB, Kumasi, Ghana

[2] Department of Energy and Environmental Engineering, University of Energy and Natural Resources, Box 214, Sunyani, Ghana

Correspondence: G.K. Anornu, Department of Civil Engineering, Kwame Nkrumah University of Science and Technology, PMB, Kumasi, Ghana. E-mail: anoprof@hotmail.com

Abstract

The Black Volta basin is one of the most important basins in West Africa that provides support for water supply, irrigation, hydropower and recreation for the communities within this basin. The Tain River basin is a sub-basin within the Black Volta. Most communities within this basin depend on groundwater resources as their main source of water supply. However, the over-growing dependence and limited assessment of the availability of groundwater resources in quality and quantity in the basin have become a worrying factor towards the sustainable management of the resources. The study therefore assessed the quality and quantity of groundwater resources in the Tain River Basin and showed that the total permanent groundwater reserve is estimated as 6.89×10^7 m^3 while the total recoverable groundwater as 2.59×10^7 m^3. The total recoverable groundwater forms 38 % of the total permanent groundwater reserve and this is the water available for abstraction and recharging of streams and springs. Both the quality and quantity of groundwater in the basin were found to be adequate according to national standards and World Health Organisation guideline values.

Keywords: quality and quantity assessment, transmissivity, specific water level, specific capacity, borehole depth

1. Introduction

Water is an essential natural resource that sustains the life of man and all living things on earth. It is central to many human activities such as industrial, domestic, animal watering, hydropower generation, transport services, tourism and recreation. There is increasing competition globally on the scarce water resources, especially surface water due to rapid population growth, urbanization and technological advancement. Current global climate change processes are expected to affect both the spatial and temporal available water resources (Arnell*et al.*, 2011; Oki & Kanae, 2006; Reddad *et al.*, 2013; Walther *et al.*, 2002). In 2004, WHO estimated that 1.1billion people lacked access to improved water sources and about 3,900 children under the age of 5 years died from water related diseases (World Health Organisation, 2004). The lack of access to potable water negatively impacts health, girl-child right to education in developing countries, and availability of water for agriculture. These collectively affect a country's urge to meet the Millennium Development Goals (MDGs) (Bakker, 2007; World Water Council, 2006).

According to the Ghana Water Policy (Ministry of Water Resources, 2007), improving water services and uses are essential for improving and increasing hygiene and sanitation services levels that affect productive lives of people, enhance enrolment and retention of girls in school, enhance women's dignity and ability to lead, reduce morbidity and mortality, reduce pre and post-natal women's risks and prevent vector and water borne diseases. In view of the concerns raised and the rate at which surface water is under threat, the only alternative source is groundwater. However, good estimation of groundwater resources of the basins in Ghana have not been carefully conducted and documented. For instance, in the Tain basin of the Black Volta, access to clean and potable water is a challenge. This is because most inhabitants depend on water supply from nearby unprotected streams and unreliable hand-dug wells constructed by individuals.

Over the years, groundwater development has been taken as the only alternative source of water to supplement the erratic surface water supply in the sub-basin. However, most of the boreholes are either dry or low yielding over time making it difficult for some communities to have easy access to potable water. Information gathered from Wenchi Municipal Water and Sanitation Agency located in the basin indicates that out of the 27 boreholes drilled in communities in 2012, eighteen (18) were successful and nine (9) were dry. Even those which were productive (successful), seven were low yielding ranging between 7 *l*/min. This is below the marginal yield of 13 *l*/min for development as stated by Community Water and Sanitation Agency (Ofori-Agyemang*etal*., 2008). This might be connected to the lack of basic information on possible groundwater potential zones and the complex nature of the subsurface geology of the area prior to the drilling exercise. There is no baseline study into groundwater resources in the River Tain basin of the Black Volta. Therefore this study examined and investigated into the groundwater resources potential in the Tain basin of the Black Volta in terms of success rate and recoverable groundwater storage. It provides a baseline for the available and current status of the groundwater resources in the basin.

2. Study Area, Materials and Methods

2.1 Study Area

River Tain basin of the Black Volta is located to the north-western part of BrongAhafo Region of Ghana within latitude 7°15′ N and 8°40′ N, and longitude 1°45′ W and 2°34′ W (See Figure 1). The basin covers four (4) administrative districts and four (4) municipalities in the region. These are Jaman North, Tain, Banda and Jaman South Districts, and Wenchi, Techiman, Sunyani and Brekum Municipalities. River Tain takes its source from the Republic of Cote d'Ivoire where it flows eastwards into Ghana through Jaman North and Jaman South, Brekum, Sunyani, and forms a boundary between Tain District and Wenchi Municipality. The river changes direction from the boundary towards the north and passes through Banda district and finally joins the Black Volta downstream of the Bui dam. The ground level elevation of the sub-basin ranges from 240 to 300 m above sea level with some few areas either undulating or rugged. Tain sub-basin of Black Volta lies in the Wet Semi-Equatorial climate region, which experiences bi-modal rainfall regime.

The major rainy season occurs between April and July with peak rainfall in June while the minor rainy season occurs between September and November, with its peak occurring in October. There is a short dry spell in August with the major one occurring between November and March. The mean annual rainfall is 1300 mm. The pattern of rainfall in the sub-basin follows the Inter-Tropical Convergence Zone (ITCZ) that shows distinction between the dry air from Sahara Desert and the moist monsoon wind from the Atlantic Ocean. The North-East trade wind, locally called the Harmattan brings in hot dry weather from December to February. The total mean evapotranspiration value for the sub-basin during the same period is 599.5 mm (Kortatsi, 1997). The mean monthly temperature is about 26 °C while average monthly humidity is 78 %.

The vegetation of the basin spans the moist-semi-deciduous forest and the Guinea Savanna woodland zones. The Guinea Savanna woodland represents an eco-climatic zone which has evolved in response to climatic and edaphic limiting factors and has been modified substantially through human activities. The original forest vegetation has been subjected to degradation, caused mainly by anthropogenic activities which include indiscriminate bush fires, slash and burn agriculture, logging and felling of trees for fuel over the years.

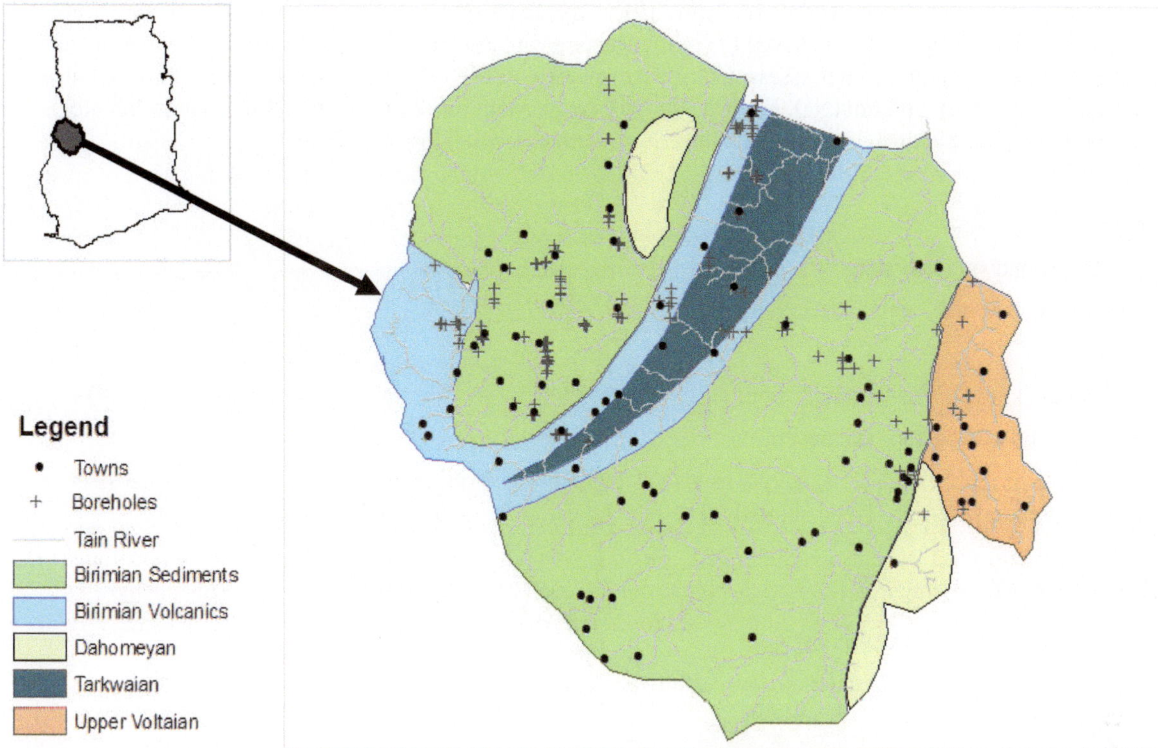

Figure 1. River Tain basin of the Black Volta showing borehole locations

2.2 Method

Geophysical data of boreholes and hydrological reports were used to identify local lineaments, the possible presence of aquifers, static water level, borehole logs, borehole yield, aquifer depth, thickness of the overburden and the aquifer zones pertaining to the basin. The study was also conducted by selecting specific communities in the sub-basin to locate all boreholes (wet and dry). The fundamental borehole parameters such as well depths, yield, static water level (SWL), dynamic water level (DWL) and geological logs were collated and analysed. The specific capacity was computed using the approach from Fetter (2001):

$$S_c = Q / s \tag{1}$$

Where

S_c = specific capacity (l/min/m);

Q = discharge in (l/mm);

s = Drawdown (m) = Dynamic water level (DWL) – Static water level (SWL);

Fifty eight (58) boreholes were selected based on the geological formations within the basin and their pumping test data analysed. For each borehole selected, a displacement-time graph was drawn and the widely used Cooper and Jacob's method employed to determine aquifer transmissivity (T) (Meier *et el.,*1998; Sánchez-Vila *et el.,* 1999).The aquifer transmissivity was then computed with.

$$T = \frac{2.303Q}{4\pi\Delta s} \tag{2}$$

Where;

T = Transmissivity;

Q = Discharge

Δs = Drawdown per log cycle

The water quality of the boreholes was assessed for their physicochemical parameters by evaluating thirty-two (32) boreholes. The distribution of the selection of the boreholes was done carefully to ensure that spatially each geological zone was represented. There were nine (9) in the Brimian Sedimentary, seven (7) in the Birimian volcanic, six (6) in the Upper Voltain, seven (7) in the Dahomeyan and three (3) in the Tarkwain formation zone. Relevant parameters for groundwater assessment were evaluated. These included calcium (Ca), Iron (Fe), Nitrate (NO_3^-), Magnesium (Mg), Sodium (Na), Flouride and pH.To ascertain the approximate total groundwater storage within the sub-basin, the approach used is as below (Asomaning, 1992; Schoeller, 1967):

$$Q_s = \omega\theta HA \tag{3}$$

Where

Q_s = Total groundwater storage;

ω = Percentage of study area underlain by groundwater zone;

θ = Porosity;

H = Mean thickness of the saturated zone;

A = Area of the hydrogeological basin.

The recoverable or usable storage capacity was also estimated to determine the amount of groundwater resource that could be withdrawn from a basin annually. The method used (Schoeller, 1967) is as below.

$$Q_r = \omega\beta HA \tag{4}$$

Where

Q_r = Recoverable groundwater storage;

ω = Percentage of study area underlain by groundwater zone;

β = Specific yield

H = Mean thickness of the saturated zone;

A = Area of the hydrogeological basin

Groundwater accessibility was assessed using 300 people per borehole threshold and a minimum yield of 13 l/min according to the Community Water and Sanitation Agency (CWSA) standard. The estimated 2012 population for the various communities were obtained from the Ghana Statistical Agency.

3. Results and Discussion

3.1 Aquifer Baseline Characteristic

The research conducted has shown that, the depths to aquifer in the sub-basin vary between 13-52 m with a mean value of 30.5 m as illustrated in Figure 2(a). Birimian Sedimentary which covers about 66.4 % of the sub-basin has depth to aquifer ranging from 16-43 m with a mean value of 27 m while Birimian Volcanic, Upper Voltaian and Tarkwaian formations, the depth to Aquifer varies from 22-52 m, 19-43 m and 25-35 m respectively. The deepest depth to aquifer values occurs in the areas where the regolith is thickest such as Botenso, Nchiraa, Seketia and Akete. The aquifer materials encountered in the Birimian Sedimentary include laterite with quartz veins, deep brown clay, and dry fine clay at a depth of between 0-55 m with high to moderately weathered black shale between the depths of 55-70 m. In the Upper Voltaian sandstone, fine-textured dry sand with clay was encountered at the depth of 0-55 m with highly weathered moist sandstone struck at the depth between 55-70 m. The data analyzed indicates that boreholes in the five geological formations ended in shale, sandstone or phyllite geological units. Groundwater occurrence in the basin depends mainly on secondary porosity through weathering, fracturing and fissuring. Fractured aquifers were encountered at Adadiem, Arkokrom, Abekwai, WalaNkwanta and Amoakokrom. In the Birimian sedimentary and volcanic formations, fissure zones were encountered underneath the thick regolith at KwabenaSuo, Kogua, Tadeakwae and Abekwai.

The aquifer transmissivity values for the Birimian sedimentary range between 1.32 and 25.61 m^2/day while in the Birimian volcanic, Upper voltaian and Tarkwaian formations, transmissivity values range between 1.019 and 16.6 m^2/day, 3.14 and 40.58 m^2/day and 1.54 and 20.34 m^2/day respectively. Table 1 summarises some borehole parameters considered in the study.

Figure 2(b) shows the specific water levels of the Tain river basin. The static water levels generally vary based on the type of the geological formation. The study indicates that Birimian volcanic recorded the deepest mean static water level of 26.1 m while the Tarkwain, Birimiansedimetary and Upper voltaian recorded the mean values of 16.4, 14.4 and 15.3 m respectively. Data on the Dahomeyan formation was not available for analysis during the study period. The shallowest static water level of 1.4 m occurred at GedengeNewsite in the

Birimianvolcanic with the deepest of 40.1 m occurred at Okyerekrom in the Birimian sedimentary formation.

From Figure 2(c), the research revealed that Birimian sedimentary has a specific capacity range of 0.37-7.23 *l*/min/m with a mean value of 2.44 *l*/min/m while the specific capacities of the Tarkwaian, Birimian volcanic and Upper voltaian range from 1.89-2.33, 0.8-1.57 and 0.61-22.82 *l*/min/m respectively for a period of six (6) hour pumping and three (3) hour recovery periods. The low value of specific capacity in the sub-basin indicates that transmissivity in the various geological formations are generally low. However, the low values may be the effect of losses in the wells, geological boundaries and partial penetration of the aquifers by the boreholes.

Figure 2(d) shows that the borehole yield is between 20 *l*/min and 113 *l*/min. However, it can be deduced that Upper voltaian recorded the highest average yield of 26.7 *l*/min followed by Birimian sedimentary with a mean value of 26.6 *l*/min while Birimian volcanic and Tarkwaian formations recorded mean values of 15 and 16 *l*/min respectively. The values suggest that most of the boreholes in the sub-basin are low yielding and this is very characteristic of sedimentary formations. It is however important to note that most of the boreholes were drilled purposely for rural water supply and therefore drilling was completed anytime yield was adequate to meet rural demands and this may justify the reason for low yielding wells. Since the purpose of the drilling was geared towards the meeting of rural water supply, the optimum yielding capacities were not attained for each borehole. This explains why the Birimian Sedimentary and Upper Voltaian have different mean transmissivities but similar yield.

Table 1. Summary of boreholes parameters in the river Tain basin

Geological formation	No. of Borehole	Depth (m)		Yield (l/min)		SWL(m)		Transmissivity (m^2/day)	
		range	mean	range	mean	range	mean	range	mean
Birimian Sedimentary	25	30-75	48.9	7-113	26.6	4.8-24.7	14.4	1.3-25.6	0.63
Birimian Volcanic	16	39-69	45.8	9-21	15	12.6-38.2	26.1	1.02-16.6	10.44
Dahomeyan*	2	40-70	55	*	*	*	*	*	*
Tarkwaian	6	40-56	47.2	12-20	16	4.9-27.8	16.4	1.5-20.3	7.29
Upper Voltaian	11	30-60	49.3	12-55	26.7	2.4-34.2	15.3	3.1-40.6	20.15

*pumping test has not been performed on the boreholes due to inaccessibility.

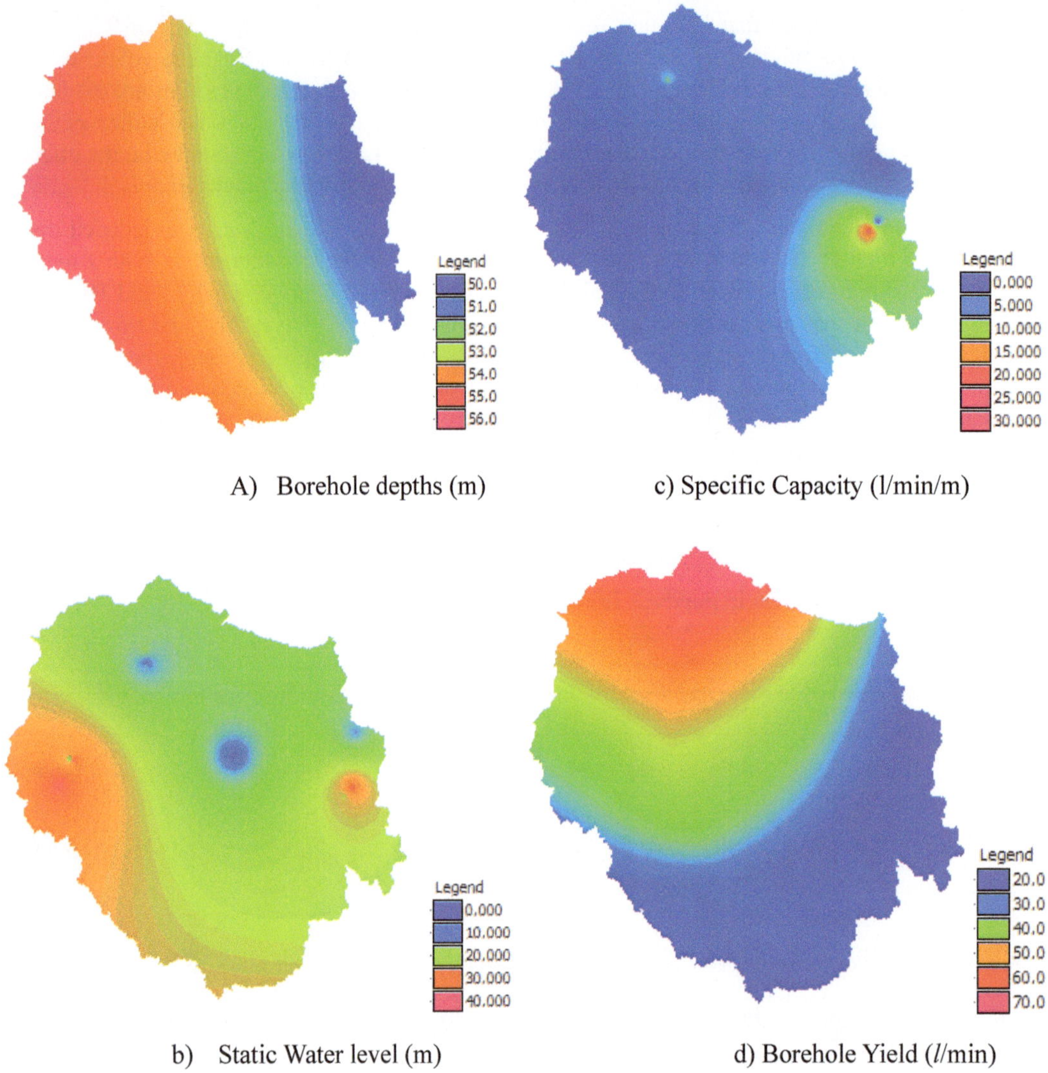

Figure 2. Spatial borehole yield, static water level, borehole depths and specific capacity

The borehole success rate in the various geological formations in the river Tain basin of the Black Volta is shown in Figure 3. It could be taken as the groundwater potential of the individual geological units which form the basin. Based on this, the Tarkwaian formation which records 94 % success rate is the richest in terms of groundwater in the study area. Birimian volcanic is next with about 89 % success rate and Birimian sedimentary which forms the largest formation in the sub-basin records about 70 % success rate. Most of the high yielding boreholes are found in the Birimian sedimentary, hence making it quite complex to conclude that it contains low yielding aquifers. Upper voltaian which is composed mainly of sandstone has a success rate of about 70 % and Dahomeyan formation recorded the highest failure rate of 50 %. This is due to the fact that Dahomeyan formation has impervious weathered zone and massive crystalline structure with limited fractured zones which limit its groundwater yielding capacity.

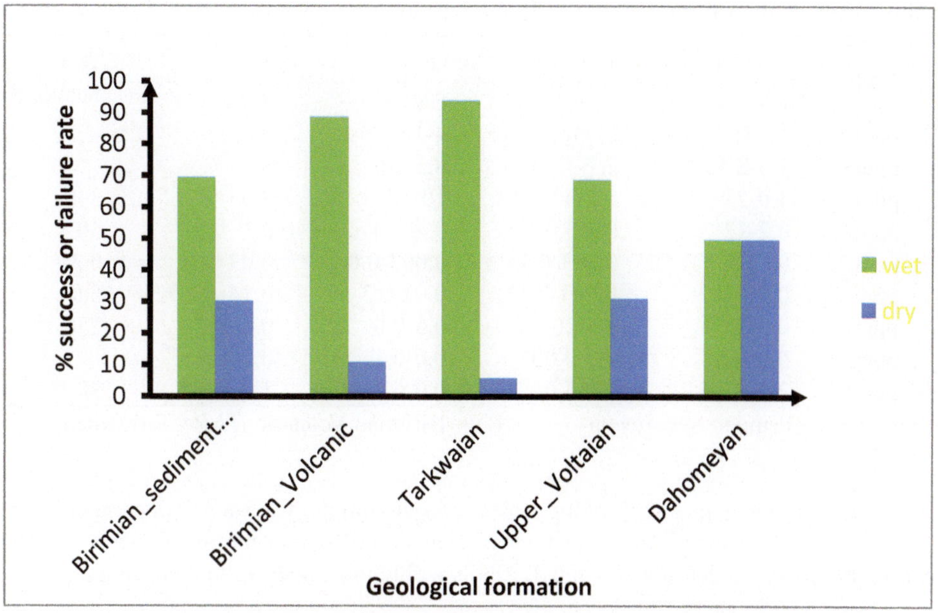

Figure 3. Borehole success and failure rate in the river Tain basin

Evaluation of the relationship between the borehole depth and yield was conducted but the study revealed no strong correlation ($R^2 < 0.2$). Therefore, there is no empirical relationship between yield and borehole depth and this can be attributed to early completion of drilling when yield is enough for water supply. The total permanent groundwater reserve was estimated at 6.89 x $10^7 m^3$ whiles the total recoverable groundwater as 2.59 x $10^7 m^3$ which is about 38 % of the total permanent groundwater reserve and this is the water for abstraction, recharging of streams and springs. The permanent groundwater reserves and recoverable groundwater storage for the various geological formations is summarized as in Table 2.

Table 2. Total and recoverable groundwater reserves in the five geological formations

Geological formation	Area covered (km^2)	Mean Specific capacity (m^3/min/m)	Percentage of saturated zone in the formation	Mean Saturated Thickness (m)	Total groundwater reserves (m^3)	Total recoverable groundwater reserves (m^3)	Percentage of recoverable groundwater as the total groundwater vol. (m^3)
Birimian Sedimentary	3865	2.44	75.9	17.32	5.08 x 10^7	1.2 x 10^7	23.6
Birimian Volcanic	731.9	1.3	88.2	15.71	1.01 x 10^7	1.3 x 10^6	12.9
Upper Voltaian	217.2	3.62	68.8	18.42	2.7 x 10^6	9.96 x 10^5	36.9
Dahomeyan	342.4	**	**	**	**	**	**
Tarkwaian	667.1	2.14	66.7	14.75	6.56 x 10^6	1.40 x 10^6	21.3

**pumping test has not been performed on the boreholes due to inaccessibility.

Available data from sampled boreholes in the river Tain basin indicates that the quality of groundwater abstracted through boreholes is generally of good physicochemical quality and therefore suitable for domestic uses including drinking, agriculture and industrial purposes. The groundwater quality of the Tain basin is within the WHO guidelines limits (WHO, 2004; WHO, 2011). Table 3 presents the various water quality parameters that were measured.

Table 3. Water quality ranges in the various geological formations in the sub-basin

Parameter	Unit	Geological formation/Quality range				WHO Standard
		UV	BS	BV	TAR	
Calcium	ppm	12-28	21-108	4-14	9.6	-
Magnesium	ppm	1.5-8.5	2.7-12.2	8.5-9.6	3.9	-
Sodium	ppm	4.0-23	5.0-30	10-17	3.0	300
Nitrate	ppm	0.2-12	0.3-3.0	0.4-7.0	0.3	50
Iron	ppm	0.01-0.2	0.003-0.15	0.003-0.01	0.03	0.3
Nitrite	ppm	0.001-0.9	0.003-0.14	0.01-0.3	0.14	3.0
Sulphate	ppm	0.0-15.0	0.5-20.0	0.5-9.0	0.0	250
Manganese	ppm	0.0-0.5	0.0-0.001	0.0-0.1	0.00	-
Potassium	ppm	2.0-13.0	0.0-5.0	0.0-6.5	1.0	30

UV= Upper Voltaian, BS=Birimian Sedimentary, BV= Birimian Volcanic, TAR= Tarkwaian.

WHO recommended pH value for domestic drinking water is in the range of 6.5-8.5. Water sampled from the boreholes in the river Tain sub-basin indicates that their pH values fall within the recommended WHO value except a borehole at Wurumpo which has a value of 6.4. It falls within the Upper Voltaian formation.

3.2 Bacteriological Quality Analysis

The results from the test indicated that out of 17 boreholes sampled in the sub-basin, 7 samples contain coliforms levels above the WHO recommended limit of 0.0 MPN Index/100 ml. The boreholes were located at Birimian Sedimentary, Shale and Sandstone formations.

3.3 Groundwater Accessibility

The study revealed that quality of groundwater abstracted through boreholes is generally of good physicochemical quality and therefore suitable for domestic including drinking, agriculture and industrial uses. However, based on the CWSA standard, majority of the population has no access to groundwater. A population of 2496 at Amponsahkrom depend on one (1) borehole instead of eight (8) with average yield of 10 *l*/min. Geological formations such as Birimian Sedimentary and Birimian Volcanic had the required number of boreholes based on their population threshold but still face water crisis due to frequent breakdown and drying of boreholes. Sixty percent of the communities have less than the required number of boreholes based on the CWSA threshold of 300 persons per a borehole, 16 % exceeded their threshold whiles 24 % have the exact number of boreholes based on their population.

4. Conclusion

Groundwater resources assessment is an important requirement for efficiently managing the quality and quantity of water resources in a basin. The study assessed the quality and quantity of groundwater resources in the river Tain basin of the Black Volta in West Africa. The study showed that the borehole yield in the basin does not depend on the depth. This is because boreholes with depth range between 20 and 30 m at Ampieni, Agubie, Kenasi, Abotereye and Ohiampeanika gave average yield of 40 *l*/min while boreholes with depth range between 50 and 80m at Adadiem, Kodua and Damaabi gave an average yield of 8 *l*/min.

The aquifers in the sub-basin are low yielding because average aquifer yield is 32 *l*/min with a range between 7 and 113 *l*/min and it varies from one geological zone to another. The geological formation with the best yield is the Tarkwaian formation with a success rate of 94% while Dahomeyan formation recorded highest failure rate of 50%. About 60 % of the communities did not have the recommended WHO standard of 300 persons per borehole leading to perennial water problems in such communities. Groundwater quality in the sub-basin falls within the WHO (2004) recommended limits for domestic water use, except for 30 % of the sampled communities. It is envisaged that findings in this study would support decision making in the integrated water resources management in the Black Volta.

Acknowledgments

We are most grateful to the critical review from Mary Magdalene Tanguo, an office assistant at the GEONETCast Centre of the University of Energy and Natural Resources-Sunyani-Ghana; and all anonymous reviewers for their time and commitment.

References

Arnell, N. W., van Vuuren, D. P., & Isaac, M. (2011). The implications of climate policy for the impacts of climate change on global water resources. *Global Environmental Change, 21*(2), 592–603. http://dx.doi.org/10.1016/j.gloenvcha.2011.01.015

Asomaning, G. (1992). Groundwater resources of the Birim basin in Ghana. *Journal of African Earth Sciences (and the Middle East), 15*(3), 375–384. http://dx.doi.org/10.1016/0899-5362(92)90022-5

Bakker, K. (2007). The "Commons" Versus the "Commodity": Alter-globalization, Anti-privatization and the Human Right to Water in the Global South. Antipode. http://dx.doi.org/10.1111/j.1467-8330.2007.00534.x

Fetter, C. W. (2001). Applied hydrogeology (Vol. 3). Prentice Hall Upper Saddle River, NJ.

Kortatsi, B. (1997). Ghana Water Resources Management study. Information Building Block.

Meier, P. M., Carrera, J., & Sánchez-Vila, X. (1998). An evaluation of Jacob's method for the interpretation of pumping tests in heterogeneous formations. *Water Resources Research, 34*(5), 1011–1025. http://dx.doi.org/10.1029/98WR00008

Ministry of Water Resources, W. and H. (2007). Ghana Water Policy.

Ofori-Agyeman, M. C., Kpordze C. S. K., & Anornu, G. K. (2008). *Viability of marginal yield boreholes in selected geological formations in Ghana.* Proceedings 33rd WEDC International Conference, Accra, Ghana, April, 2008.

Oki, T., & Kanae, S. (2006). Global hydrological cycles and world water resources. *Science, 313*(5790), 1068-1072. http://dx.doi.org/10.1126/science.1128845

Reddad, H., Etabaai, I., Rhoujjati, A., Taieb, M., Thevenon, F., & Damnati, B. (2013). Fire activity in North West Africa during the last 30,000 cal years BP inferred from a charcoal record from Lake Ifrah (Middle atlas--Morocco): climatic implications. *Journal of African Earth Sciences.* http://dx.doi.org/10.1016/j.jafrearsci.2013.03.007

Sánchez-Vila, X., Meier, P. M., & Carrera, J. (1999). Pumping tests in heterogeneous aquifers: An analytical study of what can be obtained from their interpretation using Jacob's method. *Water Resources Research, 35*(4), 943–952. http://dx.doi.org/10.1029/1999WR900007

Schoeller. (1967). Quantitative evaluation of groundwater resources. Methods and Techniques of Groundwater Investigation and Dev. *Water Reso, 33*, 21–44.

Walther, G. R., Post, E., Convey, P., Menzel, A., Parmesan, C., Beebee, T. J. C., ... Bairlein, F. (2002). Ecological responses to recent climate change. *Nature, 416*(6879), 389–395. http://dx.doi.org/10.1038/416389a

WHO (2011). *Guidelines for drinking-water quality* (4th ed.). World Health Organization. Retrieved from http://whqlibdoc.who.int/publications/2011/9789241548151_eng.pdf

World Health Organisation. (2004). Water, sanitation and hygiene links to health. Retrieved from http://www.who.int/water_sanitation_health/publications/facts2004/en/

World Water Council. (2006). The right to water, from concept to implementation (p. 8:405).

19

Development of a Charge Controller Dedicated to the Small Wind Turbine System

Ababacar Ndiaye[1,2], Cheikh M. F. Kébé[2], Vincent Sambou[2] & Papa A. Ndiaye[2]

[1] Université Assane Seck de Ziguinchor, UFR Sciences et Technologies, Département de Physique. BP: 523 Ziguinchor, Sénégal

[2] Centre International de Formation et de Recherche en Energie Solaire (CIFRES), Ecole Supérieure Polytechnique-Université Cheikh Anta Diop de Dakar (UCAD), BP 5085 Dakar-Fann, Sénégal

Correspondence: Ababacar Ndiaye, Université Assane Seck de Ziguinchor, UFR Sciences et Technologies, Département de Physique. BP: 523 Ziguinchor, Sénégal. E-mail: ab.ndiaye@univ-zig.sn

Abstract

In this paper, we present the development of charge and discharge controller of battery used in low power wind applications. This controller allows on the one hand protecting the battery against overcharging and deep discharge. On the other hand, it helps to protect the turbine against strong winds. It is controlled by PIC microcontroller 16F877A. This control function is performed using an algorithm that continuously compares the battery voltage to the charge and discharge thresholds, and tilting towards dissipation resistors. The control signals generated by the microcontroller are PWM (Pulse Width Modulation) type. The validation of the main functions of the controller is presented.

Keywords: wind power generator, microcontroller, PIC16F877A, pulse width modulation, MOSFET, battery

1. Introduction

As in photovoltaic systems, wind energy applications include most often batteries for storing excess energy and feed loads low and medium powers. The batteries are weak with the phenomena of overloading and deep discharge. This adversely affects the life of the batteries and is a weak point for renewable energies. To protect the batteries and prolong their lives, a controller should be used. For photovoltaic systems, the controller can only protect the batteries. The principle often used for photovoltaic controller consists in disconnecting the PV generator from the batteries. This disconnection occurs when the batteries are fully charged or deeply discharged (Amin et al., 2008). For wind turbine applications, the same principle cannot be used. Indeed, we cannot afford to disconnect the turbine from the batteries and keep it in a vacuum which will increase the risk of its destruction by overspeed. The principle adopted for the control of the batteries charge/ discharge for wind systems must take into account this constraint. The wind turbine must always be loaded by the batteries and/or other charges (use, discharge resistors). In addition to the batteries, the controller protects the wind turbine.

The importance of a charging / discharging controller in an autonomous system such as a photovoltaic system or wind turbine needs no more to be discussed. However, it must be done very carefully in order to meet the requirements for reliability, simplicity, portability and cost. As in the embodiment of any system, the controller also poses a number of problems related to the existence of several possible architectures for load control (Usher & Ross, 1998; Koutroulis & Kalaitzakis, 2004) (maximum charging current with control battery voltage, constant voltage charging with control current battery, charging with adjustable intensity together with control of the battery voltage, etc..) and battery discharge (Usher & Ross, 1998) (direct control of the battery voltage, check the battery voltage compensated or not the discharge current, discharge control across the state of battery charge. Several solutions are possible: analog, digital or mixed.

Additional constraints imposed on the controller such as the possibility of varying the parameters of the control algorithm and display the battery charge level, increase its complexity and lead us to opt for a smart solution based on microcontrollers.

This paper presents the different development steps of a battery controller for wind turbine applications, operating under the control of the microcontroller PIC16F877A.

2. Wind Turbine System

2.1 Operational Mode

The wind turbine is a device that converts kinetic energy from the wind into mechanical energy. The moving air passes through the section of the active surface of the airfoil. The power of the mass of air passing through this active surface is partially recovered by the turbine in the form of mechanical energy. According to BETZ law, only 59% of this power is recoverable.

It is necessary to know as accurately as possible the behavior of the wind depending on various parameters. Figure 1 below shows the power curve of a wind turbine in function of wind speed. In part (A) no power is delivered by the turbine for wind speeds lower than the starting speed V_{min}. If the maximum power is extracted from the wind turbine for each wind speed, cubic variation of wind speed is obtained in part (B). Beyond a certain limit of wind speed, the rated power of the turbine is reached. Then the latter must be stopped.

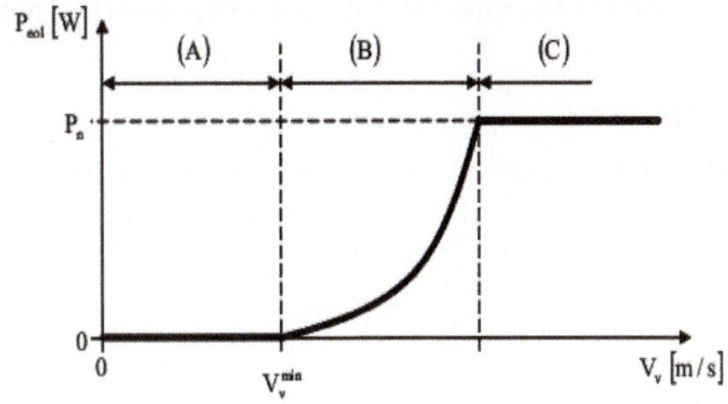

Figure 1. Power curve of a wind turbine depending on the wind speed

2.2 The Wind Turbine Control

There are occasional periods of gusty wind that inflict serious mechanical constraints to rotating parts. To prevent their destruction when the winds are of high intensity, the wind turbine must be equipped with a control board. This card is also used to operate the electrical energy delivered by the wind turbine to charge batteries. The voltage supplied by the wind turbine is a three phase system. It attacks a diode bridge to provide DC voltage. This DC voltage is used by the board to charge the batteries. When the batteries are sufficiently charged the excess of production must be consumed in order to prevent the destruction of the wind turbine. Figure 2 shows the chain formed by wind, controller, battery and load.

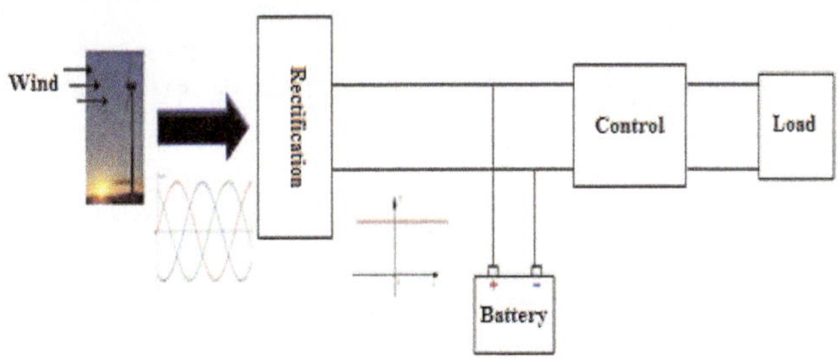

Figure 2. Wind turbine/ Controller/ Battery/ Load

3. Control Principle

The primary function of this controller is to allow the battery charging while ensuring that it is not overloaded. In fact, the battery has a high load threshold given by the manufacturer. If the battery voltage exceeds the threshold for an extended period, it can cause destruction of the batteries. To perform this control function of the load, there is a dedicated block. The principle on which the charge control block works is as follows: the voltage seen by the battery will continuously be measured. If it exceeds the high threshold, a resistor is connected in parallel with the battery to dissipate the energy excess provided by the wind turbine. This resistor must be variable in order to be adaptable to the power to be dissipated which also varies with the turbine production. The second function of the controller is to allow the discharge of the battery, while ensuring that it is not deeply discharged.

In fact, the battery has a low threshold for discharge given by the manufacturer. If the battery voltage exceeds the threshold for an extended period, it can cause the destruction of the batteries.

The control block of the discharge operates on the following principle: the voltage seen by the battery will continuously be measured. If it is inferior to the low threshold of the battery then the load is disconnected. The controller includes a third function, whose role is to inform the user on the charge level of the battery in real time. This information will be through three LEDs: green, yellow and red. Indeed, in addition to the high and low thresholds given by the manufacturer, we define a transition range between the low threshold and the high threshold.

This function is carried out according to the principle described below. The voltage seen by the battery is continuously measured. If it is below the lower threshold then the red LED is lit. This state corresponds to a level of deep discharge and therefore loads are shed. If it is greater than the upper threshold then the green LED lights. This condition corresponds to a correct level of charge and therefore the loads can be connected. If one is in the range between the two upper and lower thresholds then the yellow LED is lit. This state corresponds to the first alert.

4. Development of the Hardware Controller

The proposed controller is divided into three blocks: the control unit, the power unit (charge and discharge) and the block level meter (charge). Figure 3 shows the block diagram of the developed controller.

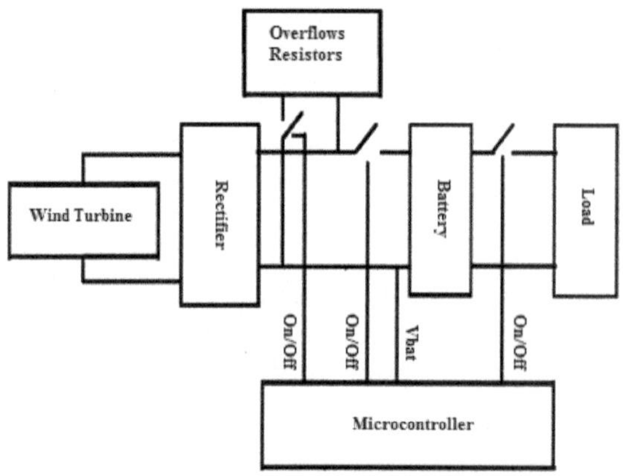

Figure 3. Block diagram of the whole Wind turbine / Controller / Load

4.1 The Control Block

Figure 4 shows the diagram of the control unit which consists essentially of the microcontroller PIC16F877A, its 20 MHz clock circuit and its initialization circuit. It also includes a 5V power supply circuit for the microcontroller and voltage divider resistors with a variable potentiometer setting. The PIC 16F877A (Datasheet PIC 16F877A, 2012) is the key control block. It allows running the program corresponding to the operation of the controller. This is a Microchip microcontroller. The clock circuit is realized around a 20MHz frequency quartz imposing the operating frequency of the PIC. The "Return To Zero" resets the microcontroller to reconfigure the controller. The battery voltage Vbat is measured through the RA0 pin of PIC configured as Analog/Digital input. The PIC only admits as analog input, voltages ranging from 0 to 5 volts. It is therefore necessary to adjust the battery voltage Vbat in the 0/5 volts range. A voltage divider resistor consisting of RV1, R14 and R15 was used. RV1 is a variable resistor for the calibration of the bridge and better accuracy. For a 24 V (1), the divider is made up of R14 and RV1. The selection between R14 and R15 is via a jumper.

$$\mathrm{Va} = \frac{\mathrm{RV1}}{\mathrm{R14} + \mathrm{RV1}} V_{bat} \tag{1}$$

$$\mathrm{Va} = \frac{\mathrm{RV1}}{\mathrm{R15} + \mathrm{RV1}} V_{bat} \tag{2}$$

The value measured through the voltage divider is then converted into a digital value Vn through an integrated internal analog to digital converter (ADC) (Ndiaye et al., 2012).

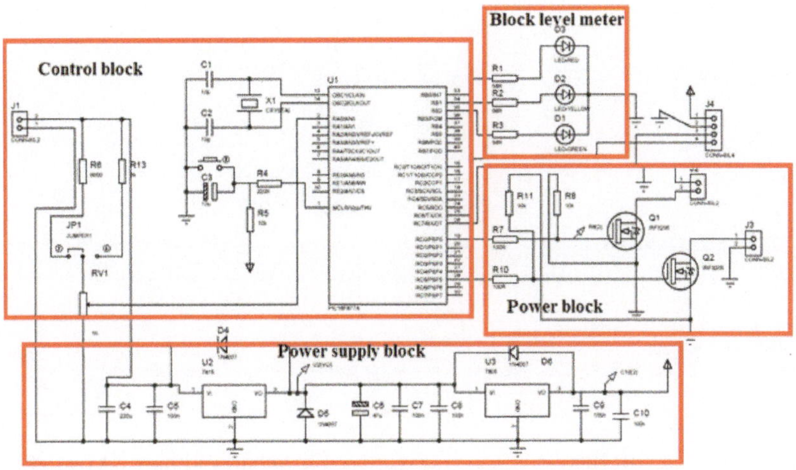

Figure 4. Electrical Schematic of the controller

4.2 The Power System

The power system consists of three blocks (Figure 4): the power supply unit (power supply block), the Power block (control charging/discharging) and the display unit of the battery charge level (block level meter).The power supply unit generates, from the battery voltage, a DC voltage of 5V for powering the microcontroller (Figure 4). It integrates two 7805 regulators (Ndiaye et al., 2010). The control blocks of the battery charging and discharging use PWM commands signals for controlling the charging and discharging of the battery. PWM signals generated by the PIC control the opening/ closing IRF3205 MOSFET transistors (Ndiaye et al., 2012).

For the control, an IRF3205 MOSFET transistor is used and connected to the overflow resistor. The resistor has been dimensioned in relation to the maximum power corresponding to excess energy production to be dissipated in order to avoid battery overcharging. According to the over-production, the overflow resistor will dissipate power variably. For discharge control, the block is composed of an IRF3205 MOSFET transistor connected to a pulling resistor (56 Ω). It is controlled by the PWM signal generated by the microcontroller. The connection of the load (use) is done by through the transistor. Thus, when the low threshold of the battery is reached, the duty cycle is very low and the load is no longer supplied. When the upper threshold of the battery is reached, the load is supplied again.

One block measures the battery voltage continuously. A voltage divider converts this voltage into the voltage range allowable by PIC microcontroller between 0V and 5V. Depending on the value of this signal, the microcontroller generates the digital signal for each transistor according to its function. A display unit of the battery charge level is integrated into the board. It consists of three LEDs of different colors. Depending on the signal generated by the PIC, the LEDs light up to indicate the battery status: Green (fully charged battery), Yellow (moderately charged battery) and Red (flat battery).

5. Development of the Software Part of the Controller

5.1 Presentation of the Operation Flow Chart

The algorithm to implement is symbolized by the main flowchart in Figure 5. This flowchart describes the general principle of the controller operation. The program is based on a comparing test of the battery voltage to the various thresholds continuously during the controller operation. The program is described in a flowchart to facilitate understanding. It has three sub-charts; each one explains the progress of each phase of the algorithm charging to be implemented:

- Flowchart of the charging phase « **Float** »,
- Flowchart of the charging phase « **Absorption** »,
- Flowchart of the discharging phase « Bulk».

Depending on the state of battery charge (V_{batt}), the program switches to the charge phase suitable.

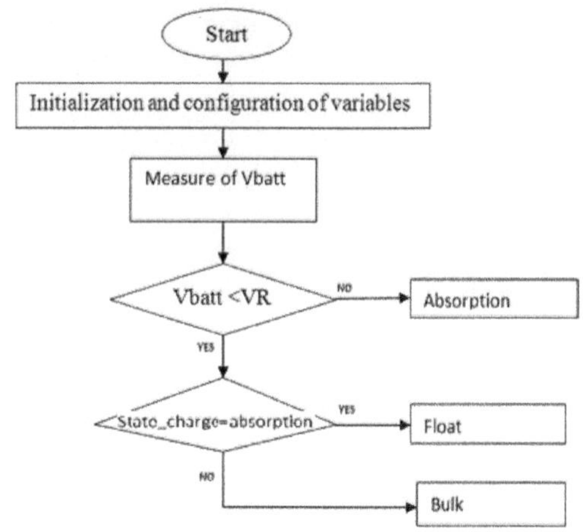

Figure 5. The main flowchart of controller

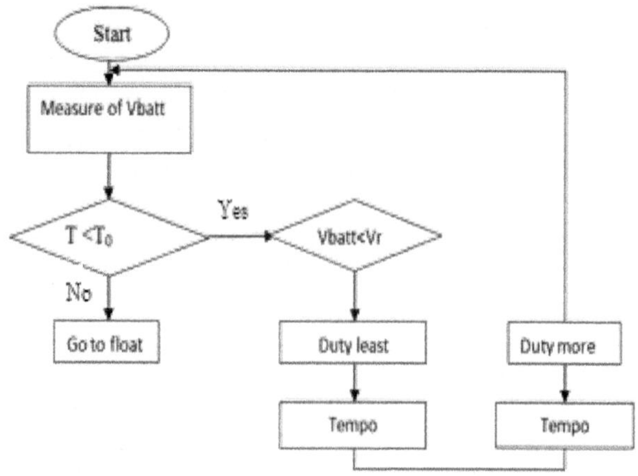

Figure 6. Flowchart of phase *"***Absorption***"*

"Absorption" Phase achieves a full charge of the battery that was previously less than 80%. During this phase, the voltage is kept constant (VR) allowing more or less dissipation (Duty more, Duty least). This step is only completed after a time T = T_0. During our tests, T_0 is set to 1hour (Huet, 1998). The corresponding flowchart is given in Figure 6.

"Float" Phase maintains the constant voltage to V_{float} value slightly lower VR. This voltage prevents self-discharge (holding voltage). Depending on the state of charge (V_{batt}), more or less dissipation is permitted (Duty more, Duty least) with a delay (tempo). In the case of our test V_{float} = 26.8 volts and tempo = 1second (Huet, 1998). Its flowchart is given in Figure 7.

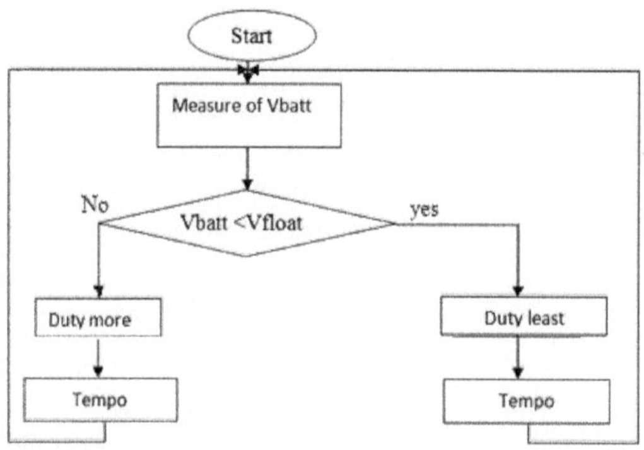

Figure 7. Flowchart of phase *"Float"*

It is in the *"Bulk"* phase whose the flowchart is given in Figure 8 that the permission or the prohibition to connect a load is taken. A voltage lower than V_{lvd}, indicates that the use of the battery is not allowed. When the battery voltage V_{batt} reaches the reconnection voltage *VAR*, the use becomes authorized.

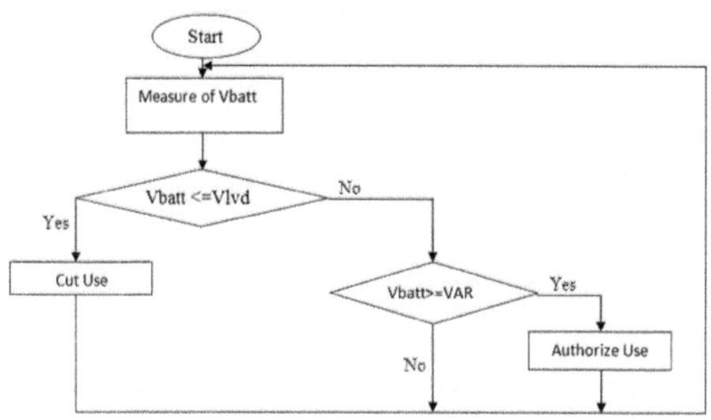

Figure 8. Flowchart of phase *"Bulk"*

5.2 The Programmer

A programmer is necessary to implement the program in the microcontroller. Several architectures are available on the literature. However, some of these make it possible to program only one *PIC* category, it is important to choose a programmer that recognize the *PIC16F877A* (Mayeux, 2002; Kirchev, 2007). The program developed in C and compiled on a computer must be sent to the PIC. But in the output of the computer we have a *RS232* connector that sends information on +12 V / -12 V. However, the PIC only supports voltages between 0V and 5V, so we must develop a device for performing this conversion function ±12 V to 0/5 V. This device shown in Figure. 9 is formed around an integrated circuit named *MAX 232*.

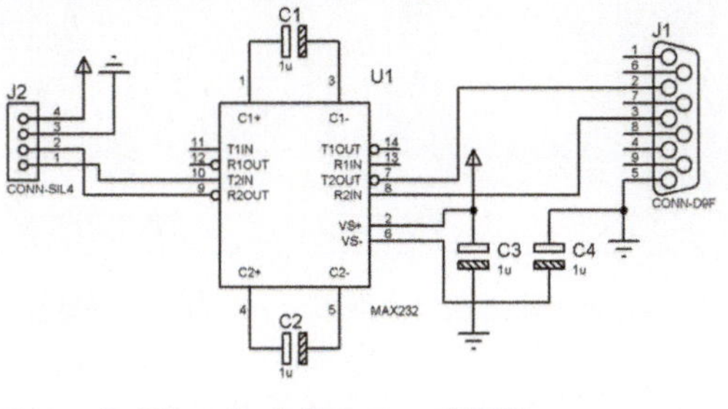

Figure 9. Electrical diagram of PIC programmer

It integrates a serial port (J1) to the computer, and allows transferring quickly and easily compiled program by *SourceBoost* tool to the microcontroller flash memory via *Tinybootloader* environment. The programming board thus produced is shown in Figure 10. The MAX 232 component around which is performed can clearly be identified, as well as links connecting the programmer to the computer on one side and to the controller board on the other hand.

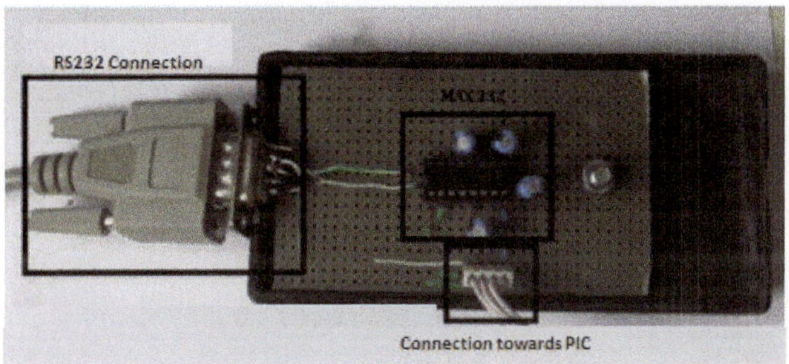

Figure 10. The programmer card

6. Tests and Experimental Results

6.1 Presentation of the Test Platform

Figure 11 shows the test platform. It consists of a power supply, a PC, an oscilloscope and the controller. The PC contains the elements of our development tool: software editing SourceBoost, and IC-PROG software for download the program to the microcontroller. Stabilized power supply is used to simulate the voltage after rectification of the wind turbine output voltage. The oscilloscope is used to visualize the individual signals such as the duty cycles and the different voltages of the controller board.

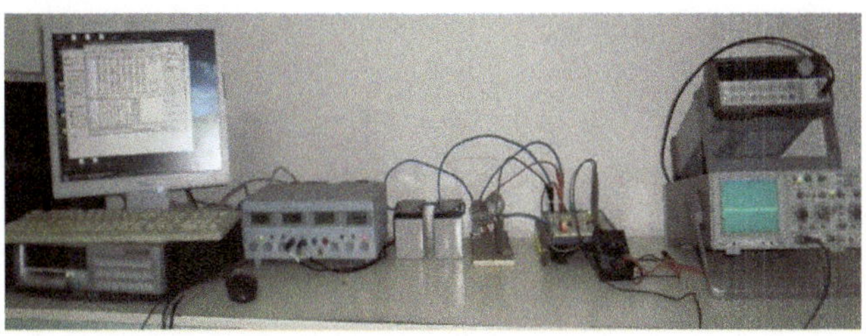

Figure 11. The test platform

6.2 Presentation of the Experimental Device

The realization of wind turbine controller took place at the International Center for Training and Research in Solar Energy (ICTRSE) of Dakar University. As mentioned earlier, this controller consists of a power unit, a display unit of battery charge level and a control block around the PIC microcontroller. Figure 12 shows the controller on the platform test.

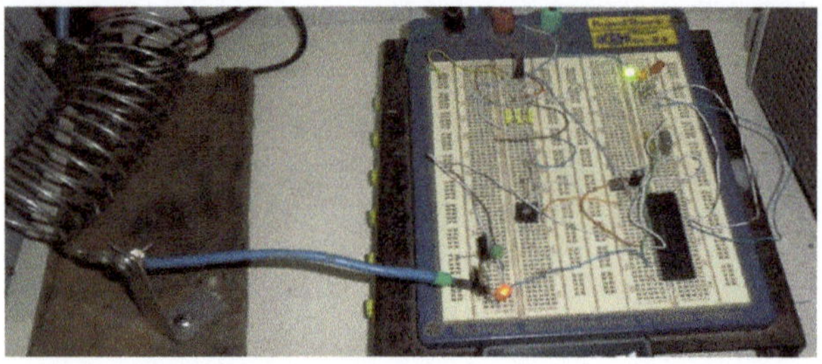

Figure 12. Presentation of the device on the test platform

6.3 Presentation of Tests Results

6.3.1 Test of the Voltage Divider

Before the voltage divider block is permanently connected to the microcontroller, it is tested to ensure proper operating condition. Measurements of the output voltage of the voltage divider (V_a) were obtained for different battery voltage V_{batt}.

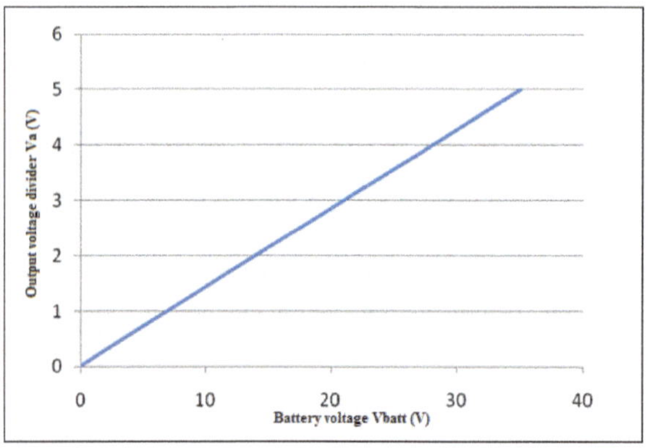

Figure 13. Output voltage divider block (V_a) in function of the battery voltage (V_{batt})

The output voltage variation as a function of input voltage V_{batt} is represented by a line whose slope is 0.139. This slope corresponds to 1/7 for the ratio of the voltage divider.

6.3.2 PWM Function Test

Figure 14 and Figure 15 show the PWM signals generated by the PIC16F877A. This is PWM signal amplitude 5V, a fixed frequency of 1.25 kHz and a variable duty cycle according to the different values of the battery voltage.

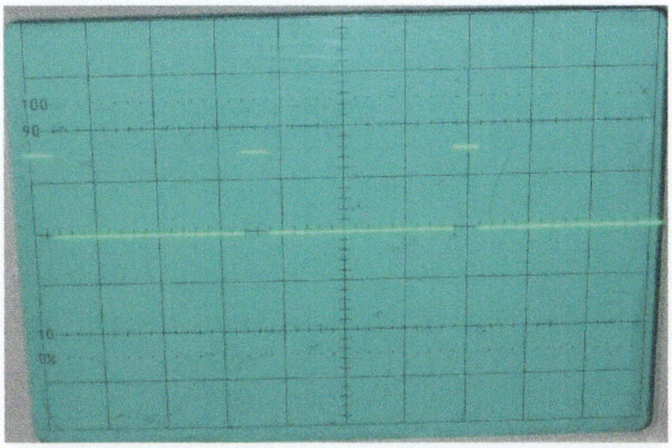

Figure 14. PWM duty cycle 11% (3V/div, 0.2ms/div)

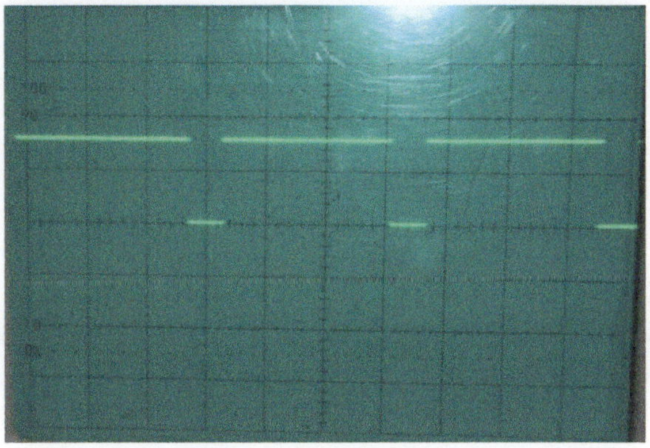

Figure 15. PWM duty cycle 83% (3V/div, 0.2ms/div)

6.3.3 Charge Control Test

To test the charge control, the variation of the power dissipated as a function of battery voltage is drawn in the Figure 16 on the one hand. On the other hand, the variation of the power dissipated as a function of duty cycle is shown in Figure 17. It is possible to notice that from dissipation threshold set at 28.3 V, a part of the energy produced (excess) is diverted to the overflow resistor. The power dissipated increases according to the battery voltage V_{batt} before reaching a maximum of 900 Watts from which it remains constant despite the increase of the battery voltage. During this phase all wind turbine production is dissipated in the overflow resistor because the batteries are full and there is no request by the loads. When the battery voltage reaches again the threshold dissipation, dissipation of power begins to decrease gradually until it becomes zero.

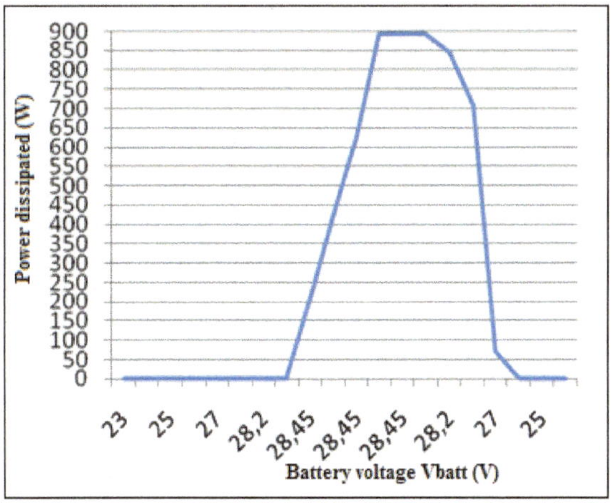

Figure 16. Power dissipated in function of the battery voltage V_{batt}

Figure 17 shows the evolution of the dissipated power as a function of duty cycle (D). We note that when the duty cycle is 100%, the total power is dissipated. For a duty cycle of 50%, half of the total production is dissipated.

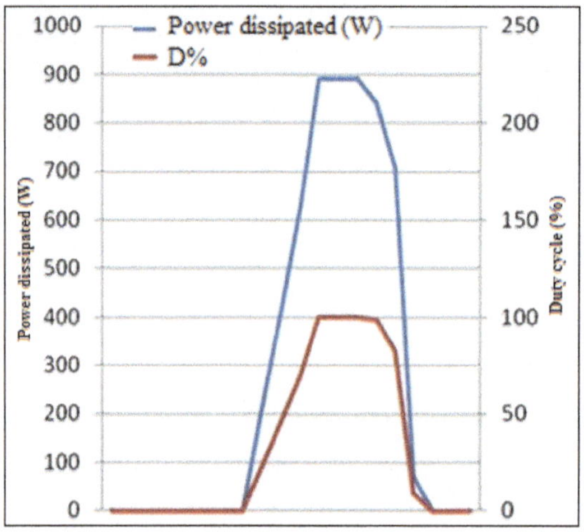

Figure 17. Power dissipated in function of duty cycle

6.3.4 Control Discharge Function

If the state of the battery charge is below the low threshold of discharge (VLVD), the controller disconnects the load (use). It is reconnected if the state of battery charge reaches the threshold of reconnection (VAR). Figure 18 shows the change of the control signal of the MOSFET which manages the charge control.

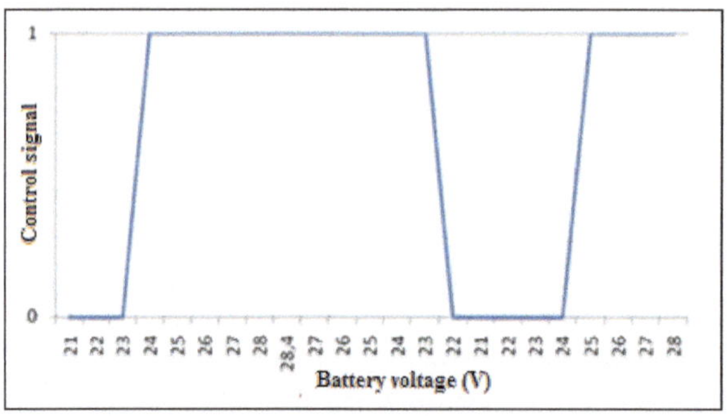

Figure 18. Evolution of the control signal of discharge

7. Conclusion

The objective of this work was to develop a controller dedicated to the small wind turbine. The first tests made in conditions of laboratory are rather satisfactory and gave us runways of improvements and optimizations. The oscillograms of the various command signals PWM generated by the PIC microcontroller 16F877 obtained were noted. The main functions were validated: deviation and dissipation of the excess power starting from the fixed thresholds, the disconnection and reconnection of the load according to the state of the battery charge. The correlation between the evolution of the dissipated power and the duty cycle was made obvious fact. In prospect, after the completion of the electronic regulator board which in printed circuit boards, tests will be performed in real conditions.

References

Amin, N., Azim, M. A., & Sopian, K. (2008). Development of cost effective charge controller with data acquisition options for PV powered sensor nodes. *Photovoltaic Specialists Conference, 2008*. PVSC '08, 33rd IEEE.

Datasheet PIC 16F877A. (2012). *Microchip Technology Inc.*

Huet, F. (1998). A review of impedance measurements for determination of the state-of charge. *Journal of Power Sources, 70.*

Kirchev, M. A. (2007). Studies of the pulse charge of lead-acid batteries for PV applications, part II. Impedance of the positive plate revisited.

Koutroulis, E., & Kalaitzakis, K. (2004). Novel Battery Charging Regulation System for Photovoltaic Applications. *IEEE Proc. Electr. Power Appl., 151*(2), 191 - 197.

Mayeux, P. (2002). Apprendre la Programmation des PIC par l'Expérimentation et la Simulation. *ETSF 2ème édition 2002.*

Ndiaye, A., Dzahini D., Ndiaye, P. A., Kébé, C. M. F., & Sambou, V. (2012). Impact of injected charges, clock noise and Operational Amplifier imperfections on the Sample and Hold (SH) overall performance. *Applied Physics Research, 4*(4), 18-25. http://dx.doi.org/10.5539/apr.v4n4p18

Ndiaye, A., Judalet, V., Kébé, C. M. F., & Ndiaye, P. A. (2010). Développement d'un régulateur de charge/décharge de batterie à seuils configurables pour des applications éoliennes. *Journal des Sciences Pour l'Ingénieur, 1*(12), 57-62. http://www.ajol.info/index.php/jspi/article/view/67976

Usher, E. P., & Ross, M. M. D. (1998). Recommended Practices for Charge Controllers. *Report IEA PVPS T3-05.*

20

CSD Type Family Solar Drying Technology and Its Potential Impact on Rural Development and Nutrition – A Case Study

Sangeeta Sinha[1,2], Anil Roy[3] & S. N. P. Singh[4]

[1] Department of Physics, B. R. Ambedkar Bihar University, Muzaffarpur, Bihar, India

[2] Centre for Renewable Energy and Env. Res., Muzaffarpur, Bihar – 842001, India

[3] DA-IICT, Gandhinagar, Gujarat, India

[4] Department of Zoology, B. R. Ambedkar Bihar University, Muzaffarpur, Bihar, India

Correspondence: Sangeeta Sinha, Centre for Renewable Energy and Env. Res., Muzaffarpur, Bihar – 842001, India. E-mail: prof.ssinha@gmail.com

Abstract

Globalization has adversely affected development of rural areas in developing countries. Rural areas are unable to absorb new technologies due to low literacy rates. Local Products, mostly from agricultural produces, are increasingly replaced by cheaper products from organized mechanized farming from one or the other corner of the globe. In this paper, family type cabinet solar dryer designs are presented in commensuration with the contemporary requirement of rural Bihar. Design objective includes simplification of technology and process, easy maintenance and low price to dry exotic produces which has significant export potential. Field trials are conducted, social surveys are undertaken and nutritional values of the products are determined in actual condition. Results indicate that simpler technology with state-of-art control of temperature by a combination of high heat capacity material and phase change material is most suitable to dry vegetables and spices, commonly called "Cash Crops". Their nutritional values remain within 80% limits even after six months. The technology is appropriate as it uses local materials; sustainable as it is cost-effective and employment generating ability and; robust as it uses very simple way for controlled drying.

Keywords: CSD, rural development, solar drying, nutrition

1. Introduction

Rural areas in developing countries are typically characterized by low literacy rates, dependence on agriculture related activities, and absence of alternative means of development, poverty and unemployment. Agricultural markets in rural areas are regulated by demand and supply rule in local context. The system had self-balancing mechanism for any drop in the quantum of production due to reasons beyond control. However, the buffering ability of local markets has taken a retreat in recent times due to globalization. Now, as and when price of a commodity rises, market supply is immediately supplemented from other places. Organized mechanized farming at a faraway place is increasingly and decisively controlling markets at local levels in rural areas. Traditional farmers with hardly any means or competency to adopt new technology get trapped in the vicious circle of high input price and low return. They are slowly alienating from agriculture as the sector depends upon several risk factors, that includes but not limited to uncertainties in government policy, energy price, labour migration, support price and even climate change. It has contributed to destabilization of traditional system and have caused mass migration of able labor forces in rural areas. In rural Bihar, villages are now mostly inhabited by children, females and old people, whose working capacity is very limited. Their ability to absorb new technologies is hence further reduced.

Arid rural areas in rural Bihar are further hit by global climate changes. Options of sustainable development have shrinked at a faster pace than expected. At the same time, arid areas are also home to priceless medicinal and horticulture produces, rich in vitamins, essential minerals and medicinal values. These produces have short life and a large part of them gets rotten before they could reach markets, where they can fetch very high values. Scattered and small scale of production is limiting factors in introducing cost-effective preservation technology at local and small scale levels. Besides, the average labour requirement for fruit production is very high, which is very important for rural development. It is approximately 900 man days per hectare per annum as against 140 man days for cereal crops. A few horticulture crops even generate much larger employment, from 1000–2500 man days per hectare per annum.

The huge employment and export potential of these agri-produces comprising horticulture, medicinal plants and vegetables have also been fully recognized at various levels by government and non-government planning agencies. Tenth Plan draft acknowledges that the "horticulture sector only contributes about 24.5 per cent towards agriculture GDP from only about 8 per cent of the cultivated area" (GOI, 2001). It acknowledges that this sector provides nutritional and livelihood security and helps poverty alleviation and employment generation. This sector supports a large number of agro-industries, which generate huge additional non-farming employment opportunities. Despite all these admissions, in practice, it is not getting the kind of priority it actually deserves. The post harvest loss of these agri-produces is very high in Bihar, India. Horticulture sector alone shows a loss of 5% to 39%, mainly during transportation. Though losses of medicinal plants and vegetables are very difficult to quantify, it is estimated to be much more due to delayed market absorption. In recent years, the demand of these agri-produces has increased considerably. Inevitably losses have also increased proportionately. The losses due to supplies from remote areas are even higher than what has been estimated.

Family size solar dryers have many advantages (Choudhary et al., 2005, 2006). It can be handled by left over rural population of women, old people and children. Besides, single day solar drying of agri-produces (medicinal and horticulture) almost eliminates degradation possibilities due to insect attack. The end product falls into the category of organic category, which has high market demand in developed countries. These dried produces can be preserved for long time in simple packaging and can be transported without any requirement of specialized vehicles. However, most of these agri-produces need to be dried in particular temperature range to preserve vitamins, color and taste. Controlling temperature variation by increasing thermal mass increases drying time and therefore losses during off sunshine hours. Latent heat storage in a Phase Change Material (PCM) is very attractive option to overcome this lacuna because of its high storage density with small temperature swing.

Phase Change Materials (PCM) is latent heat storage material. When the source temperature rises, the chemical bonds within the PCM break up as the material changes phase from solid to liquid. It is an endothermic process. Upon storing heat in the storage material, the material begins to melt when the phase change temperature is reached. The temperature then stays constant until the melting process is finished. The heat stored during the phase change process of the material is called latent heat. Latent heat storage provides the flexibility of storing large amounts of heat with only small temperature changes and therefore to have a high storage density. As the change of phase at a constant temperature takes some time to complete, it becomes possible to smooth temperature variations. It can store 5 to 14 times more heat per unit volume than sensible storage materials such as water, masonry, or rock. Phase change materials themselves cannot be used as heat transfer medium. Latent heat energy storage system has three components; (a) PCM with its melting point in the desired temperature range; (b) compatible heat exchange surface; and (3) a container compatible with the PCM.

Cabinet solar drying (CSD) technology has been evolving to meet the dynamic rural kaleidoscope. Conventional solar drying technology at various stage of complexity has failed to make an impact in Rural India despite Government support and subsidies. The reason is mostly attributed to the abysmal understanding of the socio-cultural set-up, its ability to absorb the technology and its impact on adopting the technology. In this paper, family type cabinet solar dryer designs are presented in commensuration with the contemporary requirement of rural Bihar. Design objective includes simplification of technology and process, easy maintenance and low price to dry exotic produces which has significant export potential. The state-of-art CSD with PCM and HHCM (High Heat Capacity Material) is designed and developed for single day controlled temperature drying of medicinal agri-produces (especially with anti cancerous and curative properties). Field trials are conducted, social surveys are undertaken and nutritional values of the products are determined in actual condition. The technology uses local materials. It is cost-effectiveness and has employment generating ability. It is robust as it uses very simple way for controlled drying.

2. Materials and Methods

2.1 Design of Solar Dryer

Family size passive solar dryer was designed and constructed using galvanized iron sheets. Theoretical modeling of various parameters were done using standard program with the objective of maintaining a temperature range of 50–80 degree C for 6 to 14 hours while drying of 5 kg of the agri-produce selected. The dryers's outer and inner box is made of thin Galvanized Iron sheet measuring 0.65 m × 0.55 m × 0.25 m and 0.6 m × 0.5 m × 0.2 m. 2 mm thick toughened glass is used for upper cover. 5 cm gap between the outer and inner boxes is filled with insulating material locally available (rice husk, wooden particles etc). Inner box is painted in carbon black. Inner box has a perforated tray for holding the agri-produce. The dryer achieved 90 degree C during peak summer and dried even high water and sugar content agri-produce in a single day (12–14 Hrs). However, the design lacked in temperature control.

Mechanism and needed manual intervention quite often. Thermal heat loss was also high. Hence, it is reconstructed with Fibre Reinforced Plastic (FRP) with double layer outer wall separated by 75 mm compressed wooden sheets. FRP protects the wooden structure from deformation and air leakage due to moist air, high temperature and direct sunlight. Dimension of the family size cabinet solar dryer is shown in the Figure 1a. Experimental model were constructed of clear base area 1 m × 1 m. 3 mm toughened glass is used to cover the upper portion of the dryer. A 1 inch perforated PVC pipe is placed at the bottom for controlled inlet of ambient air to supplement the moist air removal by PV operated fan from the back. A partition is created for ease of operability and to impart strength to steel mesh drawer. It can be moved outside for loading materials to be dried as shown in Figure 1b. Space below the mesh is used for storing HHCM and PCM.

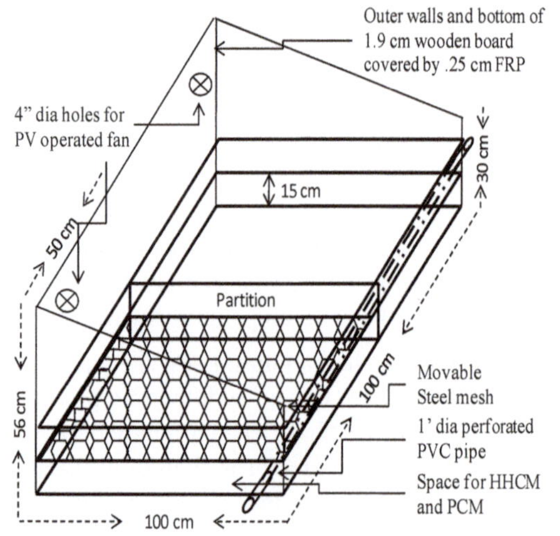

Figure 1a. Schematic diagram of family size solar dryer

Figure 1b. Family size solar dryer

Figure 1. Schematic diagram and actual family size solar dryer

At first suitable Phase Change Material (PCM) in the required temperature range for drying is selected. Latent heat of fusion, Melting Point, Density and thermal conductivity are most important parameters in selection of PCM (Dincer, 1999; Sukhatme, 1996). Theoretical modeling is used to optimize a combination of PCM and High Heat Capacity Material (HHCM) to stabilize temperature inside the dryer. Temperature range control mechanism is developed on the basis of HHCM to drying material ratio. Variation in this ratio allowed greater control of temperature for different types and quantity of materials to be dried.

2.2 Drying Experiment

A series of drying experiments were conducted in incubators to determine the optimum temperature ranges for each agri-produce. Once temperature range is fixed, theoretical modeling is used to determine the combination of PCM and HHCM for solar drying experiment. Existing literatures show that drying at temperature below 90 degree C has no impact on nutritional value of the material dried. Family size solar dryer are designed to achieve temperature less than 90 degree C. The following agri-produce with anti-cancerous properties were dried,

(a) Turmeric/Curcumin - It is one of the most important medicinal plant used as spice in India. It contains the polyphenol Curcumin, which promotes 'Apoptosis' (programmed cell death/cell suicide) that safely eliminates cancer breeding cells without posing a threat to the development of other healthy cells as in case of conventional radiotherapy and chemotherapy. It also effectively retards the growth of cancer cells causing prostate cancer, melanoma, breast cancer, brain tumour, pancreatic cancer and leukemia amongst a host of others.

(b) Fennel – It is also one of the most widely used medicinal spice. It has phyto-nutrients, antioxidants and an anti cancerous compound, 'Anethole'. Anethole is known to resists and restrict the adhesive and invasive activities of cancer cells. It is also known to suppress the enzymatic regulated activities behind cancer cell multiplication.

(c) Cayenne Pepper/Capsaicin: Capsaicin is also known for its anti-cancerous activities besides several other medicinal properties. It induces the process of apoptosis that destroys potential cancer cells and reduces the size of leukemia tumour cells considerably.

(d) Ginger: In aurvedic medicine, the medicinal quality of ginger is known to help lowering cholesterol level, boost metabolism and kill cancer cells. It is being used almost all over the world due to these medicinal properties.

(e) Oregano: Oregano is a potential agent against prostate cancer. It also consists of anti-microbial compounds. Phyto-chemical 'Quercetin' present in oregano restricts growth of malignant cells in the body, its potential as a drug against cancer-centric diseases is much more than grapes.

(f) Cumin: Cumin is called herb with anti-oxidant characteristics. Its seeds contain a compound called 'Thymoquinone' that checks proliferation of cells responsible for prostate cancer.

3. Results and Discussion

Temperature control mechanism by a combination of phase change material and high heat capacity material is achieved in harsh climatic conditions with large variation in climatic parameters. High heat capacity materials were used to offset any change in climatic parameters. On a typical day, Figure 2 shows the variation of ambient temperature and inside drier temperature of the proposed design for which temperature varies between 45 °C to 90 °C. Temperature and duration of drying is traded off by the combination of PCM and HHCM. Figure 3 shows typical variation of different combination of PCM ($NaOH\cdot H_2O$) and HHCM (boulders). It is evident that the temperature curve is flattened and thermal energy is preserved for longer period of time with use of PCM. Initial rise in temperature and maximum desired temperature can be controlled by the combination of HHCM and PCM. Spices like cumin and fennel needed slow drying. Hence, the amount of HHCM is increased to attain maximum temperature. At the same time, Cayenne Pepper is dried at high temperature for at least two hours. It is achieved by reducing HHCM in the drier. Table 1 shows the best option of PCM for drying of agri-produce known for anti cancerous properties. It also enlists weight/weight ratio of HHCM and DM (Material to be dried) for temperature control and single day drying. Dried weight is weight of DM with less than 5% moisture content except for ginger and fennel, where it is 10% and 8% respectively. The system is simple to operate and convenient for rural people particularly in developing countries.

Table 1. Selection of PCM and HHCM:DM ratio for single day solar drying of agri-produce, known for anti cancerous properties

PCM (10 kg/m^2)	M.P. (°C)	L. heat (kJ/kg)	Density (kg/m^3)	Drying Material (DM) – 5 kg /m^2	HHCM:DM (ratio)
$Fe(NO_3)_2\cdot 9H_2O$	47	170	1684	Fennel, Cumin	80
$Na_2S_2O_3\cdot 5H_2O$	48	209	1600	Fennel, Cumin	80
n- Pentacosane	54	238	769	Oregano	70
n- Octacosane	61	255	865	Cinnamon	60
$FeSO_4\cdot 7H_2O$	64	200	1893	Ginger	60
Stearic acid	69	200	848	Turmeric	50
$Ba(OH)_2\cdot 8H_2O$	78	270	1937	Cayenne Pepper	40

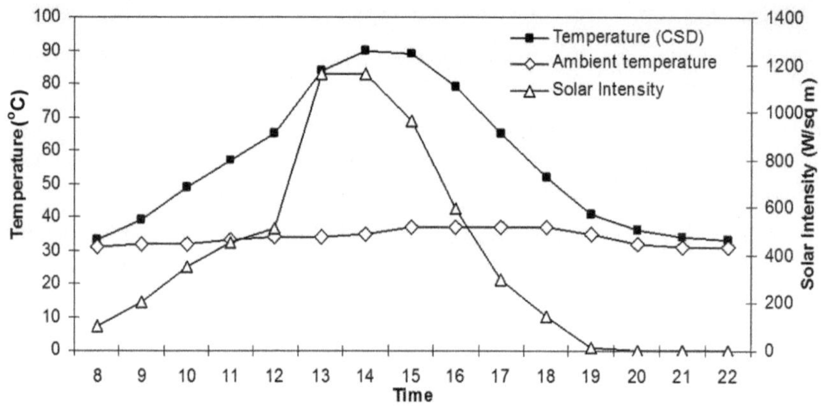

Figure 2. Variation of solar intensity, ambient temperature and temperature inside dryer on a typical day

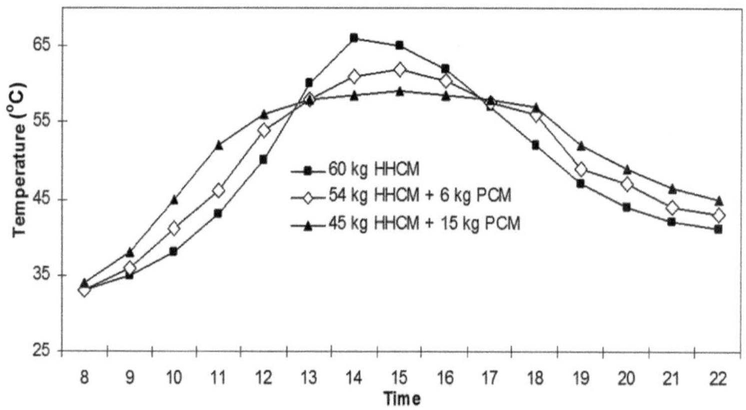

Figure 3. Variation of temperature inside dryer with a combination of PCM and HHCM

Table 2 shows the variation in drying time, drying wt. and moisture content after single day solar drying of these agri-produces. Figure 4 (a, b, c, d, e, f, g) shows the images of harvested and dried agri-produces - ginger, cumin and Cayenne Pepper.

Table 2. Variation in drying time of selected anti-cancerous produces in the proposed solar dryer

Drying Material	Drying Time (Hours)	Dried Wt. (%)	Moisture (%)
Cayenne Pepper	10	40	< 5
Turmeric	10	45	< 3
Ginger	10	50	< 10
Cinnamon	10	30	< 3
Oregano	10	10	< 3
Fennel	8	10	< 8
Cumin	8	8	< 5

(a) Harvested Ginger (b) Dried Ginger knots (c) Dried Ginger stems

(d) Harvested Cumin (e) Dried Cumin (f) Harvested Pepper

(g) Dried Pepper

Figure 4. Images of harvested and dried agri-produces - ginger, cumin and cayenne pepper

4. Conclusion

Organic farming of agro-produces with anti cancer properties has very high demand in global market. These produces are less labor intensive and most suitable for remote rural areas, where able labor forces have almost migrated to mega cities. Family size solar dryer developed provides an alternative route for economic development to rural areas. The temperature control mechanism developed can be handled by illiterate rural women. Drying in controlled low temperature range also preserves their medicinal properties, which has been investigated and documented by several scientists. Post harvest technology for these spices is based on the fact that drying below 70 degree C doesn't affect medicinal properties (Peter, 2001). Reduced weight, cheap packaging and ease of transportation from remote areas to cities, are some of the major factors, likely to drastically reduce wastage in this sector.

References

Choudhary, S., Baby, P., Sangeeta, S., & Sanjay, K. (2005). Trans-boundary Floods Induced by Global Climate Change: Development of Solar Fish Drying Technology to Combat Food Deficit in Seasonally Inundated Northern India. *Journal of Agricultural Meteorology, 60*(5), 581-584.

Dincer, I. (1999). Evaluation and selection of energy storage systems for solar thermal applications. *International Journal of Energy Research, 23*(12), 1017-1028. http://dx.doi.org/10.1002/(SICI)1099-114X(19991010)23:12<1017::AID-ER535>3.0.CO;2-Q

GOI. (2003). *Estimation Loss of Horticulture Produce due to Non-availability of Post Harvest & Food*

Processing Facilities in Bihar & UP. Socio-Economic Research, ASET, Planning Commission, Government of India. New Delhi, India. Retrieved from http://planningcommission.nic.in/plans/planrel/fiveyr/10th/10defaultchap.htm

Peter, K. V. (Ed.). (2001). *Handbook of Herbs and Spices*. Cambridge, London, Woodhead Publishing Limited.

Planning Commission. (2001). *Xth five year plan* (Vol. 2). Govt. of India.

Sanjay, K., Sinha, S., Kojima T., & Kato, S. (2006). Food Supplement and Protection in Deserts and Arid Areas by Solar Drying of Native Horticulture Products. *J. of Arid Land Studies, 16S*, 297-300.

Sukhatme, K., & Sukhatme, S. P. (1996). *Solar energy: principles of thermal collection and storage.* Tata McGraw-Hill Education.

Permissions

All chapters in this book were first published in EER, by Canadian Center of Science and Education; hereby published with permission under the Creative Commons Attribution License or equivalent. Every chapter published in this book has been scrutinized by our experts. Their significance has been extensively debated. The topics covered herein carry significant findings which will fuel the growth of the discipline. They may even be implemented as practical applications or may be referred to as a beginning point for another development.

The contributors of this book come from diverse backgrounds, making this book a truly international effort. This book will bring forth new frontiers with its revolutionizing research information and detailed analysis of the nascent developments around the world.

We would like to thank all the contributing authors for lending their expertise to make the book truly unique. They have played a crucial role in the development of this book. Without their invaluable contributions this book wouldn't have been possible. They have made vital efforts to compile up to date information on the varied aspects of this subject to make this book a valuable addition to the collection of many professionals and students.

This book was conceptualized with the vision of imparting up-to-date information and advanced data in this field. To ensure the same, a matchless editorial board was set up. Every individual on the board went through rigorous rounds of assessment to prove their worth. After which they invested a large part of their time researching and compiling the most relevant data for our readers.

The editorial board has been involved in producing this book since its inception. They have spent rigorous hours researching and exploring the diverse topics which have resulted in the successful publishing of this book. They have passed on their knowledge of decades through this book. To expedite this challenging task, the publisher supported the team at every step. A small team of assistant editors was also appointed to further simplify the editing procedure and attain best results for the readers.

Apart from the editorial board, the designing team has also invested a significant amount of their time in understanding the subject and creating the most relevant covers. They scrutinized every image to scout for the most suitable representation of the subject and create an appropriate cover for the book.

The publishing team has been an ardent support to the editorial, designing and production team. Their endless efforts to recruit the best for this project, has resulted in the accomplishment of this book. They are a veteran in the field of academics and their pool of knowledge is as vast as their experience in printing. Their expertise and guidance has proved useful at every step. Their uncompromising quality standards have made this book an exceptional effort. Their encouragement from time to time has been an inspiration for everyone.

The publisher and the editorial board hope that this book will prove to be a valuable piece of knowledge for researchers, students, practitioners and scholars across the globe.

List of Contributors

Liang Hao
Institute of Environmental and Sustainable in Agriculture, Chinese academy of agriculture science, Beijing, China
Key lab. for energy saving and waste disposal of protected Agriculture, ministry of agriculture, Beijing, China

Yang Qichang
Institute of Environmental and Sustainable in Agriculture, Chinese academy of agriculture science, Beijing, China
Key lab. for energy saving and waste disposal of protected Agriculture, ministry of agriculture, Beijing, China

Luis Alejandro
Institute of Environmental and Sustainable in Agriculture, Chinese academy of agriculture science, Beijing, China

Mohammed Saleh Al. Ansari
Department of Chemical Engineering, College of Engineering, University of Bahrain, Kingdom of Bahrain

Håkon Sælen
Center for International Climate and Environmental Research, Oslo, Norway

Hege Westskog
Center for International Climate and Environmental Research, Oslo, Norway

Tanja Winther
Centre for Development and the Environment, University of Oslo, Norway

Sophie Bouly de Lesdain
R&D, Electricité de France, associated with IIAC-LAU (EHESS-CNRS), France

Bosko Josimovic
Centre for Spatial Development and Environment, Institute of Architecture and Urban and Spatial Planning of Serbia, Belgrade, Serbia

Jasna Petric
Centre for Spatial Development and Environment, Institute of Architecture and Urban and Spatial Planning of Serbia, Belgrade, Serbia

Sasa Milijic
Centre for Spatial Development and Environment, Institute of Architecture and Urban and Spatial Planning of Serbia, Belgrade, Serbia

Chien-Li Lee
Research and Development Center for Water Resource and Conservation, National Kaohsiung First University of Science and Technology, Taiwan

Chih-Ju G. Jou
Department of Safety, Health and Environmental Engineering, National Kaohsiung First University of Science and Technology, Taiwan

Kutlu Sevinç Kayıhan
Department of Architecture, Faculty of Architecture, Gebze Institute of Technology, Kocaeli, Turkey

Seda Tönük
Department of Architecture, Faculty of Architecture, Yıldız Technical University, İstanbul, Turkey

Elizabeth Funmilayo Aransiola
Department of Chemical Engineering, Obafemi Awolowo University, Ile-Ife, Nigeria

Adrian. K. James
Environmental Science & Engineering, University of Northern British Columbia, 3333 University Way, Prince George, BC, V2N 4Z9, Canada

Steve S. Helle
Environmental Science & Engineering, University of Northern British Columbia, 3333 University Way, Prince George, BC, V2N 4Z9, Canada

Ronald W. Thring
Environmental Science & Engineering, University of Northern British Columbia, 3333 University Way, Prince George, BC, V2N 4Z9, Canada

Gurkaran S. Sarohia
Environmental Science & Engineering, University of Northern British Columbia, 3333 University Way, Prince George, BC, V2N 4Z9, Canada

P. Michael Rutherford
Environmental Science & Engineering, University of Northern British Columbia, 3333 University Way, Prince George, BC, V2N 4Z9, Canada

Joshua L. Konne
Department of Chemistry, Rivers State University of Science & Technology, Port Harcourt, Nigeria

Kenneth Okpara
Department of Chemistry, Rivers State University of Science & Technology, Port Harcourt, Nigeria

Alphonse Nahayo
Higher Institute of Agriculture and Animal Husbandry (ISAE-Busogo), Department of Forestry and Nature Conservation (FNC), Musanze, Rwanda

Isaac Ekise
Higher Institute of Agriculture and Animal Husbandry (ISAE-Busogo), Department of Forestry and Nature Conservation (FNC), Musanze, Rwanda

Alphonsine Mukarugwiza
Higher Institute of Agriculture and Animal Husbandry (ISAE-Busogo), Department of Forestry and Nature Conservation (FNC), Musanze, Rwanda

Ming Chun Hsieh
Department of Electronic Engineering, Kun Shan University, Tainan, Taiwan, R.O.C

David King Jair
Department of Mechanical Engineering, Kun Shan University, Tainan, Taiwan, R.O.C

Mohammad Aslam
Sardar Swaran Singh National Institute of Renewable Energy, India

Prashant Saxena
DR BR Ambedkar NIT Jalandhar, India

Anil K. Sarma
Sardar Swaran Singh National Institute of Renewable Energy, India

Elizabeth Locke
School of Engineering, Cardiff University, Wales, UK

Sally Hewlett
School of Engineering, Cardiff University, Wales, UK

Misri Gozan
Industrial Bioprocess Engineering Research Group, Chemical Engineering Department, Faculty of Engineering, Universitas Indonesia, Kampus UI Depok 16424, Indonesia

Evelyn Alves Nunes Simonetti
Escola de Engenharia de Lorena, University of São Paulo, Lorena, São Paulo, Brazil

Gilberto Garcia Cortez
Escola de Engenharia de Lorena, University of São Paulo, Lorena, São Paulo, Brazil

Celine Manoosingh
Georgia Southern University

James Mihelcic
University of South Florida

Manjriker Gunaratne
University of South Florida

D. F. Dramani
Department of Civil Engineering, Kwame Nkrumah University of Science and Technology, PMB, Kumasi, Ghana

G. K. Anornu
Department of Civil Engineering, Kwame Nkrumah University of Science and Technology, PMB, Kumasi, Ghana

A. T. Kabo-bah
Department of Energy and Environmental Engineering, University of Energy and Natural Resources, Box 214, Sunyani, Ghana

S. Oduro-Kwarteng
Department of Civil Engineering, Kwame Nkrumah University of Science and Technology, PMB, Kumasi, Ghana

Ababacar Ndiaye
Université Assane Seck de Ziguinchor, UFR Sciences et Technologies, Département de Physique. BP: 523 Ziguinchor, Sénégal
Centre International de Formation et de Recherche en Energie Solaire (CIFRES), Ecole Supérieure Polytechnique-Université Cheikh Anta Diop de Dakar (UCAD), BP 5085 Dakar-Fann, Sénégal

Cheikh M. F. Kébé
Centre International de Formation et de Recherche en Energie Solaire (CIFRES), Ecole Supérieure Polytechnique-Université Cheikh Anta Diop de Dakar (UCAD), BP 5085 Dakar-Fann, Sénégal

Vincent Sambou
Centre International de Formation et de Recherche en Energie Solaire (CIFRES), Ecole Supérieure Polytechnique-Université Cheikh Anta Diop de Dakar (UCAD), BP 5085 Dakar-Fann, Sénégal

Papa A. Ndiaye
Centre International de Formation et de Recherche en Energie Solaire (CIFRES), Ecole Supérieure Polytechnique-Université Cheikh Anta Diop de Dakar (UCAD), BP 5085 Dakar-Fann, Sénégal

Sangeeta Sinha
Department of Physics, B. R. Ambedkar Bihar University, Muzaffarpur, Bihar, India
Centre for Renewable Energy and Env. Res., Muzaffarpur, Bihar - 842001, India

Anil Roy
DA-IICT, Gandhinagar, Gujarat, India

S. N. P. Singh
Department of Zoology, B. R. Ambedkar Bihar University, Muzaffarpur, Bihar, India